PUBLIC PARTICIPATION
IN ENVIRONMENTAL ASSESSMENT AND DECISION MAKING

Panel on Public Participation in
Environmental Assessment and Decision Making

Thomas Dietz and Paul C. Stern, *Editors*

Committee on the Human Dimensions of Global Change
Division of Behavioral and Social Sciences and Education

NATIONAL RESEARCH COUNCIL
OF THE NATIONAL ACADEMIES

THE NATIONAL ACADEMIES PRESS
Washington, D.C.
www.nap.edu

THE NATIONAL ACADEMIES PRESS   500 Fifth Street, N.W.   Washington, DC 20001

NOTICE: The project that is the subject of this report was approved by the Governing Board of the National Research Council, whose members are drawn from the councils of the National Academy of Sciences, the National Academy of Engineering, and the Institute of Medicine. The members of the committee responsible for the report were chosen for their special competences and with regard for appropriate balance.

This project was supported by the Environmental Protection Agency, with contributions from the Food and Drug Administration and U.S. Department of Energy, Grant No. X-82873001, and the U.S. Department of Agriculture, Grant No. PNW 07-DG-1123416976-342. Any opinions, findings, conclusions, or recommendations expressed in this publication are those of the author(s) and do not necessarily reflect the views of the sponsors.

Library of Congress Cataloging-in-Publication Data

Public participation in environmental assessment and decision making / Panel on Public Participation in Environmental Assessment and Decision Making ; Thomas Dietz and Paul C. Stern, editors ; Committee on the Human Dimensions of Global Change, Division of Behavioral and Social Sciences and Education.
  p. cm.
  Includes bibliographical references.
  ISBN 978-0-309-12398-3 (pbk.) — ISBN 978-0-309-12399-0 (pdf) 1. Environmental impact analysis—Citizen participation—Evaluation. 2. Environmental policy—Decision making—Citizen participation—Evaluation. 3. Environmental policy—United States—Decision making. 4. Administrative agencies—United States—Decision making. 5. Administrative procedure—United States—Citizen participation. 6. Environmental protection—United States—Citizen participation. I. Dietz, Thomas. II. Stern, Paul C., 1944- III. National Research Council (U.S.). Panel on Public Participation in Environmental Assessment and Decision Making. IV. National Research Council (U.S.). Committee on the Human Dimensions of Global Change.
  TD194.6.P83 2008
  363.7'0525—dc22
                                        2008038571

Additional copies of this report are available from the National Academies Press, 500 Fifth Street, N.W., Lockbox 285, Washington, DC 20055; (800) 624-6242 or (202) 334-3313 (in the Washington metropolitan area); Internet http://www.nap.edu.

Copyright 2008 by the National Academy of Sciences. All rights reserved.

Printed in the United States of America

Suggested citation: National Research Council. (2008). *Public Participation in Environmental Assessment and Decision Making*. Panel on Public Participation in Environmental Assessment and Decision Making, Thomas Dietz and Paul C. Stern, eds. Committee on the Human Dimensions of Global Change. Division of Behavioral and Social Sciences and Education. Washington, DC: The National Academies Press.

THE NATIONAL ACADEMIES
Advisers to the Nation on Science, Engineering, and Medicine

The **National Academy of Sciences** is a private, nonprofit, self-perpetuating society of distinguished scholars engaged in scientific and engineering research, dedicated to the furtherance of science and technology and to their use for the general welfare. Upon the authority of the charter granted to it by the Congress in 1863, the Academy has a mandate that requires it to advise the federal government on scientific and technical matters. Dr. Ralph J. Cicerone is president of the National Academy of Sciences.

The **National Academy of Engineering** was established in 1964, under the charter of the National Academy of Sciences, as a parallel organization of outstanding engineers. It is autonomous in its administration and in the selection of its members, sharing with the National Academy of Sciences the responsibility for advising the federal government. The National Academy of Engineering also sponsors engineering programs aimed at meeting national needs, encourages education and research, and recognizes the superior achievements of engineers. Dr. Charles M. Vest is president of the National Academy of Engineering.

The **Institute of Medicine** was established in 1970 by the National Academy of Sciences to secure the services of eminent members of appropriate professions in the examination of policy matters pertaining to the health of the public. The Institute acts under the responsibility given to the National Academy of Sciences by its congressional charter to be an adviser to the federal government and, upon its own initiative, to identify issues of medical care, research, and education. Dr. Harvey V. Fineberg is president of the Institute of Medicine.

The **National Research Council** was organized by the National Academy of Sciences in 1916 to associate the broad community of science and technology with the Academy's purposes of furthering knowledge and advising the federal government. Functioning in accordance with general policies determined by the Academy, the Council has become the principal operating agency of both the National Academy of Sciences and the National Academy of Engineering in providing services to the government, the public, and the scientific and engineering communities. The Council is administered jointly by both Academies and the Institute of Medicine. Dr. Ralph J. Cicerone and Dr. Charles M. Vest are chair and vice chair, respectively, of the National Research Council.

www.national-academies.org

PANEL ON PUBLIC PARTICIPATION IN ENVIRONMENTAL ASSESSMENT AND DECISION MAKING

THOMAS DIETZ (*Chair*), Environmental Science and Policy Program, Michigan State University
GAIL BINGHAM, Resolve, Washington, DC
CARON CHESS, Department of Human Ecology, School of Environmental and Biological Sciences, Rutgers University
MICHAEL L. DEKAY, Department of Psychology, The Ohio State University
JEANNE M. FOX, New Jersey Board of Public Utilities, Newark
STEVEN C. LEWIS, Integrative Policy & Science, Inc., Washington, New Jersey
GREGORY B. MARKUS, Center for Political Studies, University of Michigan
D. WARNER NORTH, NorthWorks, Inc., Belmont, California
ORTWIN RENN, Institute of Management and Technology, University of Stuttgart, Germany
MARGARET A. SHANNON, Rubenstein School of Environment and Natural Resources, University of Vermont
ELAINE VAUGHAN, School of Social Ecology, University of California, Irvine
THOMAS J. WILBANKS, Environmental Sciences Division, Oak Ridge National Laboratory, Oak Ridge, Tennessee

PAUL C. STERN, *Study Director*
JENNIFER BREWER, *Staff Officer*
SETH TULER, *Consultant*
LINDA DEPUGH, *Administrative Assistant*

COMMITTEE ON THE HUMAN DIMENSIONS OF GLOBAL CHANGE

THOMAS J. WILBANKS (*Chair*), Oak Ridge National Laboratory, Oak Ridge, Tennessee
RICHARD N. ANDREWS, Department of Public Policy, University of North Carolina, Chapel Hill
ROBERT CORELL, H. John Heinz III Center for Science, Economics and the Environment, Washington, DC
ROGER E. KASPERSON, George Perkins Marsh Institute, Clark University
ANN KINZIG, Department of Biology, Arizona State University, Tempe
TIMOTHY MCDANIELS, Eco-Risk Unit, University of British Columbia, Vancouver
LINDA O. MEARNS, Environmental and Societal Impacts Group, National Center for Atmospheric Research, Boulder, Colorado
EDWARD MILES, School of Marine Affairs, University of Washington, Seattle
ALEXANDER PFAFF, Public Policy Department, Duke University
EUGENE ROSA, Natural Resource and Environmental Policy, Washington State University, Pullman
CYNTHIA E. ROSENZWEIG, NASA Goddard Institute for Space Studies, New York
GARY W. YOHE, Department of Economics, Wesleyan University
ORAN R. YOUNG (*ex officio*), International Human Dimensions Programme on Global Environmental Change Scientific Committee; Bren School of Environmental Science and Management, University of California, Santa Barbara

PAUL C. STERN, *Study Director*
JENNIFER F. BREWER, *Staff Officer*
SETH TULER, *Consultant*
LINDA DEPUGH, *Administrative Assistant*

Preface

This report began with two simple ideas. One was that the environmental problems of the 21st century can be effectively addressed only by processes that link sound scientific analysis with effective public deliberation. The second was that analysis and deliberation in environmental assessment and decision making can be improved by careful examination of scientific evidence.

Discussions about public participation have become especially intense in the last half century. Novel methods of public engagement have emerged to complement more venerable modes of participation, such as voting, lobbying, and protesting. In response to the new practices, a growing literature has offered theory to define and justify public participation, has proposed tools and strategies for participation, and has begun to examine what happens in participation processes. But this literature, while substantial in size and including much work of high quality, has not been cumulative. It provides no overall assessment of whether or not, in general, public participation enhances environmental assessments and decisions; those designing participation processes have trouble extracting lessons from it; and it does not reflect a consensus about the key questions requiring further research.

This study attempts to address what have been missing: to provide an overall assessment of the merits and failings of participation, to offer guidance to practitioners, and to identify directions for further research. Participation research and practice is so dynamic that our analysis is somewhat dated even as it is published, yet I believe we have made some progress in synthesizing across a diverse literature. We have found that participation can be an invaluable part of environmental assessment and decision making. Although there are no simple "best practices" that provide universal guidance in designing participation, there are principles and "*best processes*" that can enhance the effectiveness of participation. We have taken a few steps toward structuring the research literature. Our hope is that this report will prove useful for those who are assessing participation policy and practices, those who design and conduct participation, and those who study participation. We know it is not the final word, but we believe it lends some coherence to future conversations and provides a starting point for further analysis.

As one would expect of a work on participation, many have participated in creating the final product. It is, first and foremost, the work of the panel and Paul Stern, the study director. The study draws together diverse strands of literature and bridges across diverse disciplines and substantive domains. In doing so, the panel and Paul have worked very hard and exhibited great patience and a wonderful openness to synthesis.

We conducted two scoping workshops before the study began and one workshop midstream in the study. The participants in those workshops—scholars, practitioners, and nonspecialists—had a profound influence in shaping the study. We thank first the participants in our July 2001 workshop: Bonnie Bailey, Water Environment Research Foundation; Thomas C. Beierle, Resources for the Future; Mohandas Bhat, U.S. Department of Energy; Steve Blackwell, Agency for Toxic Substances and Disease Registry; Judith Bradbury, Pacific Northwest National Laboratory; Frank Clearfield, National Resource Conservation Services' Social Sciences Institute; Martha Crosland, U.S. Department of Energy; Katherine Dawes, Office of Environmental Policy Innovation, U.S. Environmental Protection Agency; Michael Donnelly, Radiation Studies Branch, Centers for Disease Control and Prevention; John Hogan, Office of Food Safety, U.S. Department of Agriculture; Debora Martin, U.S. Environmental Protection Agency; Michael Sage, Centers for Disease Control and Prevention; Michael Slimak, U.S. Environmental Protection Agency; Peter Smith, U.S. Department of Agriculture; and Elizabeth White, U.S. Department of Energy; and Susan Wiltshire, JK Research Associates.

We also thank the participants in our December 2001 workshop: Laurel Ames, Sierra Nevada Alliance; John Applegate, University of Indiana; L. Katherine Baril, Washington State University; Thomas C. Beierle, Resources for the Future; Sue Briggum, WMX Waste Management; Fred Butterfield, U.S. Department of Energy; Susan Carillo, U.S. Environmental Protection Agency; Martha Crosland, U.S. Department of Energy, Samantha Dixon, City of Westminister, Colorado; Paul Gagliardo, Metropolitan Wastewater Public Works, City of San Diego, California; Troy Hartley, RESOLVE, Washington, DC; Kenneth Jones, Green Mountain Institute for Environmental Democracy; Jeffrey Jordan, City of South Portland,

PREFACE

Maine; Marshall Kreuter, National Center for Chronic Disease Prevention and Health Promotion, Centers for Disease Control and Prevention; Mark Lubell, Florida State University; Eric Marsh, U.S. Environmental Protection Agency; Tom Marshall, Rocky Mountain Peace and Justice Center; Robert O'Connor, National Science Foundation; Dennis Ojima, Colorado State University; Kathryn Papp, National Council for Science and the Environment; Karen Patterson, Tetra Tech NUS; Trisha Pritkin, Hanford Downwinders; Beth Raps, independent consultant; Douglas Sarno, The Perspectives Group, Inc.; Michael Slimak, U.S. Environmental Protection Agency; James Smith, Centers for Disease Control and Prevention; Bruce Stedman, RESOLVE, Washington, DC; Vicky Sturtevant, Southern Oregon University; Patrice Sutton, Western States Legal Foundation; Merv Tano, Council of Energy Resource Tribes; John Till, Risk (Radiation) Assessment Corporation; William Toffey, Philadelphia Water Department; Bruce Tonn, University of Tennessee; and Chris Wiant, Caring for Colorado Foundation.

Our mid-study workshop was held in February 2005, and we thank the participants: Beth Anderson, National Institute for Environmental Health Sciences; Mitchell Baer, U.S. Department of Energy; Bonnie Bailey, Water Environment Research Foundation; Anjuli Bamzai, U.S. Department of Energy; Patricia Bonner, U.S. Environmental Protection Agency; Nina Burkardt, U.S. Geological Survey Fort Collins Science Center; Francis (Chip) Cameron, U.S. Nuclear Regulatory Commission; Joe Carbone, U.S.D.A. Forest Service; David Cleaves, U.S.D.A. Forest Service; Jim Creighton, Creighton & Creighton; Jeremiah Davis, The George Washington University; Sandra Dawson, Jet Propulsion Laboratory; Alvaro DeCarvalho, Water Environment Research Federation; David Emmerson, U.S. Department of Interior; Bruce Engelbert, U.S. Environmental Protection Agency Superfund Community Involvement and Outreach; Tim Fields, Tetra Tech EM, Inc.; Baruch Fischhoff, Carnegie Mellon University; Amy Fitzgerald, City of Oak Ridge, Tennessee; Victoria Friedensen, National Aeronautics and Spacec Administration; Elena Gonzalez, U.S. Department of the Interior; Tanya Heikkila, Columbia University; Kasha Helget, Federal Energy Regulatory Commission; Elizabeth Howze, Agency for Toxic Substances and Disease Registry; Marcia Keenan, Office of Policy, National Park Service; Jeremy Kranowitz, The Keystone Center; Linda Lampl, Lampl Herbert Consultants; Laura Langbein, American University; Charles Lee, U.S. Environmental Protection Agency; Onora Lien, Center for Biosecurity of UPMC; Mark Lubell, University of California, Davis; Tanya Maslak, U.S. Environmental Protection Agency; Katherine McComas, Cornell University; Jennifer Nuzzo, Center for Biosecurity, University of Pittsburgh Medical Center; Robert O'Connor, National Science Foundation; Lola Olabode, Water Environment Research Foundation; Suaquita (Kita) Perry,

U.S. Army Center for Health Promotion and Preventative Medicine Health Risk Communication Program; David Rejeski, Foresight and Governance Project, Woodrow Wilson Center International Center for Scholars; Anca Romantan, University of Pennsylvania; Adam Scheffler, Chicago, IL; Frances Seymour, World Resources Institute; Michael Slimak, U.S. Environmental Protection Agency; Roxanne Smith, U.S. Army Center for Health Promotion and Preventative Medicine Health Risk Communication Program; Jasmine Tanguay, CLF Ventures, Inc.; and Thomas Webler, Social and Environmental Research Institute.

We also commissioned several papers that were critical to the report by providing detailed analyses of public participation in what we call "families" of cases—cases that were similar in the environmental issues addressed and in the institutional contexts in which they were carried out. We thank the authors for their work, without which we could not have come as far as we did:

- *Evaluating Public Participation in Environmental Decisions*; Judith Bradbury, Pacific Northwest National Laboratory
- *Negotiated and Conventional Rulemaking at EPA: A Comparative Case Analysis*; Laura Langbein, American University
- *Watershed Partnerships: Evaluating a Collaborative Form of Public Participations*; Mark Lubell, University of California, Davis, and William D. Leach, California State University, Sacramento
- *Stakeholder Involvement in the First U.S. National Assessment of the Potential Consequences of Climate Variability and Change: An Evaluation, Finally*; Susanne C. Moser, National Center for Atmospheric Research

Finally, the sponsors of the study at the Forest Service of the U.S. Department of Agriculture, the Food and Drug Administration, the U.S. Department of Energy, and, especially, the U.S. Environmental Protection Agency have shown a deep commitment to effective public engagement by supporting this study at a time of budget constraints and shifting priorities.

We believe that our study has had benefits beyond this volume and that it will continue to do so. For example, it established new communication links between the National Research Council and organizations involved in addressing the practical challenges of environmental public participation. It provided educational opportunities for five Christine Mirzayan Fellows at the National Research Council during the course of the panel's work: Rebecca Zarger, Rebecca Romsdahl, Loraine Lundquist, Rachael Shwom, and Hannah Brenkert-Smith. Their insights and engagement were of great value to the project. And we hope it will help promote the continuation of the dialogue between theory and practice that was so helpful during the course of our study.

This report has been reviewed in draft form by individuals chosen for their diverse perspectives and technical expertise, in accordance with procedures approved by the Report Review Committee of the National Research Council. The purpose of this independent review is to provide candid and critical comments that will assist the institution in making its published report as sound as possible and to ensure that the report meets institutional standards for objectivity, evidence, and responsiveness to the study charge. The review comments and draft manuscript remain confidential to protect the integrity of the deliberative process.

We thank the following individuals for their review of this report: Richard N. Andrews, Department of Public Policy, University of North Carolina; Sue Briggum, Federal Public Affairs, WM Waste Management, Washington, DC; Archon Fung, John F. Kennedy School of Government, Harvard University; Jerome B. Gilbert, President's Office, J. Gilbert, Inc., Orinda, CA; Robin Gregory, Senior Researcher, Decision Research, Canada; Kathy Halvorsen, Forest Resources and Environmental Science and Social Sciences, Michigan Technological University, Houghton, MI; Evan Ringquist, Public and Environmental Affairs, Indiana University; Douglas J. Sarno, The Perspectives Group, Inc., Alexandria, VA; Mark E. Warren, Department of Political Science, University of British Columbia; and Julia Wondolleck, School of Natural Resources and Environment, University of Michigan.

Although the reviewers listed above have provided many constructive comments and suggestions, they were not asked to endorse the conclusions or recommendations nor did they see the final draft of the report before its release. The review of this report was overseen by Lorraine M. McDonnell, Department of Political Science, University of California, Santa Barbara, and Susan Hanson, School of Geography, Clark University. Appointed by the National Research Council, they were responsible for making certain that an independent examination of this report was carried out in accordance with institutional procedures and that all review comments were carefully considered. Responsibility for the final content of this report rests entirely with the authoring committee and the institution. Nonetheless, we thank the reviewers and the review coordinator for diligent analysis that greatly improved the quality of the report.

> Thomas Dietz, *Chair*
> Panel on Public Participation in
> Environmental Assessment and Decision Making

Contents

Executive Summary ... 1

1 Introduction ... 7
Defining Public Participation, 11
Dimensions of Participation, 14
Objectives and Scope of the Study, 18
Sources of Knowledge, 21
How We Conducted the Study, 27
Guide to the Report, 29
Notes, 30

2 The Promise and Perils of Participation ... 33
Historical Development: Laws and Agency Practices, 36
Purposes of Public Participation, 43
Justifications for and Problems with Public Participation, 46
Pitfalls, 51
Criteria for Evaluation, 66
Conclusion, 73
Notes, 74

3 The Effects of Public Participation ... 75
Does Public Participation Improve Results?, 76
Associations Among Results: Can You Have It All?, 86

Conclusion, 91
Notes, 92

4 Public Participation Practice: Management Practices   95
Clarity of Purpose, 96
Agency Commitment, 99
Adequate Capacity and Resources, 101
Timeliness in Relation to Decisions, 103
A Focus on Implementation, 105
Commitment to Learning, 106
Conclusion, 109

5 Practice: Organizing Participation   111
Public Participation Formats and Practices, 111
Dimensions of Participatory Process, 115
Breadth, 118
Openness of Design, 122
Intensity, 126
Influence, 132
Conclusion, 135
Note, 135

6 Practice: Integrating Science   137
Integration, 138
Challenges of Integration, 140
Meeting the Challenges, 144
Conclusion, 152
Summary: The Practice of Participation, 154

7 Context: The Issue   157
Purpose of the Process: Assessment or Decision Making, 158
Nature of the Environmental Issue, 161
The Science, 167
Conclusions, 180
Notes, 182

8 Context: The People   187
Convening and Implementing Agencies, 187
Who Participates, 192
Adequacy of Representation, 193
Differing Perspectives, 202
Polarization, 205

Power Disparities, 207
  Role of Representatives, 209
  Trust, 210
  Conclusions, 214
  Notes, 216

9 **Overall Conclusions and Recommendations**   223
  The Value of Public Participation, 226
  Management, 227
  Organizing the Process, 230
  Integrating Science, 233
  Implementation, 236
  Needed Research, 238
  Notes, 243

References   245

Appendix: Biographical Sketches of Panel Members and Staff   299

Executive Summary

Advocates of public participation believe it improves environmental assessment and decision making; detractors criticize it as ineffective and inefficient. The National Research Council established the Panel on Public Participation in Environmental Assessment and Decision Making at the request of U.S. Environmental Protection Agency, the U.S. Department of Energy, and the U.S. Food and Drug Administration, with additional support from the U.S. Forest Service, to assess whether, and under what conditions, public participation achieves the outcomes desired.

The term "public participation," as used in this study, includes organized processes adopted by elected officials, government agencies, or other public- or private-sector organizations to engage the public in environmental assessment, planning, decision making, management, monitoring, and evaluation. These processes supplement traditional forms of public participation (voting, forming interest groups, demonstrating, lobbying) by directly involving the public in executive functions that, when they are conducted in government, are traditionally delegated to administrative agencies. The goal of participation is to improve the quality, legitimacy, and capacity of environmental assessments and decisions.

- *Quality* refers to assessments or decisions that (1) identify the values, interests, and concerns of all who are interested in or might be affected by the environmental process or decision; (2) identify the range of actions that might be taken; (3) identify and systematically consider the effects that might follow and uncertainties about them; (4) use the best available knowledge and methods relevant to the above tasks, particularly (3); and

(5) incorporate new information, methods, and concerns that arise over time.

- *Legitimacy* refers to a process that is seen by the interested and affected parties as fair and competent and that follows the governing laws and regulations.
- *Capacity* refers to participants, including agency officials and scientists, (1) becoming better informed and more skilled at effective participation; (2) becoming better able to engage the best available scientific knowledge and information about diverse values, interests, and concerns; and (3) developing a more widely shared understanding of the issues and decision challenges and a reservoir of communication and mediation skills and mutual trust.

> **Conclusion 1: When done well, public participation improves the quality and legitimacy of a decision and builds the capacity of all involved to engage in the policy process. It can lead to better results in terms of environmental quality and other social objectives. It also can enhance trust and understanding among parties. Achieving these results depends on using practices that address difficulties that specific aspects of the context can present.**

The panel found that participatory processes have sometimes made matters worse. However, it also found that across a wide variety of environmental assessment and decision contexts, there are practices that can simultaneously promote quality, legitimacy, and capacity.

> **Recommendation 1: Public participation should be fully incorporated into environmental assessment and decision-making processes, and it should be recognized by government agencies and other organizers of the processes as a requisite of effective action, not merely a formal procedural requirement.**

PUBLIC PARTICIPATION PRACTICE

The panel offers four recommendations for carrying out public participation processes that embody six principles of program management, four principles for the conduct of participation, and five principles for integrating science and participation.

> **Recommendation 2: When government agencies engage in public participation, they should do so with**

1. clarity of purpose,
2. a commitment to use the process to inform their actions,
3. adequate funding and staff,
4. appropriate timing in relation to decisions,
5. a focus on implementation, and
6. a commitment to self-assessment and learning from experience.

Recommendation 3: Agencies undertaking a public participation process should, considering the purposes of the process, design it to address the challenges that arise from particular contexts. Process design should be guided by four principles:

1. inclusiveness of participation,
2. collaborative problem formulation and process design,
3. transparency of the process, and
4. good-faith communication.

In environmental assessment and decision making, special attention must be paid to scientific analysis and the uncertainty in that analysis.

Recommendation 4: Environmental assessments and decisions with substantial scientific content should be supported with collaborative, broadly based, integrated, and iterative analytic-deliberative processes, such as those described in *Understanding Risk* and subsequent National Research Council reports. In designing such processes, the responsible agencies can benefit from following five key principles for effectively melding scientific analysis and public participation:

1. ensuring transparency of decision-relevant information and analysis,
2. paying explicit attention to both facts and values,
3. promoting explicitness about assumptions and uncertainties,
4. including independent review of official analysis and/or engaging in a process of collaborative inquiry with interested and affected parties, and
5. allowing for iteration to reconsider past conclusions on the basis of new information.

IMPLEMENTING THE PRINCIPLES

There is no specific set of tools or techniques that constitute "best practices" for all contexts, or even for meeting particular difficulties. Rather, the best technique will be situation-dependent, and practices need to be sensitive to changes that occur during the process.

Recommendation 5: Public participation practitioners, working with the responsible agency and the participants, should adopt a best-process regime consisting of four elements:

1. diagnosis of the context,
2. collaborative choice of techniques to meet difficulties expected because of the context,
3. monitoring of the process to see how well it is working, and
4. iteration, including changes in tools and techniques if needed to overcome difficulties.

This process is illustrated in Figure ES-1.

NEEDED RESEARCH

Recommendation 6: Agencies that involve interested and affected parties in environmental assessments and decision making should invest in social science research to inform their practice and build broader knowledge about public participation. Routine, well-designed evaluation of agency public participation efforts is one of the most important contributions they can make. Because public participation makes a useful test bed for examining basic social science theory and methods, the National Science Foundation should partner with mission agencies in funding such research, following the model of the successful Partnership for Environmental Research of the National Science Foundation and the Environmental Protection Agency.

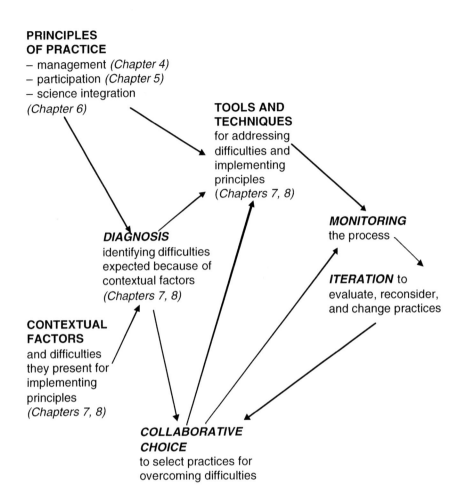

FIGURE ES-1 Elements of best process for public participation in relation to the principles of good public participation and variations in context.
NOTE: The four elements of best process are indicated in italics. Arrows indicate lines of influence: principles and contextual factors contribute to diagnosis; principles, diagnosis, and collaborative choice influence the selection of tools and techniques; the tools and collaborative choice determine what is monitored and how; monitoring leads to iteration; and iteration, via collaborative choice, feeds back to the selection of tools and techniques.

1

Introduction

Many tensions exist between the democratic aspiration of government of the people, by the people, and for the people and modern representative government with its mass electorate and elaborate bureaucracy for carrying out government functions (Finer, 1941; Schlozman and Tierney, 1986; Dahl, 1989, 1998; Morone, 1990; Held, 1996). Nowhere are these tensions more acute than in the domain of environmental policy.

On one hand, the issues are complex, laden with scientific and technical detail, and subject to change (Williams and Matheny, 1995; Fischer, 2000; Jasanoff, 2005), so that informed choices require technical expertise few of "the people" have. Advances in scientific understanding of the environment, including discoveries of new environmental phenomena; the unfolding of a variety of local, regional, and global environmental changes; and the application of new technologies ensure that environmental choices are continually evolving. Moreover, decisions that affect the environment thus present special challenges because of the need for scientific understanding of the dynamics of coupled human and natural systems. Processes of environmental change operate at large spatial and temporal scales, with linkages between processes operating at different scales; they may involve intrinsic uncertainties; they are often rapid or nonlinear; and they are sometimes irreversible (Dietz and Stern, 1998; Liu et al., 2007a,b). These characteristics of environmental decisions suggest the need for significant influence to be in the hands of experts.

But on the other hand, environmental decisions present very complex choices among interests and values, so that the choices are political, social, cultural, and economic, at least as much as they are scientific and technical. Environmental decisions have varied and uncertain effects on the values and interests of people in diverse societies, so there are rarely only two sides to a question. Furthermore, progress on environmental problems often requires changes in the behavior of a multitude of diverse groups of actors, not just corporations and governments. Citizens are now targets of policy and thus often are stakeholders in the same way that organizations are.

There typically are multiple perspectives regarding the relative importance of issues, the best courses of action, and even the right questions to ask, with strong demands from those who may be affected by policy choices to have their voices heard (Forester, 1989; Dryzek, 1990; Fischer and Forester, 1993; Ingram and Smith, 1993; Schneider and Ingram, 1997; Stone, 2002; Feldman et al., 2006; Healey, 2006). The variety of questions that must be addressed to inform a single decision is often staggering, and conflict is almost inevitable (Crowfoot and Wondolleck, 1990; Stern, 1991; Vaughan and Nordenstam, 1991; Aronoff and Gunter, 1992; National Research Council, 1996, 2005a; Dietz and Stern, 1998; Proctor, 1998; Beierle and Konisky, 2000; Lubell, 2000; Dietz, 2001; Brown et al., 2002; Campbell, 2003; Lewicki, Gray, and Eliot, 2003). Some parties may lack the power and resources to participate effectively in the policy system via traditional mechanisms: some are highly organized, and others are a more diffuse array of individual citizens (Forester, 1989; Williams and Matheny, 1995).

The high public and political visibility of many environmental issues also can add complex dynamics to the process, and the more attention is given to an issue, the more likely it will be that simultaneous opportunities to influence public decisions will exist in legislative, executive, and judicial forums, at multiple levels of government, and in the media. It is in this highly complex arena—in which those who believe that industry and development are being unnecessarily stifled contend with those who believe that the environment is being irreparably damaged and those who believe that the costs of environmental change are being unfairly distributed—that public decisions are made about making and implementing environmental policy.

The conflicts that arise in environmental policy result not only from differences in values and interests. When parties have different objectives and concerns, they need different information from science in order to consider themselves adequately informed. When they experience different parts of an environmental system, they gain different kinds of knowledge and sometimes apply different "ways of knowing" (Fischer, 2000; Feldman et al., 2006). These characteristics of environmental decisions suggest significant pitfalls in delegating too much influence to experts, because they may overlook important information or fail to analyze issues that are critical to some parties.

The degree to which nonexperts should be involved in analytic tasks typically assigned to scientists is contested among both participation theorists and practitioners. In a recent review article, Chilvers (2008) identified three different camps. One camp believes that a strict functional separation in analytic and deliberative forms of decision making is essential to avoid a muddling of facts and values. Another camp proposes that the limit of public involvement in the scientific analysis should be determined by the extent to which nonscientists possess "contributory expertise" that can complement or enhance certified scientific expertise. The third camp believes that the scientific and political dimensions cannot be separated and emphasizes the need to negotiate public meanings embedded in science as an integral part of decision making. All these camps do agree that regulatory decisions cannot be based on technical expertise alone, but need refinement by stakeholder or public involvement. However, some critics of public participation practice have suggested that participation is too expensive and slow for what they contend are the minimal benefits it provides and that participation can degrade rather than improve decisions (e.g., Graham, 1996; Rossi, 1997; Sanders, 1997; Sunstein, 2001, 2006; Collins and Evans, 2002; Campbell and Currie, 2006). We detail these concerns in Chapter 2.

Given these tensions, it is not surprising that since the 1960s, U.S. environmental policy has come under fire from different quarters and for several reasons. Environmentalists, advocates for disadvantaged communities, resource user groups, Native American tribes, and others have criticized policy makers as being out of touch with public desires and as having made too many bad environmental decisions (e.g., Bullard, 1990; Pellow, 1999; Durant, Fiorino, and O'Leary, 2004). Others, expressing concerns with efficient decision making, have criticized existing policies as having produced "environmental gridlock" (e.g., Van Horn, 1988; Kraft, 2000)—excessive delay due to continuing conflict and litigation over decisions and proposed decisions.

The criticisms concern both the legitimacy and the quality of decisions. Administrative decisions by bureaucratic agencies have been criticized for failing to follow basic principles of good policy making, for example, by failing to pay attention to legitimate interests and to take their concerns into account and sometimes short-circuiting standard administrative process. The result has been a loss of legitimacy in the eyes of some parties. Agencies have also been criticized for failing to follow basic principles of good decision making, for example, by artificially narrowing the set of choices to consider, failing to take important values into account in analyses, and making unrealistic assumptions in the face of scientific uncertainty (Shannon 1991; Office of Technology Assessment, 1992). Criticisms on grounds of inefficiency are rooted in part in these other criticisms, when conflict and litigation result because parties are seriously critical of the quality and legitimacy of agency decisions.

Broader and more direct participation of the public and interested or affected groups in official environmental policy processes has been widely advocated as a way to increase both the legitimacy and the substantive quality of policy decisions (e.g., Dietz, 1987; Shannon, 1987; Fiorino, 1989, 1990; Renn, Webler, and Wiedermann, 1995; Williams and Matheny, 1995; National Environmental Justice Advisory Committee, 1996; National Research Council, 1996; Liberatore and Funtowicz, 2003; Renn, 2004; Stirling, 2004, 2008). Such arguments have had political success in some situations. As Creighton (2005:1) pointed out, "Public participation requirements have been embedded in virtually every important piece of environmental legislation in the United States and Canada since the 1970s," and "more than thirty-five European countries are signatories to the 1998 Aarhus Convention," which commits their governments "to ensure public participation and access to information in all environmental decision making." Proponents claim that increased public participation will inform the decision-making process in ways that lead both to more informed and reasoned discussion of these complex issues and to better and more widely acceptable decisions. Others, however, raise concerns about hazards of public participation, such as the accountability and representativeness of self-appointed public participants, the inability of nonexpert communities to understand and process complex scientific relationships, the unlikelihood of reaching a meaningful consensus among conflicting interests, the effects of misdirected pressure to achieve consensus at the expense of achieving other important societal goals, and manipulation of outcomes either by those who frame the questions to be addressed or by those who get a "seat at the table" (Cupps, 1977; Abel, 1982; Graham, 1996; McCloskey, 1996; Coglianese, 1997; Rossi, 1997; Pellizoni, 2001; Sunstein, 2001, 2006; Ventriss and Keuntzel, 2005; Bora and Hausendorf, 2006; Abels, 2007).

Assessing these claims is central to the aims of this study. The Panel on Public Participation in Environmental Assessment and Decision Making was established in response to a request from the U.S. Environmental Protection Agency (EPA), the U.S. Department of Energy, and the Food and Drug Administration to make such an assessment, and it also received support from the U.S. Forest Service. Its task was to "undertake a study of public participation processes in environmental assessment and policy making" that would focus on "indicators of success and variables that may influence these indicators; lessons from experience concerning which approaches work well under which conditions; testable hypotheses that would allow verification or refinement of such lessons; and ways that government agencies can learn systematically from their own experience and the experience of others." The panel was charged with writing "a consensus statement about the implications of current knowledge for public participation, practice and research."

The panel includes researchers and practitioners with expertise in environmental assessment, public participation, risk analysis, adaptive management, group process, decision making, environmental policy, evaluation research, and related fields. We were selected to provide the study with a range of knowledge and expertise across these fields and over a wide variety of environmental and biomedical policy issues. We sought additional input from other researchers and practitioners as we conducted the study, as described later in this chapter.

DEFINING PUBLIC PARTICIPATION

It is necessary to be clear at the outset about what we mean by public participation and to describe how we assess the evidence about it. In one sense of the term, all decisions in a democracy involve public participation. People participate through voting, expressing opinions on public issues and governmental actions, forming interest groups or holding public demonstrations to influence government decisions, lobbying, filing lawsuits to contest government actions, physically interfering with the execution of objectionable policy decisions, acting in partnership with government agencies, and even producing films, songs, and artistic events to mobilize public attention to issues. Defined broadly, public participation includes all of these forms. For example, Creighton (2005:7) defines it as "the process by which public concerns, needs and values are incorporated into governmental and corporate decision making." Indeed, public participation may be defined even more broadly to include citizens making and implementing decisions on matters of public concern directly and in ways that are largely or even entirely independent of government (Fung and Wright, 2001; Boyte, 2004). In the United States, citizens engage directly in environmental stewardship through a host of watershed councils and "stream teams," through "bucket brigades" that monitor air quality, through land trusts and forest councils, and in dozens of other ways (Knopman, Susman, and Landef, 1999; Sabel, Fung, and Karkkainen, 2000; O'Rourke and Macey, 2003; Weber, 2003; www.bucketbrigade.net).

Our focus is narrower. We are concerned with organized processes adopted by elected officials, government agencies, or other public- or private-sector organizations to engage the public in environmental assessment, planning, decision making, management, monitoring, and evaluation. These processes supplement the traditional forms of public participation noted above by adding direct involvement in executive functions that, when they are conducted by government, are traditionally delegated to administrative agencies. Often the role of the public is advisory, but increasingly there are experiments with shared governance and ongoing collaboration (e.g., Sabatier, 2005).[1] These processes may engage people at the earliest stages of environmental assessments, but they are most common as an immediate precursor to decision making. In some cases, public participation is focused on providing input to ongoing decisions about implementation. The focus in this study, then, is on participation that takes place in institutionalized decision processes. We recognize that when such processes fail to incorporate public concerns adequately, people can and do participate by going outside these organized venues. Indeed, the evolution of official mechanisms of participation is at least in part a response to participation outside the system.

The term "public participation," as used in this study, includes any of a variety of mechanisms and processes used to involve and draw on members of the public or their representatives in the activities of public- or private-sector organizations that are engaged in informing or making environmental assessments or decisions. Our interest is in mechanisms and processes other than the traditional modes of public participation in electoral, legislative, and judicial processes. These processes are mainly used in bureaucratic agencies charged with administering policies, although they may also be used in policy development.[2] "The public" may consist of organized interests, sometimes referred to as stakeholders; people selected by a systematic process to create a representative sample, as is done in survey research; people selected purposively to represent particular perspectives, knowledge bases, or interests; or individuals who themselves choose to engage in processes that are open to all. Which of these versions of "the public" are the actual participants can make a difference to participatory processes, because different selections are likely to represent different sets of interests or concerns in the process.[3]

A concrete example of how public participation is defined in agency regulations is the definition used by the EPA in its regulations related to the Resource Conservation and Recovery Act, the Safe Drinking Water Act, and the Clean Water Act:

> 40 CFR§25.2(b) Public participation is that part of the decision-making process through which responsible officials become aware of public attitudes by providing ample opportunity for interested and affected parties to communicate their views. Public participation includes providing access to the decision-making process, seeking input from and conducting dialogue with the public, assimilating public viewpoints and preferences, and demonstrating that those viewpoints and preferences have been considered by the decision-making official. Disagreement on significant issues is to be expected among government agencies and the diverse groups interested in

INTRODUCTION   *13*

and affected by public policy decisions. Public agencies should encourage full presentation of issues at an early stage so that they can be resolved and timely decisions can be made. In the course of this process, responsible officials should make special efforts to encourage and assist participation by citizens representing themselves and by others whose resources and access to decision-making may be relatively limited.

40 CFR§25.2(c) The following are the objectives of EPA, State, interstate, and substate agencies in carrying out activities covered by this part:

(1) To assure that the public has the opportunity to understand official programs and proposed actions, and that the government fully considers the public's concerns;

(2) To assure that the government does not make any significant decision on any activity covered by this part without consulting interested and affected segments of the public;

(3) To assure that government action is as responsive as possible to public concerns;

(4) To encourage public involvement in implementing environmental laws;

(5) To keep the public informed about significant issues and proposed project or program changes as they arise;

(6) To foster a spirit of openness and mutual trust among EPA, States, substate agencies and the public; and

(7) To use all feasible means to create opportunities for public participation, and to stimulate and support participation.

It is interesting to compare the EPA language with the regulations of the U.S. Department of Agriculture's Forest Service regarding its land and resource management planning processes:

36 CFR§219.9 Public participation, collaboration and notification. The Responsible Official must use a collaborative and participatory approach to land management planning, in accordance with this subpart and consistent with applicable laws, regulations, and policies, by engaging the skills of appropriate combinations of Forest Service staff, consultants, contractors, other Federal agencies, federally recognized Indian Tribes, State or local governments, or other interested or affected communities, groups, or persons.

(a) Providing opportunities for participation. The Responsible Official must provide opportunities for the public to collaborate and participate openly and meaningfully in the planning process, taking into account the discrete and diverse roles, jurisdictions, and responsibilities of interested and affected parties. Specifically, as part of plan development, plan amendment, and plan revision, the Responsible Official shall involve the public in developing and updating the comprehensive evaluation report, establishing the components of the plan, and designing the monitoring program. The Responsible Official has the discretion to determine the methods and timing of public involvement activities.

As these regulatory definitions illuminate, the language of public involvement varies with the history, purpose, and culture of an agency. In the case of EPA, public involvement is within the framework of carrying out specific statutes regulating use and protection of the environment. In contrast, the Forest Service has a broad multiple-use mandate and must seek to satisfy a broad range of perspectives and uses of natural resources and environmental qualities.

DIMENSIONS OF PARTICIPATION

For the purpose of assessing public participation processes across a large range of types of agency activities, it is important to distinguish several dimensions along which assessments and decisions can be participatory. In a classic paper, Arnstein (1969) defined a ladder of participation with eight "steps" that ranged from manipulation of the public through consultation, placation, and partnership to citizen control. Similarly, the International Association for Public Participation offers a matrix that describes a "spectrum" of processes commonly labeled public participation (http://www.iap2.org/displaycommon.cfm?an=5). It emphasizes "increasing level of public impact" as the key dimension and identifies five levels: inform, consult, involve, collaborate, and empower. Fung (2006) articulates three dimensions of participation: who participates, how participants communicate with one another and make decisions together, and how discussions are linked to policy or action.

In our work we elaborated on Fung's approach, identifying five dimensions:

1. who is involved;
2. when—at what points—they are involved;
3. the intensity of involvement, that is, the degree of effort made by the participants to be involved and by the government agency or other convener to keep them involved;
4. the extent of power or influence the participants have; and
5. the goals for the process.

We also considered how these five dimensions relate to the design of public participation processes.

Who Is Involved

"The public" in public participation normally refers to individuals acting both in their roles as citizens and as formal representatives of collective "interested and affected parties"—people, groups, or organizations that may experience benefit or harm or that otherwise choose to become informed or involved in an environmental decision (National Research Council, 1996).[4] These may include particular ethnic groups, children, affected neighborhoods, occupational categories, or other categories of individuals, groups, or organizations, some of which are inadequately represented in traditional policy forums. Although the label "public" often refers to individual citizens or relatively unorganized groups of individuals, our definition of public participation includes the full range of interested and affected parties, including corporations, nonprofit educational or advocacy organizations, and associations, and it also considers the roles of public officials, agencies, and scientists, the last acting as individuals or on behalf of organizations. The "who" dimension includes the variety of kinds of participants as well as their number, which may range from a handful to thousands in any single process.

Dewey (1923) defined the public as all those who would be interested in or affected by a decision. In the context of environmental decision making it is useful to make distinctions among these publics (U.S. Environmental Protection Agency Science Advisory Board, 2001; Renn and Walker, 2008):

- *stakeholders*—organized groups that are or will be affected by or that have a strong interest in the outcome of a decision;
- *directly affected public*—individuals and nonorganized groups that will experience positive or negative effects from the outcome;
- *observing public*—the media, cultural elites, and opinion leaders who may comment on the issue or influence public opinion; and
- *general public*—all individuals who are not directly affected by the issue but may be part of public opinion on it.

As discussed in Chapters 4-8, how much attention should be paid to involving each of these publics depends on the context. Often it is sufficient to include only stakeholders, but for some issues it is crucial to use involvement processes that integrate stakeholders and other segments of the public to ensure that the process is not, and does not appear to be, captured by organized interests that may not raise the full range of public concerns. As we note in later chapters, the breadth of involvement must be matched to the issue. Indeed, diagnosing who should be involved often requires more content-specific characterizations of the public than these four heuristic categories provide. It would be inefficient and a waste of time and money to include the full scope of public actors in all environmental controversies. But substantial financial, organizational, and institutional resources can also be wasted if the involvement process falls short of the expectations of the general public or of organized groups. Later in the report we discuss approaches that help diagnose what is appropriate in a particular context.

Points in the Policy Process

The public can be involved to different degrees in different aspects of a policy process. The schema developed in *Understanding Risk: Informing Decisions in a Democratic Society* (National Research Council, 1996) provides a heuristic that is useful in structuring our discussion, although it will not apply exactly to all processes; see Figure 1-1. The schema identifies nine points in the policy process: five stages or elements that precede and inform decisions, the decisions themselves, two activities that follow decisions, and a learning process that uses the consequences of past decisions as input to future ones.

It is much more common for government agencies to invite public involvement at some points in the process than at others. It is normal, and sometimes required by law, for federal agencies to invite public involvement in gathering information for making environmental decisions and in commenting on draft documents that synthesize that information (e.g., the National Environmental Policy Act of 1969, http://www.nepa.gov/nepa/regs/nepa/nepaeqia.htm) or in commenting on proposed decisions (e.g., the Administrative Procedure Act, http://www.archives.gov/federal-register/

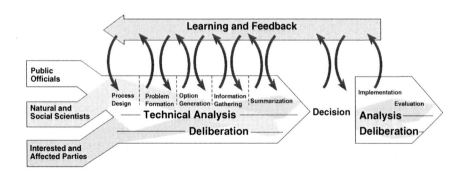

FIGURE 1-1 A schematic representation of environmental decision processes. SOURCE: National Research Council (1996:28).

laws/administrative-procedure/). It appears to be less common for them to invite broad public involvement in formulating the problem about which information will be gathered. Processes may be considered more participatory along the "when" dimension if they involve the public earlier in the policy process or at more points in the process.

Intensity of Involvement

Public participation activities can vary greatly with respect to levels of involvement. They can range from minimal opportunities to express an opinion verbally or in writing in open meetings, focus groups, or surveys that act as inputs to a process that lacks subsequent public involvement, to the highly intensive interaction, dialogue, contribution of information, and participation in shared analyses characteristic of regulatory negotiations or advisory committees. Similarly, convening organizations may exert greatly varying amounts of effort to solicit and maintain public input over time, with some processes consisting of a single meeting and others continuing over many months or years and including contributions to or ongoing learning during implementation.

Influence of Participants

The degree of public influence may vary from negligible, when public hearings are conducted only to fulfill a legal public comment requirement; to moderate, such as information exchanges or option development in a workshop setting; to an explicit requirement for consensus on recommendations, as for the decision phase under regulatory negotiation procedures (Arnstein, 1969; Fung, 2006).

Goals for Participation

Public participation processes vary in their goals. Some seek consensus on a policy choice, for example, in a regulatory negotiation process. Others have much more modest goals, such as identifying public values and concerns, gathering information for assessing environmental conditions, or shaping environmental analyses that will inform an administrative decision that is not likely to please all the interested groups. Participation processes may be convened as a way to educate or empower the public, or only to elicit information and concerns. As discussed further in Chapter 2, there is considerable dispute about what the goals of public participation should be, both among observers of the processes generally and among participants in particular processes.

Designing Participation Processes

The five dimensions of public participation define a sort of "space" of possible forms of public participation. The challenge facing those designing public participation processes is to find the appropriate "place" in this space for a particular process. Chapters 4 through 8 review the evidence about how the process and context of participation influence outcomes and offer some guidance on best practices for designing processes. As the evidence shows, there is no single "ideal" process. For example, if broad directions for policy are being set and trade-offs are being made among public values, it makes sense to have very broad engagement of all elements of the public. But if the policy issues are narrower and affect only a definable group, a less inclusive list of participants may be advisable. As another example, a process to set the agenda for issues to be considered in planning can be less intensive than one that is considering complex value trade-offs or seeking a policy consensus. Such differences are sometimes reflected in the laws and regulations that define the responsibilities of government agencies.

OBJECTIVES AND SCOPE OF THE STUDY

The Panel on Public Participation in Environmental Assessment and Decision Making was established to assess whether, and under what conditions, various forms of public participation achieve the promise of better and more acceptable assessments and decisions. In this book we identify the major challenges in achieving those goals and evaluate available evidence on ways to meet them. The book concludes with the panel's set of evidence-based recommendations for best practices in designing public participation and for future research.

Specifying the desired results is critical to evaluation. For example, it is easier to achieve the goal of a broadly informed debate, and easier to document that achievement, than it is to know that public participation has led to a better decision. Indeed, it can be quite difficult to define what is meant by better environmental decisions. Dietz (2003) has proposed six criteria that capture much of what is meant by a good environmental decision. Ideally, good decisions improve human well-being and environmental quality, are competent in the use of both facts and values, are fair with regard to both process and outcome, rely on processes that avoid errors in cognition and decision making, provide a chance to learn how to do things better, and are efficient in the use of scarce resources. Of course, there can be trade-offs among these criteria, and the relative importance of and feasibility of achieving each will depend on the context. Any public participation process is likely to look better in terms of some criteria than others.

In the nearly 40 years since the adoption of the National Environmental

Policy Act, federal, state, and local government agencies, Native American tribes, private corporations, and nongovernmental organizations have put significant effort into various public participation practices in the hope that these efforts would lead both to better and to more widely acceptable environmental choices (e.g., Bingham, 1986; Chess, Tamuz, and Greenberg, 1995; Caldwell, 1998; Chess, 2001). A growing but still fragmented body of experience and research has accompanied these efforts. In this book, we compile and assess the evidence from diverse areas of research to summarize the current state of knowledge and to build a more comprehensive framework for the future accumulation of knowledge.

This study is timely because of the confluence of several phenomena. First, many federal agencies and other organizations have been trying new forms of public participation, indicating their willingness to search for more effective tools of engaging the public. At the same time, in recent years a number of scholars and agency officials have expressed concern that participatory processes may degrade the quality of environmental assessments and decisions. Second, additional empirical studies about public participation have made it possible to build on both academic studies and practitioners' clinical knowledge. Third, the research, though of increasing quality, remains scattered. Knowledge has not been cumulative because researchers have not addressed a common set of theoretical and methodological issues, there is no shared vocabulary, evaluative criteria are poorly specified, and different streams of research do not cite each other. So there is both scientific and practical value in taking a broad, though necessarily not exhaustive, view of the literature. And finally, the need for effective methods of arriving at wise environmental policy decisions through fair and widely trusted processes continues to grow as the issues become more complex and the consequences of the decisions more significant. Many people hope and believe that sound public participation will be a key method for meeting this need, but others are skeptical. A synthesis of available knowledge can both advance basic understanding and provide practical advice to those who carry out environmental public participation processes.

Environmental public participation is a matter of concern nearly everywhere in the world. Indeed, the Aarhus Convention makes effective public input into environmental decision making an element of international law (United Nations Economic Commission for Europe, 1998). Although the panel drew on insights from other countries as appropriate, we have focused on the United States for three practical reasons. First, even this restricted scope is quite broad: research on environmental public participation in the United States is extensive and not well integrated. Second, it is unwise to presume that findings generalize across countries when the countries have very different formal systems of governance and political cultures. Third, this study was requested by several U.S. federal agencies that are concerned with conducting public participation within their own contexts and constraints. We believe our detailed assessment of research on the U.S. case can guide practice in this country while contributing to scientific discourse on participation that spans national boundaries. Integrating across domains of participation and across national contexts is an important challenge for future research on public participation.

This study focuses on environmental assessment, planning, evaluation, management, and decision making. We do not exhaustively review the excellent work on public participation in other policy arenas, such as urban policy (e.g., Berry, Portney, and Thomson, 1993; Sager, 1994; Fung, 2006) or biomedical policy (e.g., Abelson et al., 2003; Fleck, 2007; Furger and Fukuyama, 2007). We also acknowledge that there is a huge body of research, much of it in the gray literature of technical reports, on public participation in development project planning and implementation. Although we have considered some materials from each of these domains of study, we have not extensively reviewed them.

Our scope is nevertheless quite broad: the study covers issues ranging from toxic contamination to forestry and from local watershed management to global climate change. It covers a spectrum of activities ranging from comprehensive scientific assessments, such as the U.S. National Assessment of Climate Variability and Change and the Millennium Ecosystem Assessment, which did not deal with policy directly, to processes intended to produce specific policies, such as negotiated regulations, federal land management plans, and local permitting decisions, with many processes falling in between. Our coverage ranges from the global (climate and ecosystem change assessments), through the national (dialogues to build consensus on national legislation or to negotiate federal regulations), to the regional (deciding flow regimes for major river basins and developing forest plans), and the local (toxic waste problems). And it includes processes with very different goals and processes and that are in the purview of a diversity of agencies.

A wide range of approaches to structuring participation have been proposed (for overviews, see Renn, Webler, and Wiedemann, 1995; Rowe and Frewer, 2000; Kasemir et al., 2003; Creighton, 2005; International Association for Public Participation, 2006; Abels, 2007; Renn, 2008). This spectrum of approaches has its origins in analyses from the 1960s that described a "ladder" of increasing intensity and influence of public participation processes (Arnstein, 1969; see also Fung, 2006). When the stated objective is information exchange, the process is appropriately designed to merely elicit information about the perspectives of all relevant segments of the public. Sometimes the objectives are more ambitious, such as to produce recommendations or develop agreement among parties. Our review considers all of these goals and much of the diversity of mechanisms that have been deployed. However, the existing research does not always allow for empirical comparison of various closely related methods of public participation because the details of process are too varied. We thus focus on general aspects of the context and process of public participation and on a search for general principles that can be used in the design of specific processes.

SOURCES OF KNOWLEDGE

Environmental public participation has been a topic of increasing interest among scholars for decades. Some early writings were atheoretical, prompting the criticism that public participation was a practice in search of a theory (Wengert, 1976). Since at least the 1970s, however, researchers have been working to develop theories of environmental public participation (e.g., Barber, 1984; Dietz, 1987; Benhabib, 1992; Dryzek, 1994a; Renn, Webler, and Wiedemann, 1995; Sclove, 1995; Webler, 1995; Bohman, 1997; Hajer, 1997; Brulle, 2000; Fischer, 2000; Florig et al., 2001; Morgan et al., 2001; Hajer and Wagenaar, 2003; Renn, 2004).

There are at least three streams of theoretical and empirical work that have converged to form the current literature on environmental public participation. One stream (e.g., Dietz, 1987; Shannon, 1987, 1991; Dryzek, 1994a; Sclove, 1995; Webler, 1995; Brulle, 2000; Renn, 2004, 2008; Chilvers, 2005) flows from the ideas of Jürgen Habermas (1970, 1984, 1987) and his philosophical predecessors. Habermas in particular has emphasized the quality of deliberation as a key to successful decision making and has influenced ideas about how to evaluate the participation process (Dietz, 1987; Webler, 1995; Kruger and Shannon, 2000; Shannon and Walker, 2006). This line of research has also emphasized that deliberative processes can lead to public value changes in the face of novel policy challenges (Shannon, 1987).

Virtually all public decisions require dealing with differences among varied constituencies. Thus, a second stream that contributes to current research has its headwaters in conflict resolution. Conflict resolution theory offers additional lenses through which to understand the dynamics of environmental public participation and to arrive at useful prescriptions for practice. The conflict resolution field rests on the premises that conflict is inevitable and that it can be a positive force in human interactions (Simmel, 1955; Coser, 1956; Deutsch, 1973; Fisher and Ury, 1981; Susskind, Bacow, and Wheeler, 1983; Nicholson, 1991; Shannon, 1992b). Conflict resolution theory has influenced critical assumptions in current participation practice, such as that most conflicts originate in competing interests (Raven and Rubin, 1983) and that the ways in which individuals and organizations pursue their differences can affect conflict dynamics (Simmel, 1955; Coleman, 1957; Deutsch, 1973; Kriesberg, 1973; Wehr, 1979; Felstiner, Abel, and

Sarat, 1980-1981). Related lines of research have had significant influence on how public participation practices, including research on bargaining and negotiation and from game theory (e.g., Bartos, 1974; Gulliver, 1979; Fisher and Ury, 1981; Lewicki and Litterer, 1985; Lax and Sebenius, 1986; Bazerman et al., 2000; Raiffa, 2007), on the ways that human beings create meaning and misunderstandings in conflict and conflict-handling processes (Mather and Yngvesson, 1980-1981; Cobb and Rifkin, 1991; Ross, 1993), on issues of procedural justice and the dimensions of satisfaction sought (Thibaut and Walker, 1975; Lind and Tyler, 1988), on participation as a political process (Wondolleck, 1988; Cortner and Shannon, 1993; Cortner, 1996), and on the effects of interventions by third parties (Walton and McKersie, 1965; Deutsch, 1973; Bercovitch, 1984; Pruitt and Rubin, 1986; Donohue, 1991; Dingwall and Greatbatch, 1993; Kolb, 1994).

A third stream flows from the practice of environmental public participation and the need to draw lessons from that practice. Government agencies have increasingly implemented procedures to broaden public input to environmental decisions. During the 1970s and 1980s, federal, state, local, and tribal government agencies organized many hundreds of public participation processes (Bingham, 1986, 2003). Since the 1990s, the number of participation efforts has increased into the thousands. In many of these efforts, agencies experimented with methods to improve participation, often engaging scholars interested in evaluation, natural resources management, or risk management. The research literature that has developed around these efforts is a major source of evidence for this study. The earliest and most common type of analysis involves case studies examining one or a few specific applications, and the literature has grown to include studies of multiple related cases. Private corporations and nongovernmental organizations have also attempted to engage the public in environmental decision making, for example, in relicensing hydroelectric power plants, corporate social responsibility efforts, forest certification and forest management planning (Brun and Buttoud, 2003; Shannon, 2003), and efforts convened by nongovernmental organizations on climate change and control of invasive species, although some of these efforts are not well documented. These developments have made it possible to test theoretical arguments and proposals against experience.

These sources of insight and experience, combined with the judgment of experts or panels of experts, have provided the basis for numerous handbooks, guidelines, and other prescriptive documents (e.g., Pritzker and Dalton, 1990; Society of Professionals in Dispute Resolution, 1992; Canadian Round Tables, 1993; National Environmental Justice Advisory Council, 1996; World Bank, 1996; Presidential/Congressional Commission on Risk Assessment and Risk Management, 1997a,b; Western Center for Environmental Decision Making, 1997; U.S. Environmental Protection

Agency, 1998, 2000a,b, 2001; Creighton, 1999, 2005; Policy Consensus Initiative, 1999; Susskind, Thomas-Larmer, and Levy, 1999; Susskind et al., 1999; Institute for Environmental Negotiation, 2001; Organisation for Economic Co-operation and Development, 2001; International Association for Public Participation, 2006; International Finance Corporation, 2006). A set of principles recently issued by a federal interagency task force (Office of Management and Budget and President's Council on Environmental Quality, 2005) reflects current understanding drawn from theory, practice, and case studies; see Box 1-1. This excellent summary of current advice can be thought of as a series of hypotheses or research questions. They are among the hypotheses from previous syntheses that we have tested against evidence from a variety of sources in developing our conclusions.

The emerging data have not yet been organized within a common conceptual framework that allows for the consistent measurement of variables and formal testing of hypotheses that are desirable for scientific analysis. Enough progress in that direction has been made in recent years, however, to make it possible in this study to take a significant step toward conceptualizing public participation and its intended results and in developing evidence-based guidance that can improve practice over time through systematic empirical investigation.

This study draws on six sources of evidence regarding public participation:

1. theories of participatory democracy, public discourse, and conflict resolution;
2. basic social science knowledge on phenomena directly related to public participation (e.g., small-group interaction, public understanding of science);
3. experience of public participation practitioners;
4. case studies of individual instances of environmental public participation;
5. research comparing multiple public participation processes focused on similar environmental issues, similar mechanisms, or a single convening organization ("families" of cases); and
6. studies of multiple cases that cut across families.

The majority of environmental public participation efforts in federal agencies and most existing handbooks for practitioners have drawn mainly on the third and fourth forms of knowledge, with some reliance on the first. This is appropriate as, until recently, those were the best sources of knowledge available about public participation. Now, however, analyses based on the last two approaches have also become available (some conducted specifically in support of this study). In addition, we have looked further

> **BOX 1-1**
> **Basic Principles for Agency Engagement in Environmental Conflict Resolution and Collaborative Problem Solving**
>
> **Informed Commitment** Confirm willingness and availability of appropriate agency leadership and staff at all levels to commit to principles of engagement; ensure commitment to participate in good faith with open mindset to new perspectives.
>
> **Balanced Representation** Ensure balanced inclusion of affected/concerned interests; all parties should be willing and able to participate and select their own representatives.
>
> **Group Autonomy** Engage with all participants in the developing and governing process, including choice of consensus-based decision rules; seek assistance as needed from impartial facilitator/mediator selected by and accountable to all parties.
>
> **Informed Process** Seek agreement on how to share, test, and apply relevant information (scientific, cultural, technical, etc.) among participants; ensure relevant information is accessible and understandable by all participants.
>
> **Accountability** Participate in the process directly, fully, and in good faith; be accountable to the process, all participants, and the public.
>
> **Openness** Ensure all participants and public are fully informed in a timely manner of the purpose and objectives of process; communicate agency authorities, requirements, and constraints; uphold confidentiality rules and agreements as required for particular proceedings.
>
> **Timeliness** Ensure timely decisions and outcomes.
>
> **Implementation** Ensure decisions are implementable consistent with federal law and policy; parties should commit to identify roles and responsibilities necessary to implement agreement; parties should agree in advance on the consequences of a party being unable to provide necessary resources or implement agreement; ensure parties will take steps to implement and obtain resources necessary to agreement.
>
> NOTE: These principles were derived from discussions held in 2004 among senior staff from 16 federal departments and agencies at the request of James L. Connaughton, chair of the President's Office of Environmental Quality. These principles are consistent with collective professional experience and research in interest-based negotiation, consensus building, collaborative management, environmental mediation, and conflict resolution.
> SOURCE: Office of Management and Budget and President's Council on Environmental Quality (2005).

INTRODUCTION   25

into certain areas of behavioral and social science research than has typically been done in studies of public participation. Thus, it is now possible to deploy the wider range of methods, using contrasts and comparisons to expand and make more robust the understanding of environmental public participation. Since we use all six forms of knowledge, it is useful here to review the merits and limits of each.

Practical experience, case studies, and theory all are well suited for proposing factors that matter in public participation. The first two of these are also valuable for understanding the nuances of public participation processes and the ways that such processes develop over time. But these forms of knowledge are not readily codified, which makes it difficult to assess general hypotheses that are thought to hold across a variety of cases or to evaluate systematically the plausibility of explanations that differ from those offered by experience or case studies.

Case studies, usually of one or a few instances of public participation, are of great value for demonstrating that certain phenomena can occur, for understanding particular instances of participation, and for drawing comparative conclusions across a small range of contexts (Ragin, 1987; Ragin and Becker, 1992; McKeown, 2004; George and Bennett, 2005). Although case studies of public participation can provide "existence proofs" of relevant phenomena, they have not yet been well connected to common theoretical concerns, research questions, concepts, or methods of measurement. As a result, they generally provide less guidance for future research or for the practice of public participation than would be ideal. Moreover, studies based on single cases or a small number of cases are of necessity limited in the variation they exhibit across key variables. Strong conclusions from case studies require the analysis of many cases (or many repeated observations through time) that exhibit substantial variability in key factors.

Theory is useful for conceptualizing the contexts, processes, and outcomes of public participation, for identifying factors that should be considered as explanations of the outcomes, and for developing explicit hypotheses about relationships among contexts, processes, and outcomes. It is not useful for drawing conclusions until the theories are examined in light of empirical data.

In recent years, other sources of knowledge and insight about public participation have become available. These sources, in addition to a continued expansion of case reports, make it possible to check past lessons learned against information from new studies, new bodies of knowledge, and more sophisticated multivariate research methods. One can now "triangulate" in looking for robust findings verified by more than one method. The new and emerging knowledge complements existing knowledge and provides a more solid basis for advice to public participation practitioners.

Basic social science knowledge has been advancing on a number of topics of obvious relevance to environmental public participation, including individual judgment and decision making, group process, conflict management, and civic participation. Much of this knowledge has not yet been brought to bear on the design of environmental public participation processes. The panel examined several of these lines of research for their implications for environmental public participation. Panel members and staff prepared papers summarizing these implications, which were discussed at a public workshop held on February 3-5, 2005. Workshop materials are available at http://www7.nationalacademies.org/hdgc/Workshop%20Materials.html.

Comparative studies of families of cases—that is, cases with similar content or purpose—are a relatively recent development in environmental public participation research. They now include studies of watershed partnerships (Duram and Brown, 1998; Sommarstrom and Huntington, 1999; Leach, Pelkey, and Sabatier, 2002), forest management (Gericke and Sullivan, 1994; Williams and Ellefson, 1996), land use conflicts (Lampe and Kaplan, 1999; Rauschmayer and Wittmer, 2006), and cleanup of toxic sites (Aronoff and Gunter, 1994; Henry S. Cole Associates, 1996; Carnes et al., 1998; Ashford and Rest, 1999; Bradbury, Branch, and Malone, 2003). At least one study considers a family of cases defined by a similar participation format—regulatory negotiation (Langbein, 2005). Multiple cases provide for replication and for comparison of cases that vary on some dimensions while others remain constant. In addition, the use of common concepts across cases reduces the potential for ambiguity in findings.

Finally, we can draw on multicase, multifamily databases. By the end of the 1990s, a sufficient body of data on single cases was available to allow Beierle and Cayford (2002) to identify 276 documents describing environmental public participation in sufficient detail to be included in a database.[5] Beierle and Cayford coded these case reports on a large set of variables presumed to be important for assessing and explaining the outcomes of public participation, using common definitions for variables and a transparent coding system. Although ambiguity certainly exists in the case reports, databases such as that of Beierle and Cayford provide an invaluable resource for seeking generalities about public participation that cut across particular decision contexts.

It is worth noting that the evidence base for the present study does not include experimental field research involving case-control studies in which environmental assessment or decision processes are randomly assigned to two or more conditions (e.g., an experimental participatory process and a less participatory standard practice) and the results are compared. Such research is often considered the "gold standard" in policy evaluation because it can provide the strongest possible evidence of the causal efficacy of an intervention. Such field experiments could, in principle, be conducted, but the panel has not identified any in environmental public participation (although a few exist in the broader literature on public participation, e.g., Fishkin and Luskin, 2005).

However, experimental studies have their own limitations. The experiment, and especially the laboratory setting, may create a context for decision making different from that of participation processes in practical settings. This altered context may in turn alter the ways in which participants interact and make decisions (Lopes, 1983; Fischhoff, 1996a,b). Experiments in field settings reduce such concerns about what researchers call external validity, but they have other limitations. It is difficult in the field to hold constant all factors extraneous to the public participation process being implemented, and it is also difficult to conduct a large enough number of field trials to give confidence that these factors are randomly distributed across experimental conditions. Thus, there is almost always room for legitimate dispute about the import of results from field experiments in complex social settings. It is also worth emphasizing that given the state of knowledge in this field, it is not yet clear which variables are most important to investigate with rigorous research designs.

Because of the limitations of all methods of evaluation, social scientists draw inferences about complex social phenomena by triangulation across multiple methods of data collection, which can together provide robust evidence not vulnerable to the flaws of any single method. We have followed this strategy by seeking a convergence of evidence from multiple sources, some of which have not to our knowledge been included in previous assessments of environmental public participation.

This study examines all six of the sources of knowledge and insight identified, reconsiders the conclusions stated in past guidance documents for public participation, and presents a set of conclusions and recommendations based on our assessment of currently accumulated knowledge.

HOW WE CONDUCTED THE STUDY

The basic strategy of this study has been to consider possible conclusions and guidance for environmental public participation in light of all the available sources of knowledge and insight. Our presumption is that conclusions that are robust across various methods and sources of knowledge provide a stronger basis than previously available for offering the science-based guidance that government agencies and others need in order to improve the practice of public participation. A multimethod approach can also move knowledge forward by testing current beliefs against the best available evidence, a point on which we expand below.

The National Academies began by soliciting knowledge and insights based on practical experience. Two workshops conducted before the panel was formed invited practitioners from various levels of government, professionals in environmental dispute resolution and public participation, citizens with substantial experience in public participation, and researchers. Their ideas were solicited about the most important issues for the study to address. Some of the participants in the early workshops were appointed as members of the study panel when it was formed. Input was also sought from the original participants and other outside parties throughout the study via an electronic mailing list, a website, and open invitations to the panel's meetings and to the major workshop held in February 2005.

In order to seek broad public input, the panel was provided with internal funds and approval from the National Research Council (NRC) for efforts to elicit input beyond what is typical for the NRC. We think the quality of our work was greatly improved by the input we received; nevertheless, we wish we had been more successful in eliciting input from citizens with experience with environmental public participation. Understandably, most such individuals are not familiar with NRC studies and are unlikely to attend open meetings in Washington; our resources for supporting any travel to our meetings were limited. Despite reasonable efforts to make the study widely known and a public commitment that the panel would discuss all materials submitted via our interactive website, we received relatively little input. So, ironically, the question of how to effectively engage the public at appropriate stages in an NRC study remains an open one.

We commissioned a series of papers to synthesize the many sources of available data and discussed them within the panel and at the 2005 workshop. One set of papers included a draft conceptual framework for consideration by the panel (Stern, 2003), a review of practitioner handbooks (Zarger, 2003), and a summary of the findings of several existing case-family papers (Tuler, 2003).

A second set of papers sought insights for environmental public participation from basic social science knowledge on such topics as civic engagement and political participation (Markus, Chess, and Shannon, 2005), conflict resolution (Birkhoff and Bingham, 2004), interpersonal processes in decision-making groups (Stern, 2005b), decision analysis (North and Renn, 2005), and individual judgment and decision processes (DeKay and Vaughan, 2005). This social science knowledge is seriously underrepresented in past writing on environmental public participation.

A third set of papers examined selected families of cases. We invited researchers who had already synthesized knowledge about particular families of public participation cases to reexamine those cases in relation to a common set of issues and concepts, so as to make it possible to draw comparisons both within and across case families. These papers examined public participation in watershed management (Lubell and Leach, 2005), regulatory negotiation (Langbein, 2005), remediation of Superfund sites associated with nuclear weapons production (Bradbury, 2005), and regional and sectoral assessments under the U.S. National Assessment of Climate Change (Moser, 2005). In addition, we conducted a partial reanalysis of the Beierle and Cayford (2002) dataset (Dietz and Stern, 2005) to address important questions for this study. Finally, we drew on other independently produced case-family analyses (e.g., Ashford and Rest, 1999; Leach, 2005; Mitchell et al., 2006; National Research Council, 2007a) that were similar in scope and purpose to the ones we commissioned, using these as additional sources when evaluating hypotheses. The case families selected were the subset of all possible case families that the panel thought would best clarify key issues, given our resources and time constraints. More work comparing case families is certainly possible and is likely to be fruitful.

This report is the synthesis of all these sources of knowledge and insight. Comparing and synthesizing knowledge from these diverse sources and methods increases confidence in results, allows for testing of tentative conclusions from one approach for consistency with evidence of other types, and creates a stronger basis for developing practical guidance. It can also improve the basis for future research, moving toward a science of public participation that is increasingly cumulative and that contributes both to theoretical understanding of democratic governance and to future public participation practice.

GUIDE TO THE REPORT

Following this introduction, Chapter 2 considers the history of public participation in U.S. environmental policy and discusses the major justifications that have been offered for broad public participation in environmental policy decisions as well as the major arguments that have been proposed against it. These justifications and arguments provide hypotheses about the effects of participation that are examined in the remainder of the report. The chapter also considers when in a process evaluation is appropriate and identifies the three types of results that are used in this study as criteria of success: the quality of assessments or decisions, the legitimacy of those assessments or decisions, and improvements in the capacity of those involved to make good, legitimate assessments and decisions in the future.

Chapters 3 through 8 examine and summarize evidence relevant to hypotheses about the consequences of environmental public participation, focusing especially on the factors inside and outside the process that determine those consequences. Chapter 3 considers the most basic evaluative questions about environmental public participation: the overall degree to which public participation efforts succeed, and whether tradeoffs among the desired consequences are necessary, so that some can only be achieved at the expense of others.

Chapters 4, 5, and 6 consider how the practice of public participation affects the outcomes. Chapter 4 examines the effects of aspects of program management (e.g., setting goals, providing resources and organizational commitment, developing a realistic timeline). Chapter 5 considers alternative ways of organizing participation. It reviews the effects of such factors as breadth of participation, openness of design, intensity of participation, and influence of participants on the results of the process. Chapter 6 discusses the effects of the ways scientific analysis is integrated with public participation, which is always a special challenge in environmental assessment and decision making. It identifies the key challenges in achieving this integration and identifies a number of mechanisms and tools that have been used for meeting the challenges. These chapters identify evidence-based basic principles of good practice for participation.

Chapters 7 and 8 examine aspects of the context of participation. They show how particular contextual factors may make effective participation difficult and identify specific practices that have been used to help overcome these difficulties. Chapter 7 reviews the characteristics of the issue under consideration, including the nature of the environmental problem and the state of the science available to understand the problem. Chapter 8 considers the constraints faced by agencies in conducting participation and the characteristics of those who might participate.

In Chapters 3 through 8, we examine all the available sources of evidence, compare and weigh the evidence of various types, consider the limitations of each type of evidence, and identify conclusions supported by a convergence of evidence.

Chapter 9 summarizes our conclusions and presents our recommendations. It offers a "best process" for diagnosing participation contexts, choosing practices to address the difficulties they present, and improving the process over time. The chapter also presents our recommendations for further research.

NOTES

[1] In such cases, public participation begins to blend with commons management (National Research Council, 2002a; Dietz, Ostrom, and Stern, 2003) and with "citizen science" (Irwin, 1995). However, the literature from these two fields of research have not been integrated into the literature on public participation, and their implications for public participation are not clear. We discuss the implications of work on the commons further in Chapter 5.

[2] Environmental public participation processes may also be convened outside of government, for example, by a business or nonprofit nongovernmental organization or even by a previously unorganized group of affected individuals. We sometimes use the term agency to refer broadly to any

INTRODUCTION

entity or group of entities that may convene a public participation process, provide the resources for it to proceed, or take action based on its results.

[3]We recognize that public participation is sometimes organized merely for appearance or to comply with external requirements and without the intent to make use of public input, but the use or nonuse of public input is separate from the definition of participation.

[4]Some researchers make a sharp distinction between "stakeholder involvement" and "public participation" (English et al., 1993; Yosie and Herbst, 1998; Ashford and Rest, 1999). When this distinction is made, public participation generally connotes processes that do not "differentiate among different members of the public" (Ashford and Rest, 1999:1-3), and stakeholder involvement refers to processes that define participants in terms of the interests or organized groups they represent (English et al., 1993; Ashford and Rest, 1999); we do not follow this distinction.

[5]An additional 255 such documents lacked sufficient detail to be used in the analysis.

2

The Promise and Perils of Participation

Why should agencies engage the public as part of environmental assessment and decision processes? What is to be gained? And what are the costs and risks associated with public participation? In this chapter, as a basis for assessing the effects of public participation, we examine the arguments for and against public participation, including the U.S. legal mandates for participation. These arguments and expectations identify the results that people desire, expect, or fear from public participation and thus imply criteria for evaluation. The ideas reviewed in this chapter provide a framework for considering the evidence about participation, which we review in subsequent chapters.

Some arguments for participation rest on normative theories of democracy and collective action, some are based on ideas of what constitutes a high-quality decision, and some are grounded mainly in considerations of improving agency practice and the policy process. Several arguments critical of public participation question the basic logic of citizen participation in complex science-based issues (Rossi, 1997; Sanders, 1997; Collins and Evans, 2002, Campbell and Currie, 2006). Most of the critiques of participation, however, are grounded in the practical. Critics worry that participation in practice may not achieve the lofty goals articulated in theory and may actually impede good decision making. They offer three basic arguments: that the costs are not justified by the benefits, that the public is ill-equipped to deal with the complex nature of analyses that are needed for good environmental assessments and decisions, and that participation processes seldom achieve equity in process and outcome. Others argue that participatory processes tend to experience a set of pathologies that range from paralysis by endless deliberations to reaching only trivial results when trying to accomplish a consensus among stakeholders with conflicting values and interests (e.g., Sunstein, 2001, 2006). As Ventriss and Kuentzel (2005:520) state: "a consensus in the public sphere is like a transitory mirage, contingent on the constellation of actors who happen to rise to the surface of ongoing public conflict and debate."

It is useful to recognize at the outset that decision making on matters of environmental policy is intrinsically and appropriately a political process (Cortner and Shannon, 1993; Landy, 1993; Williams and Matheny, 1995). Environmental decisions always involve both public and private interests. Furthermore, the decisions are typically backed by governmental authority, so environmental policy always involves power relations in society. Such relations shape environmental policy, and environmental policy in turn reshapes power (Stirling, 2008). This recognition provides a context for understanding participation processes, the motivations for public participation, and the challenges to it.

Science plays a special role in public participation in environmental issues. Environmental policy decisions therefore should be—and in the United States by statute typically must be—informed by the best available scientific information and judgments. Because they are matters of public policy they should—and, again by statute, typically must—also take into account the knowledge, values, and preferences of interested and affected parties. Ideally, public input and good information and judgment are complementary. Interested parties can bring critical factual information and scientific analyses to the process, whether as scientists themselves, by employing scientists, or by contributing experiential, observational or traditional knowledge. Similarly, scientific analysis can be made more decision relevant when public values and concerns frame the questions being asked and the methods deployed. Ideally, thoughtfully structured public participation can make these choices explicit and examine their implications for public decisions. Scientific analysis on its own is an inadequate guide to determining how the risks, costs, and benefits of environmental decisions ought to be balanced or how they should be distributed across the public. Such decisions depend not only on factual information, but also on values and preferences and on interpretations of factual information (e.g., National Research Council, 1983, 1994, 1996). Even setting the policy agenda—deciding which environmental matters deserve public consideration and which do not—requires the integration of scientific analysis and public input. In a democracy, such decisions cannot legitimately be made without consulting the many groups in society. When the issues are of great significance and complexity, a democracy would be foolish to forego good science.

However, the best ways to pursue the ideal of integrating scientific analysis, values, and judgment and the extent and manner in which the public should be directly involved in doing so remain matters of debate. As noted above and discussed in more detail below, the challenges of public participation in administrative processes are so great that some have questioned the value of the enterprise. But, if public concerns are not adequately addressed in such processes, people can and do become politically involved outside them, through elections, lobbying, social movements, and judicial actions. The issue for policy, and for this report, is whether public involvement in these processes can be organized in ways that provide net benefits at acceptable costs. Our goal is to review what is known about public participation and to extract lessons from that knowledge that can guide such effective participation.

In this volume, we apply social science to the task of informing the continuing discussion about methods of public participation. Although simple prescriptions cannot be found, we think that choices of methods for participation can be usefully informed by empirically and theoretically grounded analysis of how approaches to public participation, deployed in different contexts, influence the results. Our assessment, like other scientific analyses, requires context-specific diagnosis and judgment before being translated into policy. There is no escape from values and judgment in making what are fundamentally political decisions. Consequently, any reasonably comprehensive examination of public participation in environmental decision making must take into account the political context and consequences of such decisions.

Thus we emphasize that the design of any public participation process reflects value choices and the political power of the players to influence those choices, beginning with the decision about what questions are the focus of analysis and deliberation (Thomas, 1995; Schneider and Ingram, 1997; King, Feltey, and O'Neil Susel, 1998; Walters, Aydelotte, and Miller, 2000; Feldman and Khademian, 2002; Wynne, 2005). Those design choices have the potential to advantage some interests over others, empower some and disempower others, and lend differential credence to some values, preferences, and beliefs over others (e.g., Bingham, 1986; Dietz, Stern, and Rycroft, 1989; Forester and Stitzel, 1989; Stirling, 2006, 2008). The advantage of grounding the design of public participation processes in lessons from scientific analysis of public participation is that it can help avoid unintended consequences and make more transparent the implications of the choices made. As subsequent chapters show, the research literature on public participation, while rapidly evolving, already provides sound guidance for the design of effective participation processes.

This chapter begins with a brief overview of the historical development of public participation in U.S. environmental policy management at the federal level. This history shows that public participation has been proposed to serve a variety of purposes, that there is a long record of contestation over what the proper purposes should be, and that disagreements about public participation continue. We then summarize the most commonly offered justifications for public participation in environmental decisions. These include normative justifications, derived from democratic theory and considerations of fairness, as well as substantive and instrumental justifications (Fiorino, 1990; Laird, 1993; Fung, 2006) for public participation. Many of these justifications are reflected in statutes, executive orders, and official practices. We then review arguments that public participation has adverse consequences that are rarely acknowledged in official statutes or pronouncements that advocate broader participation. We consider some of the most trenchant concerns about public participation: that it may fail to handle scientific information adequately, particularly about uncertainty; that it may fail to achieve objectives of fairness; that it leads to trivial results based on a weak consensus among stakeholders with conflicting interests and values; and that its costs outweigh its benefits. The final section discusses the kinds of results of public participation that have been considered important and the feasibility of measuring those results. This discussion sets the stage for our analysis of what happens in public participation processes and of which factors influence the results.

HISTORICAL DEVELOPMENT: LAWS AND AGENCY PRACTICES

In the United States, the tradition of direct public involvement in policy making traces back at least to the New England town meeting (Bryan, 2004). Public involvement in aspects of federal environmental policy is often traced to the new organizations and programs created in the 1930s under President Franklin Roosevelt's New Deal. For example, organizations of farmers took part in the development and implementation of agricultural policy (Daneke, 1983) and in the development projects of the Tennessee Valley Authority (Rossi, 1997). Both of these early processes confirmed the dangers of participation without standards or rules to govern it. Philip Selznick's classic book, *TVA and the Grassroots* (1949), charted how the TVA co-opted and manipulated local organizations to create the appearance of public support for agency policies, many of which were contrary to the interests of many people living in the region. Such early efforts, despite their limitations, pioneered institutionalized public participation in federal agency decisions (Acheson, 1941).

The Administrative Procedure Act (APA), enacted in 1946, set forth general procedures that all agencies must use in developing policy, promulgating rules, notifying the public and other agencies of their intentions, requesting public information and disseminating information to the public, and receiving comments from the public and other agencies (5 U.S.C.§§551 to 559, 701 to 706). This act specified in some detail the processes by which federal agencies should make decisions depending on the type of decision at issue (Section 553). Although the APA did not call for direct public participation at the point of decision, it recognized the right of the public to know about, contribute to, and monitor the actions of agencies (Section 552). The "notice and comment" requirements of the APA and the creation of the Federal Register set in place some of the fundamental requirements for active participation: knowledge of what kinds of decisions agencies intend to make, an opportunity to give information to agencies prior to their final decisions, the opportunity to comment on proposed agency actions, and the opportunity to seek judicial review if informal appeal to the agency for reconsideration of its actions was unsatisfactory.

Although the APA was an important milestone because it officially mandated norms for agency conduct (Daneke, 1983), it contained only the "notice and comment" understanding of the role of the public in government decision making. In terms of the decision schema of Figure 1-1 (in Chapter 1), the APA established requirements and procedures for public participation in the information gathering and feedback phases of the process, but it did not provide for participation in the other phases.

A more active model of public participation in government decision making was encouraged by Congress with the Revised Housing Act of 1954 and later by the Economic Opportunity Act of 1964, which sought "maximum feasible participation" in community development. In reality, however, such participation was feasible only for organizations or individuals with sufficient time, money, and other resources to enable them to participate in often-distant federal processes (Fiorino, 1989). Moreover, even when the government-organized participation in decision making occurred locally, as in the Citizen Action Programs created under the Economic Opportunity Act, public officials demonstrated a readiness and a capacity to constrain, obstruct, or derail participation initiatives they perceived to be incursions on their power (Kramer, 1969; Piven and Cloward, 1971; Strange, 1972; Greenstone and Peterson, 1973; Berry, Portney, and Thompson, 1993). However, the program has also been criticized as one that bypassed the "institutions of electoral representation" leading to "maximum feasible misunderstanding" (Moynihan, 1969; see also Walinsky, 1969). The debate about public participation in antipoverty and community development programs was and is intense. It is an important element of the context in which environmental public participation evolved.

Increased public involvement in the decision process of federal environmental agencies was required beginning with the passage of the National Environmental Policy Act (NEPA) in 1969 and thereafter was mandated in nearly all other environmental and land management statutes (Fiorino, 1989). These laws were passed with the belief that participation could lead to better decisions that improved the environment and lead to a more just and prosperous society (Cramton, 1972; Fischer and Forester, 1993). NEPA required agencies to inform one another and the public of the expected environmental, social, and economic consequences of proposed actions (NEPA Section 102C(v)). President Nixon, by Executive Order 11514 (March 7, 1970, 35 F.R. 4247), expanded NEPA's public notice requirements by requiring agencies to:

> Develop procedures to ensure the fullest practicable provision of timely public information and understanding of Federal plans and programs with environmental impact in order to obtain the views of interested parties. These procedures shall include, whenever appropriate, provision for public hearings, and shall provide the public with relevant information, including information on alternative courses of action. Federal agencies shall also encourage State and local agencies to adopt similar procedures for informing the public concerning their activities affecting the quality of the environment (Section 2(b)).

The APA required agencies to make relevant documents available to the public, whereas NEPA assured access to public information from federal agencies and the opportunity to be heard after receiving this information and before decisions have been made. These requirements made it possible for members of the public to make their informed judgments known to agencies before decisions were made and thus potentially to have an influence on the decisions. However, they did not require agency decision makers, for example, to use the public input or explain why they did not.

The Council on Environmental Quality, in 1978, required agencies to engage in "scoping" processes early in an agency's assessment of the environmental impacts of options to ascertain what issues the public wished to see addressed in that assessment. Nicholas C. Yost (1979), then general counsel at the Council on Environmental Quality, stated:

> Every major affected group in the nation—from business to environmentalists to state and local governments—applauded the new regulations.
>
> Why this universal praise? I suspect it was, in part, because of the stress in the regulations, as in the process of their development, on seeking consensus. . . . [T]he new NEPA regulations will involve all those who are interested. The regulations make them part of the process. If all are part of the process, the Council believes, the process will be better. The results will be both more environmentally sensitive and less subject to disruptive conflicts and delays. . . .
>
> Don't wait, the new regulations say, until positions harden and commitments have been made to focus on the important issues and alternatives. Instead, involve all the necessary people from the beginning to see that the impact statement analyzes the information most significant to the ultimate decision. If the important issues receive attention at the outset,

later squabbles about the need for more study and new information can be avoided, along with increased costs and substantial delay.

The scoping process, often including a scoping meeting, will provide a forum for using consensus-building techniques to insure that all essential information is gathered before the ultimate decision is made. Real opportunities exist for those skilled in facilitating consensus to aid diverse participants in exploring the issues and agreeing on those to be studied. Then, when a decision is made on a particular proposal, it can at least be agreed that the analytical groundwork was complete and developed fairly.

The idea of involving the public in early scoping of a problem is often seen as one of the most important contributions of NEPA to public participation. As we note in later chapters, there is great value in engaging the public in problem formulation. This can sometimes broaden the range of alternative actions considered in ways that lead to better decisions. Of course, the participation can also identify approaches to a problem outside the scope of the convening agency, which can be frustrating for all involved. But considering a full range of options is often noted as a first principle of effective decision making. Administrative and judicial decisions under NEPA and other environmental laws have also broadened both the scope of government actions that are considered environmentally consequential and broadened the basis for the public to have "standing" to participate in both the courts and in administrative processes.

The Freedom of Information Act (FOIA) gave citizens stronger legal authority for meaningful participation by establishing the public's right to obtain information from federal government agencies, with nine exemptions, including national security. Enacted originally in 1966, FOIA states that "any person" can file an FOIA request, including U.S. citizens, foreign nationals, organizations, associations, and academic institutions. In 1974, the act was amended following the Watergate scandal to force greater agency compliance (5 U.S.C. Section 552, as Amended by Public Law No. 104-231, 110 Stat. 3048). It was also amended in 1996 to incorporate electronic information. These pieces of legislation all either specifically require certain forms of public participation or provide the public with access to information or opportunities to be heard. Agencies must comply with these requirements or face lawsuits. As is discussed below, however, agencies can interpret and implement the requirements differently and have done so.

Since the 1970s, laws have made concerns with fairness and balance explicit considerations in decisions by all federal agencies. The Federal Advisory Committee Act (FACA) of 1972 mandates standards and uniform procedures to ensure that advisory committees serve public rather than private interests. Under FACA, federal advisory committees must be "fairly balanced" in terms of points of view and have a formal charter that is reviewed by federal officials outside the agency creating the committee. At about this time, administrative law generally underwent a reformation in which fairness and equity were asserted to protect new classes of interests under an expanding government (Stewart, 1975). By the end of the 1970s, 80 percent of all federal programs and federal granting authority required some form of "public participation" (Rosenbaum, 1978; Advisory Commission on Intergovernmental Relations, 1979). However, these developments, which were meant to encourage transparency and openness, could also act as constraints on both the process and the outcome of public participation, a point to which we return in Chapter 4.

Legislation on environmental protection and federal natural resource management, beginning with the Clean Air Act of 1970, continued to expand the role of the public (including a citizen's right to sue under some statutes), and citizens for the past three decades have organized themselves to actively participate in federal environmental policy processes. These pressures have undoubtedly encouraged increased agency interest in the more intensive mechanisms of public participation. While the more passive "notice and comment" and "inform and involve" approaches to public participation often remain the official stance of federal agencies, several agencies have gone beyond the letter of the law in involving the public. For example, the collaborative licensing process of the Federal Energy Regulatory Commission is an exceptionally strong approach to empowering citizens in agency decision making. The public often has been actively involved in formulating policy, in making and implementing decisions, and sometimes in enforcement by filing "citizen suits" (Boyer and Meidinger, 1985).

The burst of enthusiasm for public participation in environmental assessment and decision making of the 1970s continued and spread to nearly every agency involved in environmentally significant activities. However, this expansion has not been monotonic and has led to expressions of concern about the value of public participation. We review such concerns and criticisms of participation later in this chapter. Here we note that in some agencies, the complexity of public participation and consultation has come to be seen as burdensome and possibly an obstacle to effective action.

Perhaps the clearest example has been around the federal management of land and ecosystems. The extension of legal requirements for participatory processes did not address issues that arose when agency responsibilities and jurisdictions overlapped, as is frequently the case for environmental assessments and decisions. Public participation processes often crossed the boundaries of agency-specific mandates. For example, decisions regarding forests, water, and wildlife inevitably require several agencies to coordinate their responsibilities toward crafting a joint decision (e.g., the

Interior Columbia Basin Ecosystem Management Project; see www.icbemp. gov). Such decisions usually engage multiple levels of governmental and nongovernmental actors in scientific assessment, planning, decision making, and implementation. Each actor has specific substantive as well as procedural duties to meet as well as distinct constituencies to engage and satisfy (Johnson et al., 1999; U.S.D.A. Forest Service, 2002). However, most people are unfamiliar with the boundaries of agencies' mandates, and this can be a source of frustration in participatory processes.

One response to this complexity is to increase the scale of the assessment and decision processes so as to allow for a broad level of agreement on strategic goals, implementation objectives, and evaluation criteria. Conflicts around how to use federal public lands in the 1990s provide an instructive example. Multiple agencies have made efforts to work at a landscape scale through bioregional assessments and thereby craft a broad policy that can guide agency-specific decisions as well as coordinate the actions of other landowners and resource users (Johnson et al., 1999). However, working with constituencies ranging from local governments to international environmental organizations in a single, multiagency process taxed the capacity of the administrative agencies involved, especially at a time when many of them were experiencing significant losses of personnel and resources (U.S.D.A. Office of General Counsel Natural Resources Division, 2002; Shannon, 2003). The Northwest Forest Plan, a landscape-scale, multiagency policy that affected the management of federal lands in western Washington, Oregon, and Northern California is the most prominent example of the approach. When the agencies relied on the plan in making more localized decisions, the courts nullified the approach by demanding that the agencies specifically consider localized and short-term consequences for each decision. They further admonished agencies with regulatory responsibilities that they had to affirmatively carry out these responsibilities for each decision rather than assume that an activity proposed by a land management agency that was consistent with the bioregional plan automatically complied with regulatory policy.

An ebb and flow of concern with the efficacy of large-scale public participation has been one consequence of the difficulties perceived with landscape-level processes. A series of reports providing guidance to the U.S. Forest Service exemplifies this trend. In 1998, the Secretary of Agriculture charged a committee of scientists to craft a new conceptual framework for planning for the 21st century. The committee's report, *Sustaining the People's Lands*, proposed participatory processes that were highly collaborative with other governmental and nongovernmental stakeholders in order to improve both the quality of the decision and its implementation capacity (U.S.D.A. Forest Service, 1997). The Forest Service wrote new planning regulations based on the report that were published in November

2000 (36 CFR Part 219), just before the George W. Bush administration took office. Although everything in these new regulations was a part of current practice of the agency, many agency and nonagency observers were concerned that an increased emphasis on public participation would not yield timely decisions and effective planning and policy.

In November 2001, a new report, *Reflecting Complexity and Impact of Laws on a USDA Forest Service Project*, documented the legal complexity of project and operational planning (U.S.D.A. Forest Service Inventory and Monitoring Institute and Business Genetics, 2001). This report noted that there are hundreds of individual activities needed to make decisions and dozens of process interaction points. Agency actions are governed by regulations requiring public participation along the way, but the public can choose not to get involved until the very end or not at all. The chief of the Forest Service convened a team to examine these issues of legal and regulatory complexity. Its report, *The Process Predicament: How Statutory, Regulatory, and Administrative Factors Affect National Forest Management* (U.S.D.A. Forest Service, 2002:5), called out three problems as critical:

1. Excessive analysis—confusion, delays, costs, and risk management associated with the required consultation and studies;
2. Ineffective public involvement—procedural requirements that create disincentives to collaboration in national forest management; and
3. Management inefficiencies—poor planning and decision making, a deteriorating skills base, and inflexible spending rules, problems that are compounded by the sheer volume of the required paperwork and the associated proliferation of opportunities to misinterpret or misapply required procedures.

The Forest Service published new planning regulations in December 2005. They categorically exempt bioregional assessments and national forest integrated land and resource management plans from the NEPA process on grounds that no decisions about action are made through those processes. Thus, NEPA compliance with its requirement for public participation now rests at the project planning—operational—level where decisions directly affecting the land and resources are made. The 2005 rules (36 CFR Part 219) still contain much the same language regarding public participation and the need for a collaborative approach. Under these rules, national forest-level planning processes could still be highly participatory and collaborative, but freed from attention to detail and therefore less costly, more timely, and more flexible. However, the new regulations could also lead to a substantial reduction in public participation. As yet, there have been no studies of the new regulations.

Since 2000, some agencies seem to have retained or even increased their commitment to public participation, while in others, formal requirements and institutional mechanisms such as advisory councils remain in place but are given little funding and attention by decision makers. As the discussion above indicates, every agency, with its unique culture, leadership, and current challenges, may alter its responses over time to the challenges of effective and efficient public participation.

PURPOSES OF PUBLIC PARTICIPATION

Within the shifting legal context, agencies have considerable discretion concerning whom they involve, when they are involved, the type and intensity of involvement, the influence of participation on decision making, and the goals they seek from public involvement. Few studies have examined how agencies exercise this discretion or what determines the level of participation they choose (Yang and Callahan, 2007).

The issue of the purposes of public participation deserves highlighting. Public input can serve many purposes in a decision or assessment process and can be used at many stages in the process. Some purposes relate to improving the quality of assessments or decisions, some relate to increasing their legitimacy, and some relate to improving the decision-making capacity of the public and the agency. Public participation can in principle improve an assessment or decision in various ways. Box 2-1 presents several of these, organized around the two general objectives of quality and legitimacy and linked to the phases of the idealized decision process presented in Figure 1-1 (in Chapter 1) at which participation may be helpful.[1]

It is worth noting that, depending on the purpose that public input is serving, different inputs may be needed from different people. For example, the people needed to provide information on environmental conditions in a managed forest are not necessarily the ones needed to assess the value of various ecosystem services provided by the forest or to agree on a process for making management decisions. Thus, what is required for a good participatory process may vary with the purpose that the process is intended to serve. We return to this issue in Chapter 4.

Agencies may, within their discretion, be restrictive about public input, inviting it only as applicable laws require, or expansive, inviting and using public input at every point in the process if doing so is not legally prohibited (e.g., where it would delegate statutory responsibility). In exercising this discretion, agency officials may or may not be explicit in stating the purposes they intend public input to serve. This situation leaves considerable room for ambiguity, misunderstanding, and contestation over who should participate, how, when, and with what kind and degree of influence. For example, an agency may invite public input to a decision with the implicit understanding that the choice will be among three defined options. But

BOX 2-1
Possible Functions of Public Participation Within an Idealized Decision Process

Text in parentheses refers to steps in the idealized decision process of Figure 1-1 (in Chapter 1).

Improving Decision Quality

- Clarify the nature of the problem or problems to be addressed (problem formulation)
- Identify the set of possible decision alternatives (selecting options and outcomes)
- Identify the set of outcomes of concern (selecting options and outcomes)
- Gather information on the state of environmental systems (information gathering)
- Gather information on how environmental conditions are affecting outcomes of concern (information gathering)
- Gather information on how each decision alternative might affect outcomes (information gathering)
- Evaluate the credibility and certainty of the information gathered (synthesis)
- Consider the implications of available information for decisions at hand (synthesis)
- Assess the decision against its objectives (decision)
- Develop methods for evaluating results of decision (evaluation)
- Monitor results of decision (evaluation)

Improving Legitimacy

- Seek consensus on the problem to be addressed (problem formulation)
- Seek consensus on a process for conducting an assessment or informing a decision (process design)
- Identify and consider the outcomes that parties want to achieve or prevent (selecting options and outcomes)
- Identify the range of decision alternatives that parties want to consider and consider them if the agency has the discretion to do so (selecting options and outcomes)
- Gather information from the parties relevant to how each decision alternative might affect outcomes of concern to them (information gathering)
- Gather information from the parties on the credibility and certainty of decision-relevant information (synthesis)
- Seek broadly based agreement on decision (decision)
- Seek public acceptance of decision (implementation)
- Seek agreement on implementation strategies (implementation)
- Seek agreement on evaluation methods (evaluation)
- Monitor results of decision (evaluation)

some participants may favor an option that the agency does not consider to be on the table or even one that is outside its legal authority.

Because agencies and participants may come to a participatory process with diverse and sometimes contradictory goals for the process, the purposes of a process can be as much the foci of conflict as the environmental issue under consideration. Agencies need to acknowledge this source of conflict and be prepared to deal with it effectively (Niemeyer and Spash, 2001). Moreover, conflicts about the character of and goals for the participation process may require a different approach than conflicts about the substance of the environmental issue under consideration. For example, conflicts over substance can be addressed with methods for representing and comparing values, including argumentation and discussions of trade-offs among values. Conflicts about purposes, while also often reflecting deeply held values, are often best addressed by deliberations about the essential nature of the issue, the agenda of the meetings that will be held, and the potential influence the results can and should have on policy. We return to issues of process in Chapter 5.

Conflict about the purposes of public participation may arise and play out in many different ways. For example, invited public participants may not understand the legal limits to which an agency can delegate authority or its willingness to share responsibility in its range of discretion, so the participants may assume that their input will have more influence than is possible. Agency officials may not be clear about their purposes, and so may convey ambiguous messages to public participants with regard to which of the many steps in a decision process are being opened to public input and influence. They may be deliberately vague about the intended use of public input in the hopes of increasing acceptance of agency decisions while promising little in return. Or the agency officials who are convening a process may offer clear statements of their intended purposes, but may be overruled at the end of the process by higher-level officials who do not share those purposes.

Sometimes conflict among social groups or between social groups and agencies is an important element of the context in which the participatory process is taking place. Indeed, as we note below, a common justification for such processes is the hope that they can be effective in finding resolutions to such conflicts without the need to resort to other political processes for addressing conflict, such as demonstrations, lobbying, and litigation. Furthermore, when more than one administrative agency has an interest in and responsibility for an environmental assessment or decision, interagency differences in perspective and mission may lead to conflicts that influence participatory processes and the extent to which they influence decisions.

JUSTIFICATIONS FOR AND PROBLEMS WITH PUBLIC PARTICIPATION

Normative Justifications

As Fiorino (1990:239) put it, "the case for participation should begin with a normative argument—that a purely technocratic orientation is incompatible with democratic ideals." Public participation is intrinsic to democratic governance. However, there is no single theory of democracy and so no unitary theoretical basis for public participation. Rather, there is an array of such theories (e.g., Dewey, 1923; Barber, 1984; Habermas, 1984, 1987, 1996; Held, 1987; Dryzek, 1994a; Dahl, 1998; Shapiro, 2003). Taken together, they lay out various conceptions of what democracy is, articulate its justifications as a system of government and indeed as a way of life, explicate its shortcomings and inherent tensions, and consider how it should proceed in principle and in practice.

Some authors (most notably Schumpeter, 1942) argue that, in the modern nation-state, "democracy" means little more than regular, competitive elections involving institutionalized opposition parties and universal franchise. But many other theorists call for more extensive and meaningful engagement of the public in decisions that affect them. Indeed, Dewey (1923) defined "the public" as those who will be affected by a decision and thus should have some say in it. Many scholars have pointed critically to the gaps (some would say chasms) between the promises or aspirations of democracy and the facts on the ground (e.g., Bobbio, 1987). Despite their many important differences, however, most theories of democracy tend to converge on a few fundamental ideas, which may be subsumed under three broad headings: political equality, popular sovereignty, and (somewhat more controversially) human development (Dewey, 1923; Pateman, 1970; Rosenbaum, 1978; Dahl, 1989, 1998; Habermas, 1991, 1996; Sen, 1999; Young, 2000; Warren, 2001; Shapiro, 2003). Although major theorists use slightly different terms—for example, Gastil and Levine (2005), following Dahl (1989), emphasizes inclusion, effectiveness, and enlightened understanding—there is a remarkable consensus about these three elements of democracy.

Political equality is the tenet that every citizen possesses an inalienable right to participate on even terms in the making of public policies. Although this principle is unfulfilled in practice, it remains a broadly shared aspiration. Equality entails not only equal consideration in terms of "one person, one vote," but also equal opportunity to express preferences throughout the process of decision making on public matters and to shape the public agenda. It entails, as well, what Dahl (1989:141; see also Habermas, 1991) called the criterion of enlightened understanding: "adequate and equal opportunities for discovering and validating, in the time available, what his or her preferences are on the matter to be decided—opportunities for the acquisition of knowledge of ends and means, of oneself and others." Because of the importance and complexity of the issue of equality and the related normative issue of fairness, we return to this topic below.

Popular sovereignty, or self-government, is the principle that the authority for making and enforcing laws and rules under which citizens live derives from the consent of the governed themselves (Richardson, 2002). Taken together, equality and autonomy imply that democratic governance is a means of managing power relations so as to minimize domination (Shapiro, 1999). That virtually all systems of modern government that are classified as being democratic (including those at the federal, state, and local levels in the United States) involve elaborate divisions of political labor and responsibility among elected representatives and elected and appointed executives and judges complicates but does not abrogate the foregoing foundational ideas of democracy (e.g., Dahl, 1989).

The idea of human development follows a line of argument that stretches (at least in some important respects) from Aristotle through Jean-Jacques Rousseau, Alexis de Tocqueville, John Stuart Mill, and John Dewey. A number of modern democratic theorists propose that participation in democratic governance is not only a method by which citizens can advance their interests, but also an important means through which they come to understand their interests in the first place and how those interests relate to and depend on those of other citizens (Kaufman, 1960; Habermas, 1970; Pateman, 1970; Barber, 1984). The claim is essentially that participation in public life, as with other forms of experiential learning, is a means of human development; it is a means through which private individuals become public citizens (Pateman, 1970). Thus, democratic participation is a key element in personal development (Warren, 1992; Young, 2000). Sen (1999) moves this argument to the societal level. In his view, when a society provides widespread opportunities for people to exercise agency in shaping their shared future, not only is that conducive to development, it *is* development. "[T]he liberty of political participation," Sen wrote, is "among the *constituent components* of development" (1999:5, italics in original).

In practice, of course, everyone cannot participate in every decision, and few would care to try. Time and energy are finite resources for even the most avid citizens. Furthermore, there are some compelling reasons for citizens to be wary of official invitations to "get involved." Even so, the working premise of democracy is that the burden falls on those who would seek to delimit public participation to justify such limits, rather than on those who advocate more participation (Shapiro, 1996).

Theories of democracy have in turn led to theories of public participation. Renn and Schweizer (in press) and Renn (2008) provide extensive reviews of these and the theories of society in which they are embedded. They categorize the literature into six theoretical approaches, each linked to key ideas from substantial theoretical literatures. Table 2-1 summarizes them, along with the methods that might be deployed to implement them (Renn, 2008).[2] It illustrates that the process of participation may differ depending on the goals for the process. Thus, differing views about the goals can lead to conflicts about how to conduct the process, a point we have noted.

A functionalist approach emphasizes the importance of participation for strategic planning and adaptive social change. Neoliberal theories see participation as a way of eliciting public preferences and finding optimal compromises among interests. What has been called the anthropological or pragmatist approach emphasizes individuals articulating their preferences as citizens and reaching consensus based on those preexisting preferences. The deliberative or Habermasian approach seeks normative consensus via discourse and thus moves beyond the pragmatist view in positing that new norms and shared preferences can emerge from a participatory process. The emancipatory view envisions participation as a process by which the least powerful in society gain a voice in specific decisions and increase their capacity to have influence in the future. The postmodern perspective argues that deliberative processes should reveal power relations and thereby help reframe decisions.

These perspectives overlap in many respects. However, they offer somewhat different views of what can and should be expected of a participatory process and imply different approaches to and goals for the practice of participation. This taxonomy demonstrates the rich and complex character of normative arguments about participation. Advocates of different perspectives may find themselves in opposition about the likely value of a participatory process because of their differing priorities about the purposes of such processes.

Substantive and Instrumental Justifications

Officially sanctioned opportunities for direct citizen participation in governance proliferated markedly over the last half of the 20th century in the United States—and indeed throughout much of the world—even as participation in elections and related activities stagnated (Franklin, 2004; Franklin, Lyons, and Marsh, 2004; Geys, 2006). It has been argued (Roberts, 2004:315) that such opportunities will continue to expand "as democratic societies become more decentralized, interdependent, networked, linked by new information technologies, and challenged by "wicked problems" (Rittel and Webber, 1973). There are many substantive and instrumental justifications offered for public participation: for an entrée into the great variety of such justifications, see the reviews by Mendelberg (2002), Delli

TABLE 2-1 Six Concepts of Public Participation

Concept	Main Objective	Rationale	Models and Instruments
Functionalist	Improvement of quality of decision output	Representation of all knowledge carriers; integration of systematic, experiential, and local knowledge	Delphi, workshops, hearing, inquiry, citizen advisory committees
Neoliberal	Representation of all values and preferences in proportion to their share in the affected population	Informed consent of the affected population; Pareto-rationality plus Kaldor-Hicks (improvements)	Referendum, focus groups, deliberative polling, Internet-participation, negotiated rule-making, mediation, etc.
Deliberative	Competition of arguments with respect to criteria of truth, normative validity, and truthfulness	Inclusion of relevant arguments, reaching consensus through argumentation	Discourse-oriented models, deliberative round tables, citizen forums, deliberative juries
Anthropological	Common sense as ultimate arbiter in disputes (jury model)	Inclusion of noninterested laypersons representing basic social categories such as gender, income, and locality	Consensus conference, citizen juries, planning cells
Emancipatory	Empowerment of less privileged groups and individuals	Strengthening the resources of those who suffer most from environmental degradation	Action group initiatives, town meetings, community development groups, tribunals, science shops
Postmodern (reflexive)	Demonstration of variability, plurality, and legitimacy of dissent	Acknowledgment of plural rationalities, no closure necessary, mutually acceptable arrangements are sufficient	Open forums, open space conferences, panel discussions, public fora

SOURCE: Based on Renn (2008). Permission to reprint from Ortwin Renn, *Risk Governance: Coping with Uncertainty in a Complex World*, 2008, London: Earthscan Publishers, http://www.earthscan.co.uk.

Carpini, Cook, and Jacobs (2004), and Stirling (2008). The most important of these can be grouped under three categories, improving quality, enhancing legitimacy, and building capacity.

Improving Quality

Sound public policy must be based on both an accurate assessment of facts and an accurate assessment of public values.[3] Scientific or technical analysis is usually seen as the arbiter of facts in environmental assessment and decision making. But in dealing with environmental issues, both local context and the behavior of individuals and organizations matter substantially. There are more than a few examples of local knowledge in the hands of those who would not normally be called experts (i.e., those lacking official credentials) serving as a corrective to a scientific or technical analysis that misrepresented the local context in which it was being applied (e.g., Peterson and Stunkard, 1989; Wynne, 1989; Vaughan, 1993; National Research Council, 1996). Nor is the public's ability to strengthen the scientific and technical underpinnings of a decision always limited to local knowledge (Shannon and Antypas, 1996). Thus, public engagement can be essential for "getting the science right" (National Research Council, 1996:6). Scientific methods, such as surveys and economic valuation techniques can be of use for assessing public values and concerns, but public participation has been held to provide an essential complement to these methods (e.g., Gregory et al., 1993; Dietz, 1994; Dietz and Stern, 1998; Niemeyer and Spash, 2001; Ackerman and Fishkin, 2004; Rauschmayer and Wittmer, 2006). One reason is that the meanings of people's responses on contingent valuation surveys and other value elicitation instruments are rarely self-evident (e.g., Dietz and Stern, 1995; Brouwer et al., 1999; Clark, Burgess, and Harrison, 2000; Svedsater, 2003; Dietz, Stern, and Dan, in press).

Enhancing Legitimacy

Many federal agencies and other mission-oriented organizations that convene public participation processes see them as a means of making their decisions more broadly acceptable to the public and thus of helping them move forward with their missions. Moreover, with many governmental decisions, people expect to be consulted, or at least to have the opportunity to be heard. Ideally, public participation provides a mechanism for obtaining the consent of the governed in more specific ways than are possible with elections. In the ideal case, public participation is a form of democracy in action, and its results are likely to be widely accepted as legitimate (Nonet, 1980). Of course, it is possible for an organization to convene a participation process that has no effect on its subsequent actions. When corrupted, misrepresented, or insincerely applied, public participation may function as a form of co-optation rather than a democratic practice—a problem we discuss in more detail below.

Building Capacity

Many federal agencies operate with a vision in which agency personnel are continuously engaged with the public not only in making decisions, but also in implementing, assessing, and revising them. Ongoing relationships of this kind require building a certain level of mutual understanding and trust among all the parties engaged, and, conversely, building understanding and trust makes continuing engagement operate more smoothly. Thus it is argued that participatory activities, if done well, strengthen and improve ongoing relationships, with benefits to future decision making, assessment, and implementation activities. It is also argued that participation increases public understanding of science and scientists' and agency officials' understanding of public concerns, thus enabling future participatory processes to proceed more efficiently (Schwarz and Thompson, 1990).

Beyond offering opportunities for acquiring a better understanding of relevant information, participatory processes, if they are designed to do so, can provide a space in which participants develop their capacities to articulate their interests and concerns and also come to understand how their interests and concerns relate to those of others (Kruger and Shannon, 2000). That is, public participation need not be viewed exclusively as a means by which interested parties express their already-held views; it can also be a means through which those parties develop and refine their views and perhaps articulate, discover, and create shared interests (Fishkin, 1991; Gutmann and Thompson, 1996; Shannon and Walker, 2006). We discuss this as a normative justification above, but such changes also build capacity in a way that is of instrumental value.

PITFALLS

As just discussed, public participation is often described as a tool for enhancing democratic practice and improving the quality of environmental assessments and decisions. Although participation often fulfills these goals (see the following chapters), it is important to remember that participation can have a number of other consequences that are unintended and sometimes unwanted, even by its advocates (see, e.g., Mansbridge, 1983; Webler and Renn, 1995; Sanders, 1997; Schudson, 1997; Mendelberg, 2002; Mutz, 2002a,b; Sunstein, 2003; Delli Carpini, Cook, and Jacobs, 2004). Indeed some scholars have argued that public participation may degrade environmental decision making more often that it improves it (van den Daele,

1992; Breyer, 1993; Dana, 1994; Rose-Ackerman, 1994; Coglianese, 1997, 1999; Rossi, 1997; Sanders, 1997; Cross, 1998; Löfstedt, 1999; Sunstein, 2001, 2006; Durodie, 2003; Ventriss and Keuntzel, 2005).

These criticisms offer a strong cautionary note to the literature advocating public participation. The concerns fall into four broad categories, which we consider in turn. First, public participation can devolve to little more than political manipulation. Second, public participation may degrade rather than improve decision quality and especially the handling of scientific information. Third, public participation is too often unfair, inequitable. Fourth, public participation may yield trivial or even undesirable results at substantial costs in time, effort, and funds.

Political Manipulation

From the standpoint of participation advocates, some of the problems we review in this section may be considered as outcomes of processes gone awry. Some of these consequences, however, may be desired by some participants, such as an agency that is pursuing its own ends or participants who wish to slow decision making and weaken agency power (Ventriss and Keuntzel, 2005). There are four lines of argument regarding political pitfalls of public participation.

First, the perception that participation confers legitimacy to policies may lead an agency to initiate a participatory process simply for that purpose, with no intention of affecting its decisions. Contemporary public participation practices are often seen as more legitimate than the traditional "decide, announce, defend" approach to fulfilling the minimal requirements of the Administrative Procedure Act, perhaps because of greater trust in processes that are seen as more democratic. From the standpoint of democratic theory, building legitimacy and trust among all participants, including public trust of a decision-making agency, are appropriate normative goals if the agency is making active use of public input in shaping its assessments and decisions.

Agency officials are sometimes divided in their stance on public participation, with some officials sincerely committed to using public input and others viewing such input as having little value, even as they orchestrate the public participation process to gain the desired legitimacy. However, such orchestration usually does not hold for long (Burgess and Clark, 2006). Participants may come to trust an agency when a participation process is conducted with the best of intentions by those officials directly responsible for it, even if the officials who will make decisions ignore what is learned from the process. If agency assessors and decision makers elicit public input but the agency does not take it seriously, there may be a short-term gain in public acceptance at the expense of legitimacy in the longer run (Nonet,

1980; Löfstedt, 1999; Abels, 2007). In Chapter 4, we identify agency commitment to taking the results of participation seriously as a key determinant of successful results.

Second, in part because public participation is widely perceived as legitimate, participatory processes can be used to insulate an agency from legitimate external challenges, such as from the legislative and judicial branches as well as the public. For example, lawsuits may be successfully defended, or avoided entirely, if an agency has fulfilled the formal requirements for participation, even if the participation had no real influence. In addition, involving the public in a participatory process can keep people from acting outside the agency's control by absorbing their time and energy and even by building unwarranted trust (Selznick, 1949; Modavi, 1996; Murphree, Wright, and Ebaugh, 1996; O'Toole and Meier, 2004). People so involved may be less likely to lobby with legislative bodies, bring lawsuits, pursue their goals through other agencies, or otherwise deploy strategies of influence outside the formal public participation process.

It is reasonable for an agency to expect to avoid lawsuits, legislative action, and countermoves by other agencies if participation processes are used to shape an agency's assessments and decisions. But when a participation process is done pro forma, with no influence on the agency, it can subvert the legitimate influence of the public on government decisions (Bora and Hausendorf, 2006). If an agency ignores the public input it receives, the fact that it has organized a participation process can disempower and delegitimate public opposition, both by those who participated ("they've had their say") and those who did not ("they've had their chance"). In these ways, public participation has the potential to erode the ability of the public to recognize, articulate, and effectively advocate for their own interests (see Shannon, 1991, for a critical analysis of public participation in U.S.D.A. Forest Service forest planning). In some circumstances, it can co-opt, localize, and contain or channel conflicts that would otherwise influence agency actions and thus function as a way for an agency to exert control and engage in hollow public relations, rather than being truly responsive (e.g., Rosener, 1982; Mouffe, 1999).

Third, public participation may create or increase conflict rather than decrease it and entrench differences rather than resolving them. Many people may have given little thought to a public policy issue before a participation process begins, so differences of opinion or in willingness to accept a policy choice, such as those revealed in opinion polling, may not be based on careful reflection (Shannon, 1991). This is particularly likely for novel, complex issues like those common in environmental decisions and assessments. Theorists have long noted that participation can help shape people's values, beliefs, preferences, and opinions through discourse with others and by elaborating their knowledge (Dewey, 1923; Habermas, 1984,

1987, 1991; Dietz, 1987; Chambers, 1996; Dietz and Stern, 1998; Abelson et al., 2003; Ackerman and Fishkin, 2004). As participants think through the implications of an agency action and consider others' viewpoints, they may develop diverging rather than converging views of what is best (e.g., Sunstein, 2001, 2003; for reviews of the research, see Mendelberg, 2002; Delli Carpini, Cook, and Jacobs, 2004). Naïve differences of opinion may, through public participation processes, develop into thoughtful, well-argued disagreements (Stirling, 2008). This outcome can be very frustrating even to an agency that is open to being influenced by the public, because a lack of consensus gives no clear direction as to how to proceed. By informing the public, a participatory process may also generate opposition to possible agency choices, although before the participation process there was no opposition. This result may be seen as negative because it interferes with agency action, but from a broader perspective the effect may be simply to articulate conflicts so that they appear before a decision rather than afterward. Thus, the emergence of disagreement in a public participation process does not necessarily reflect an invalid or poorly conducted process (e.g., Morgan et al., 2001; Webler, Tuler, and Krueger, 2001). But there is no doubt that a poor process can lead to unnecessary and futile conflicts.

Fourth, public participation can decrease an agency's autonomy and control, thereby making outcomes less predictable. Organizations have imperatives to fulfill their external mandates, such as statutory requirements, as well as to meet internal organizational imperatives, such as keeping to timetables for action. Public participation takes time, requires an investment of resources, and often produces results that are messier than what might emerge from a purely internal agency process. These attributes of participation constitute perceived disadvantages from the perspective of the sponsoring organizations.

Decision Quality

One of the most critical concerns about public participation processes is that they may reduce the overall quality of assessments and decisions by introducing poor-quality thinking or reducing the effective use of science in public decision making. These issues are closely connected.

Concern with the quality of public understanding of information about environmental problems, particularly information about uncertainty, motivated one of the most influential calls for the use of formal risk analysis as a basis for technological and environmental policy. Starr (1969) argued that the views of the public should not be given much weight in technological and environmental policy because the average citizen does a very poor job of handling probabilities and contingencies, yet such probabilities and contingencies are central to societal decisions about environment and technology.

Influential psychological research on "heuristics and biases" in human decision making (e.g., Tversky and Kahneman, 1972; Kahneman, Slovic, and Tversky, 1982) suggests that, especially in the face of uncertainty, people may deviate substantially from the kinds of thinking that normative decision theory assumes. Deviations can be the product of errors in quantitative or probabilistic thinking, or they may reflect alternative ways of construing the problem that make other reasoning strategies seem more relevant (Epstein, 2000; Stanovich and West, 2000). Related work shows that public estimates of the risk of injury or death associated with various technologies do not match the estimates of experts or actuarial and epidemiological studies (Fischhoff et al., 1978; Covello, 1983; Fischhoff, 1985, 1989; Borcherding, Rohrmann, and Eppel, 1986; Slovic, 1987, 1992; Boholm, 1998; Rohrmann, 1999; Sjöberg, 1999; Rohrmann and Renn, 2000; Slimak and Dietz, 2006). If the public incorrectly interprets or misuses complex technical information or analyzes an environmental problem with different reasoning strategies than experts, the concern is that processes resulting from engagement with the public will reflect errors in reasoning (Cross, 1992, 1998; Breyer, 1993; Dana, 1994; Okrent, 1998; Campbell and Currie, 2006) or the adoption of less than optimal decision strategies (Futrell, 2003). Indeed, professionals working on risk policy often see "public ignorance" as a major source of conflict in environmental policy (e.g., Dietz, Stern, and Rycroft, 1989; Futrell, 2003).

However, several qualifications are appropriate for interpreting the meaning for public participation of the substantial literature on risk perception and decision making under uncertainty. First, the inability to handle information about uncertainty was initially demonstrated among highly educated subjects whose quantitative skills may be closer to those of scientists than to those of the average member of the public. Thus, difficulty in dealing with probabilities and contingencies is not restricted to the nonspecialist public (Fischhoff et al., 1981; Kraus, Malmforms, and Slovic, 1992; Barke and Jenkins-Smith, 1993; Slovic et al., 1995; Pultzer, Maney, and O'Connor, 1998; Sterman and Sweeney, 2002; Bramwell, West, and Salmon, 2006; Slimak and Dietz, 2006; Silva, Jenkins-Smith, and Barke, 2007). Second, a growing body of literature suggests that the chance of errors in interpreting probabilistic information depends on how the problem is framed (e.g., Gigerenzer and Hoffrage, 1995; Cosmides and Tooby, 1996; Gigerenzer, 1998). Third, recent social science research has demonstrated that people have a repertoire of reasoning strategies and routinely use automatic and rule-based processes for judgments under uncertainty. Both analytical and heuristic processes may make independent contributions to judgments for any given task (Stanovich and West, 2000; Ferreira et al., 2006). These lines of evidence suggest that analyses involving uncertain information require procedures to help people avoid the pitfalls of flawed reasoning, whether they are scientists or not. We return to this point below. Thus, it seems that for public participation processes to contribute to high-quality decisions, they need to be clear about what kinds of input to obtain from whom and how to use this input (Stirling, 2008).

The argument that the public has problems in dealing with uncertainty is relevant to decision quality only to the extent that improved knowledge is essential for coping with uncertainty. This may be true for characterizing the degree of uncertainty, but it is less obvious for making decisions under uncertainty. If the consequences are uncertain, science cannot provide a "rational" unambiguous answer about what options to choose (Keeney, 1996). This is particularly true for consequences for which robust estimates of probability are missing (Wynne, 1992; Aven, 2003). In these cases, environmental managers have to find a balance between too much and too little caution, that is, make a prudent judgment on the acceptable level of uncertain consequences (International Risk Governance Council, 2005). This judgment can be informed by technical knowledge, but is essentially dependent either on a trade-off between immediate benefits and future risks or on an explicit negotiation between those who benefit and those who are at risk (De Kay et al., 2002; Pellizoni, 2003). With uncertain science, participation may help to find a fair balance or a mutually acceptable trade-off between the extremes of too much and too little caution in environmental protection. Technical expertise is neccessary, but not sufficient, for confronting this dilemma.

In contrast to these arguments that participation may degrade decision quality, two types of argument claim that participation can improve decision quality. One emphasizes that locally grounded, contextually sensitive factual information that is often essential to apply scientific analysis to a specific context often comes from nonscientists. There are many examples of how public insights have been essential for accurate analysis of environmental issues (e.g., Wynne, 1989; Brown and Mikkelsen, 1990; Nordenstam and Vaughan, 1991; Brown, 1992; Coburn, 2005; Brown et al., 2006; McCormick, 2006, 2007a,b; Metzger and Lendvay, 2006). Public input can prompt the generation of new, essential technical information or questions and thus improve the technical adequacy of decisions (e.g., Jasanoff, 1996; Futrell, 2003). In many situations, such information is essential to developing a sound scientific analysis. Thus, to be fully informed, an assessment or decision may need ways to link scientists with members of the public who possess needed information in ways that enhance the quality of information available while preserving the integrity of the scientific method.

The second, and more commonly invoked, rationale for public input is that environmental problems always involve complex value trade-offs. Although scientific methods, especially if well linked to local understandings, offer the best way to establish facts, these facts will always be somewhat uncertain and their implications for action more so. Weighing the various costs, benefits, and risks associated with a choice always involves value trade-offs, including consideration of the proper weights to be given to uncertainties and to future impacts and benefits. This is a common theme in the literature on public participation (Corrigan and Joyce, 1997; Hagendijk and Irwin, 2006). Scientific analysis can facilitate understanding of the views of a broad segment of the public, helping individuals and groups estimate the consequences of different choices for different values (e.g., Gregory, Lichtenstein, and Slovic, 1993; McDaniels, 1998). But scientific analysis cannot resolve the value problems inherent in environmental assessment and decision making. So public participation can also help articulate value concerns and the consideration of trade-offs (Langton, 1978; Barber, 1984; Dietz, 1987; Ethridge, 1987; Burns and Überhorst, 1988; Dryzek, 1994a,b; Renn, 2004, 2008). Handling value concerns and trade-offs around a complex environmental issue is fraught with as much difficulty as handling information about uncertainty. One cannot expect individuals, whether or not they have scientific training, to perform well in examining these matters unaided. So the criticism that public participation may make a muddle of the facts carries some weight—and it can make a muddle of values as well. However, there is little evidence to support a belief that excluding the public will resolve these difficulties. On the contrary, experts and public policy makers experience the same problems of separating factual information from value judgments that participatory bodies do (Hammond, Harvey, and Hastie, 1992). This problem is by no means unique to participation (Horlick-Jones et al., 2007).

The arguments about difficulties in human information processing have been described by drawing a distinction between "System 1" and "System 2" thinking (Stanovich and West, 2000; Kahneman, 2003). In System 1, judgments are reached quickly, are mindful of the context of specific decisions, use simplifying rules (e.g., generalizations from past experience), and can be strongly influenced by the heuristics and biases described in the cognitive psychology literature. By contrast, in System 2 thinking, factors that are clearly related to the values at stake are carefully considered and weighed, and those that logically should be considered extraneous are held at bay. Everyone makes use of both systems, although for most choices, people do not engage in the great effort that System 2 thinking often requires.

The challenge for participation processes is to ensure that the collective assessment or decision process benefits from the experiential knowledge that influences System 1 and yet approximates the ideals of careful thinking characteristic of System 2—even if individual participants fall short of those ideals. Some methods of eliciting information from individuals, such as surveys, seem prone to elicit System 1 thinking (e.g., Dietz and Stern,

1995; Slimak and Dietz, 2006). Group processes are also often subject to biases, even when the decisions are important enough to justify considerable cognitive effort (e.g., Janis, 1972). The strategy for meeting the challenge of high-quality group thinking is to deploy processes that compensate for human weaknesses in both individual cognition and in group deliberation. Evidence from experimental studies of small-group processes (Delli Carpini, Cook, and Jacobs, 2004) and from case study evaluations (Roch, 1997; Vorwerk and Kämper, 1997; Beierle, 2000; U.S. Environmental Protection Agency Science Advisory Board, 2001; Rowe, Marsh, and Frewer, 2004; Rauschmayer and Wittmer, 2006) suggests that this can sometimes be accomplished and cognitive biases avoided.

Understanding Risk provides guidance for the overall public participation process that we think is still core guidance (National Research Council, 1996). It argues that rather than separating public deliberation and scientific analysis by a "firewall," they should have permeable boundaries, mutually inform each other, and be iterative throughout an assessment or decision-making process. This approach has come to be termed an "analytic-deliberative" process, or "analytic deliberation." *Understanding Risk* does not give detailed guidance on how to conduct analytic-deliberative processes. However, the public participation literature has been attentive to the complex issues involved in bringing together groups of citizens and having them interact in ways that minimize the threats to high-quality thinking posed by cognitive shortcuts and group process issues (Dietz, 1987, 1988; McDaniels, 1996; Renn, 1999, 2004). The issue is a major theme in handbooks for the practice of participation. In Chapter 5 we review what is known about how the practice of participation affects decision quality, drawing on the best information from case studies, accumulated practitioner knowledge, and insights from related fields, such as decision science and the study of small-group processes.

On one hand, public participation processes can indeed result in inept handling of information on uncertainty, misunderstanding of science, and clumsy assessment of public values. On the other hand, public participation may be essential to ground scientific analysis in local contexts, calibrate the treatment of uncertainty to reflect public preferences, and both inform decisions about public values and reform those values themselves (Chilvers, 2008). The devil lies in the details. We return to the challenge of integrating science and public participation in Chapter 6.

Fairness

One of the central normative goals of participation is to enhance the fairness of decision making. The notion that rational collective decision making must be both fair and competent derives from the theories of

Jürgen Habermas (1970, 1984, 1987; see Renn, Webber, and Wiedemann, 1995). Fairness, in this view, requires that all those who will be affected by a decision have a say in the decision (either directly or by representation); competence requires that the process take full account of all available information about facts and values. Public participation processes have been criticized on the grounds that they may not be competent, as discussed in the previous section. In this section, we discuss the criticism that they may not be fair.

Although there is nearly universal acceptance that in a democratic society decisions should be fair, there are myriad views on what constitutes fairness. Indeed, the question of what constitutes fairness has been a major theme of debate by ethicists in recent decades, inspired in part by the book *A Theory of Justice* (Rawls, 1971). We cannot resolve all these issues here. However, a few key points about fairness and participation inform our subsequent discussions.

It is important to distinguish fairness in participatory process, procedural fairness, from fairness in the effects of the ultimate decisions, distributional fairness (Young, 1993). From a normative standpoint, democracies should seek both fair processes and fair outcomes. In practice, though, it is much easier to specify fair processes than fair outcomes. It is difficult to specify fairness in outcomes because nearly every environmental decision produces winners and losers and because outcomes are multidimensional, so that a party that seems to be a winner from one perspective or over one time horizon may be a loser from a different perspective or over a different time horizon. Thus, a decision that looks fair to some participants can look unfair to others, and a decision that seems fair when it is made may look unfair in retrospect. Theories of justice attempt to provide guidance regarding the circumstances under which unequal distributions of benefits, costs, and risks may be considered fair (Baumol, 1986). However, theoretical guidelines, such as Rawls' (1971) proposal to judge outcomes ex ante under a "veil of ignorance," are difficult to implement with practical multiparty decisions, a point we return to below. It is not unusual for the parties to environmental decisions to disagree about whether the outcomes are fair and even about which of the many aspects of the outcomes are most important for evaluating fairness.

The panel cannot resolve such debates by recommending a particular method that agencies should use in determining how to balance the complex outcomes of their decisions. Experience shows that whatever calculus is used, plausible claims of unfairness may still be made. One reason fair processes are important is that they can help consider and perhaps resolve such claims. We recognize fairness in the outcomes of environmental decisions as a central societal objective, but our emphasis in this study is on the role of fair procedures in helping to achieve it. We seek participatory processes that the parties accept as fair and that help them consider the fairness of outcomes. At the same time, we are attentive to the principle that perceptions of fairness will vary widely. Indeed, perceptions of fairness in participation processes have been a more important area of research than fairness in outcomes (Keeney, 1980; Keller and Sarin, 1988; Linnerooth-Bayer and Fitzgerald, 1996).

Generally, we consider a fair decision process to be one in which all those affected by a decision have an opportunity to participate meaningfully (either directly or via representatives) and in which those empowered to decide take participants' views seriously. However, one would like more specific guidance on what this means if one is to organize a process of public participation. Psychological research has illuminated several characteristics of decision processes that people generally consider to be fair. These include the opportunity to voice opinions and concerns, neutrality of the forum, trustworthiness of authorities, and quality of treatment by authorities both formally and informally (Thibaut and Walker, 1975; Lind and Tyler, 1988; Tyler, 2000; Tyler and Blader, 2000; Blader and Tyler, 2003; Clayton and Opotow, 2003; Besley and McComas, 2005; McComas, Trumbo, and Besley, 2007).

In addition to being normatively desirable, fairness may also affect various practical consequences of public participation. For example, when decision processes are judged as fair, participants are more likely to see the outcomes as fair or just, or at least to accept them (e.g., Lind and Tyler, 1988; Tyler, 2000; Tyler and Blader, 2000; McComas, Trumbo, and Besley, 2007). Acceptance of the decision as resulting from a fair process is closely related to legitimacy, one of the major criteria we identify for assessing the results of public participation. In Chapter 5 we discuss how process attributes that are associated with perceived fairness influence legitimacy and the other key results of public participation. As the evidence shows, procedurally fair processes do quite well in terms of outcomes: they tend to produce better results in terms of the quality and legitimacy of decisions and in terms of the capacity of the participants for future decision making.

As we argue, public participation inevitably has a political dimension in the sense that it reflects and can alter distributions of power and influence. Indeed, for many advocates of public participation, the main goal to be achieved by participation is to enhance the power within the process of those who might otherwise have limited influence on agency assessments and decisions (e.g., Brulle, 1994; National Environmental Justice Advisory Committee, 1996; Fung and Wright, 2001; Fischer, 2005). From this perspective, participation processes are successful in part if they empower the disempowered and thus make the political process more fair. However, fairness of political process in itself does not ensure either high-quality decision making or the distributional fairness of the decisions the process advises.

From a different public policy perspective, the hope for public participation is that it will produce a distribution of power and influence among interested and affected parties that is more broadly accepted by the interested and affected parties and that contributes to improved democratic capacity. This may imply a standard of fair process as one in which the distribution of influence in the process is more even than the distribution of power in society at large.

The effectiveness of public participation processes in leveling the playing field has been questioned. Most participation processes make a point of involving "stakeholders." As Gastil (2008:192) notes, "in the public participation and management literatures, the term ['stakeholder'] has come to refer ... to persons who represent organizations, communities, or alliances that have a particular stake in a decision." In contrast, "the public" typically refers to the broader, relatively undifferentiated collectivity of unorganized individuals who may have some interest or be affected relatively indirectly by a decision. Clearly, this distinction is a matter of degree. Nevertheless, a focus on stakeholders creates the potential for well-resourced parties to use participatory processes as one more venue in which to overwhelm the broader public.

The extent to which this threat to fairness is realized depends to some degree on how well the interests of the broader public are "represented" in the process. But "representation" is more complicated than merely statistical or demographic representativeness. As Parkinson (2006:29) points out, "representation" turns on whether nonparticipants have themselves, through one means or another, *authorized* individuals to represent them, and not merely on whether the process organizers have selected a sample of participants that the organizers deem to be "representative." Selecting participants to "represent" the broader public is not in itself adequate reason for nonparticipants to regard the outcomes of participatory processes as being legitimate. In light of this, Parkinson (2006:34) concludes that a process that involves a (random) sample of participants "is only legitimate when the aim is information-gathering, or when it is part of a wider deliberative decision-making process that involves the people more generally." So again, assessing a participatory process requires careful thought about the goals of the process.

Some critics argue that a focus on stakeholders and officially selected representatives gives focused interests more weight than the more diffuse interests of the public (Cupps, 1977; Reagan and Fedor-Thurman, 1987; Lijphart, 1997; Joss, 2005; Bora and Hausendorf, 2006). In this view, participation by stakeholders distributes representation and influence disproportionally to the size of the affected populations or the importance of the interests. Analysts of pluralist societies show that the relative power of interest groups typically does not match the relative importance of their interests in society, but depends on such factors as exclusiveness of representation, availability of power and resources, and potential for social mobilization (Downs, 1957; Olson, 1965, 1984; Breyer, 1993). Some interests nearly always have greater influence on the decision-making process than others and use the opportunity of deliberation to influence the opinion-forming process and advance their specific interests on the agenda. Thus, public participation processes can easily mirror the power distribution in society rather than level it (Waller, 1995). This result seems particularly likely with such procedures as negotiated rule-making or mediation, in which identified stakeholder groups are given access to a decision-making process without further public engagement beyond the formal accountability required by The Administrative Procedure Act and other legal requirements (see Schoenbrod, 1983; Edwards, 1986; Baughman, 1995). A corollary problem is that many important interests that are widespread and important in the aggregate are not strongly held or advocated by any particular organized group. This is the classic problem of public goods or collective action (Olson, 1965; Ostrom, 1990). Such interests are likely to have little voice in a process that emphasizes engaging stakeholders and organized groups.

Clearly, there are limits to the degree to which a participatory process can overcome larger structural inequalities in a society. Historically embedded inequalities make it much harder for some groups to engage in participation, to be effective in expressing their views, to analyze the implications of alternative decisions for their values and interests, and to have their views taken seriously. A well-structured process may be able to help compensate for some of these inequalities, but it cannot make them disappear, and in some cases the compensation may be very limited.

All policies have unintended and unanticipated consequences. Existing inequalities make some groups more vulnerable than others to adverse consequences. Fair and competently conducted participation processes can improve the ability to identify such adverse consequences and devise strategies to mitigate them, but such anticipation and mitigation will always be imperfect. Such mitigation means paying attention to the matter of participant motivation, a point we discuss in Chapter 5. It is an axiom of politics that a "special interest" that stands to benefit greatly from a policy decision can typically prevail over the much larger, but unorganized, general public, even if the cost to the public outweighs the benefit to the special interest. The motivation for members (or representatives) of the special interest is substantial, while the motivation of each person is negligible in relation to the costs of becoming informed and participating.

There are various views of what constitutes a fair process in addition to those based on the distribution of power. In one view, all those interested in or affected by a decision should have a voice in the discussion. There is evidence that this standard is sometimes achieved in environmental dispute resolution, in which such balance is especially important (Cormick, 1980; Bingham, 1986; Moore, 1996). An alternative and more difficult standard is that a representative sample of the public should be included (Hagendijk and Irvin, 2006). This might be accomplished by true random sampling, electronic participation, or quota selection. Such an approach can be used to implement Rawls' (1971) concept of justice if a group of participants representative of the general public is asked to examine the issue at hand under a "veil of ignorance," as if they did not know how the decision would affect them. However, in trying to implement such an approach it is important to remember that public participation rarely engages policies that are wholly *de novo*. Rather, the issue at hand and the agency nearly always have a history, and that history has often led to advantages for some interests and groups and disadvantages for others. It is difficult and perhaps inappropriate to attempt to ignore those histories. A number of theoretical and ethical concerns with the Rawlsian approach to deliberation have been raised (Macedo, 1999). We are not aware of well-documented efforts to implement it in environmental public participation, so it is not possible to assess its implications in practice.

Research demonstrates that, even under excellent conditions, a true representation of the public is never accomplished, but it is possible to achieve high heterogeneity and diversity (List, 2006). Indeed, some processes are designed specifically to exclude stakeholders in order to allow latent, broadly held but seldom articulated public interests to emerge. When a representative sample of the public is engaged, keeping them engaged and getting them to learn about the issues at hand may require substantial effort and expense in comparison with engaging stakeholders.

In yet another view, a fair process is one that comprehensively represents the arguments about the issues at hand, not one that proportionally represents the population (see Habermas, 1984, 1987, 1989; Webler et al., 1995; Chambers, 2003). In this view, powerful actors belong in the discourse but should have no privileged status apart from the arguments they present. It depends largely on the concept of participation itself whether fairness is seen as representing common sense, diversity on viewpoints, all relevant arguments, or a proportional sample of the affected public (Renn, 2008).

Under any of these standards of fairness, fair participatory practice probably depends on improving the competence to participate of those who do not have the background or financial resources to discuss issues on an equal stance with well-informed and powerful people (Moore, 1986; Fischer, 2005). This is especially the case if empowerment is also a goal of participation (Forester and Stitzel, 1989). Responsibility for improving the competence of those with traditionally low levels of influence does not rest only in the hands of the organizers of the participation process. Although it is difficult to mobilize some parts of the population to participate in collective action, including engaging in public participation, such mobilization does takes place. A very substantial literature examines the ways in which the collective action problem is overcome in low-influence groups (e.g., Gamson, 1990, 1992; Diani and McAdam, 2003). While this literature has not been deployed to inform the design of public participation processes, it might offer some useful insights.

Trivial or Undesirable Results

The last major criticism of participation concerns the results. The claim is that the results of highly participatory processes are too often trivial, overprotective of certain interests or values, or lead to actions that are inefficient or disproportional to the threat or problem (Sanders, 1997; Cross, 1998; List, 2006). Many critics claim that people are either unable or unwilling to accept trade-offs and to search for efficient or cost-effective solutions (Zeckhauser and Viscusi, 1996). Participatory processes, in this view, tend to favor solutions that violate rules of efficient or cost-effective spending of public money (Cupps, 1977; Rosenbaum, 1978; Graham and Wiener, 1995; Viscusi, 1998). Critics argue that participation may aggravate environmental damage or impacts on human health because it focuses on a single issue and does not take into account that minimizing the impact of one problem can increase the impact of related problems (Perry, 2000). By pursuing priorities that the public demands, regulators are likely to spend time and effort on environmental threats that are relatively benign but highly visible in the public eye and neglect those threats that are not well known to the public but very potent in their consequences (Coglianese, 1999). In the long run, in this view, more people will suffer from future damages than necessary since the funds for safety and risk reduction are spent inefficiently.

A related argument is based on the costs of participatory processes (Rossi, 1997). This line of criticism expresses a concern that participation may disrupt the normal operation of agencies or representative bodies, consuming time and money and resulting in delay, immobility, and stalemate (Aron, 1979; Cross, 1998). Some analysts have claimed that the European style of closed-shop negotiation has been much more effective in regulating environmental risks than the adversarial and open style of the United States (Coppock, 1985; Weidner, 1993). The more people are asked to take part, the more time it will take to come to any conclusion. This argument holds that effective government rests on a limited opportunity to participate and that in the long run, participation does more to harm than help the environment.

Another criticism is that public deliberative processes lead to trivial results (Coglianese, 1999). This argument holds that the more public input is allowed to enter the process, the more likely that "window-dressing"—superficial outcomes—will occur. If all the participants have to find a common ground of agreement, the language of that agreement is likely to remain vague and the outcomes will lack specificity and clear direction. This argument is directed, of course, against deliberative procedures that require consensus.

Finally, one of the most sustained criticisms of public participation is that it requires a commitment of time and money so large that the costs far outweigh the benefits of participation (Krutilla and Haigh, 1978; Aron, 1979; Coppock, 1985; Weidner, 1993; Cross, 1998), especially given concerns with the outcomes reviewed above. From an agency's perspective, unless the input gained through the process is of high quality, the funds might better be spent on other activities. From the perspective of the public, unless the outcome of the participation is influential, other mechanisms of influencing the agency may be more cost-effective (Shannon, 1987). These concerns have led some researchers to explore methods of participation that provide added value in a timely fashion with a reasonable expenditure of resources (e.g., Chess, Dietz, and Shannon, 1998; Renn, 1999).

These criticisms are an antidote to naivete with regard to participation by identifying plausible ways that participation may go astray. It is useful to think of these criticisms in two ways. First, they raise the issue of what can or should be expected from a participatory process, considering that participation may have various purposes. For example, if the goal of a participatory process is for the agency to identify the concerns of the public but not to propose solutions or reach a consensus, concerns with fairness are appropriate, but issues of political manipulation, scientific competence, and efficiency would seem less germane. However, if the goal is to recommend policy, and especially if the recommendations will be influential, these latter concerns must be given serious consideration. Different goals enhance or reduce the importance of these problems. If the main objective is to identify the range of public concerns, all that is required is to survey the views of interested and affected parties: representation matters, but other concerns are more muted. If the objective is to reach a consensus, more stress must be placed on the ability of the participants to learn from each other and weigh arguments (Wynne, 1992; Tuler and Webler, 1995; Daniels and Walker, 1997; Beierle, 2000; Webler, Tuler, and Krueger, 2001; Welp and Stoll-Kleemann, 2006). Many experienced observers claim that given the right structure and facilitation process, a rational exchange of arguments and a balanced and efficient assignment of trade-offs can be and has been achieved (Webler, 1995, 1999; Renn, 1999, 2004). Ultimately, it is an empirical question whether these claims can be validated. The limited systematic evidence indicates that consensus-seeking participation processes tend to be more time-consuming and intense (Sherington, 1997; U.S. Environmental Protection Agency Science Advisory Board, 2001; Abelson et al., 2003). They also fail more often than processes that only measure public preferences or display the diversity of opinions. However, consensus processes seem to be better than inventory-oriented processes at meeting the expectations of the participants and the users (Beierle and Cayford, 2002; Hagendijk and Irwin, 2006; Abelson et al., 2007). Furthermore, evaluations of case studies of deliberative processes provide rather convincing evidence that the results of well-designed processes range far beyond the trivial or inefficient (Rowe et al., 2004; Rauschmayer and Wittmer, 2006).

When a process is intended to empower the participants and to bring attention and consideration to the needs and interests of those who are normally neglected in the public policy arena, the results are intended to be different from the general public's preference structure. Although this objective may not be widely shared, it would be unfair to criticize such processes if the result is inefficient or disproportionate from the standpoint of the entire society (Koopmans, 1996; Fung and Wright, 2001; Fischer, 2005).

The theoretical arguments extolling participation and the cautionary literature are both sources of hypotheses regarding the outcomes of participatory processes. Chapters 3-8 examine these hypotheses by considering in detail the scientific and experiential evidence regarding the outcomes of participatory processes and what shapes them.

CRITERIA FOR EVALUATION

Our discussion so far makes obvious the diversity of expectations—positive and negative, hopeful and cautionary, normative, substantive, and instrumental—that have been expressed for public participation processes. Each of these expectations is an implicit hypothesis about the effects of public participation processes or about how attributes of the processes or their contexts determine those effects. To make sense of the evidence, it is necessary to distill the very large number of variables and hypotheses in the literature down to a manageable set to use in our assessment. This section identifies a few key types of results that stand out in the literature and that we use as evaluative criteria in the chapters that follow.

When to Evaluate: Evaluating Across Stages of Implementation

Public participation processes go through many stages, from problem formulation and process design through decision making and implementation, to the ultimate effects of decisions on the environment and society. Thus, evaluation could potentially be done at many points in the process.

A previous report of the National Research Council (2005b) distinguished five classes of metrics for evaluation—input, process, output, outcome, and impact—that are roughly ordered in relation to the point in a process at which they can first be assessed. They have a plausible causal ordering in the sense that the attributes of an assessment or decision that appear earlier on the list can influence the later ones, but not vice versa. After sufficient passage of time, environmental assessments and decisions can be judged in terms of any of these classes of metrics.

The ultimate concerns of public policy are with *impacts* on socially important values, such as environmental quality, economic activity, the distribution of the benefits and costs across the population, and public faith in government. There is also serious concern with *outcomes* that depend on implementation, such as whether responsible agencies make new commitments or decisions; whether laws, regulations, or policies change; and whether actions are taken on the ground. These outcomes were identified by Beierle and Cayford (2002) as stages of implementation that intervene between the outputs of public participation processes and the ultimate impacts of decisions on environmental quality and other social goals.

Public participation processes do not influence such ultimate outcomes and impacts directly. Environmental impacts, for example, though potentially influenced by public participation processes, are also affected in very significant ways by many other factors. These include the implementation of policy decisions, as well as events in the natural and social worlds that affect environmental quality independently of any decision resulting from a specific public participation process. Because so many factors influence environmental conditions and other impacts, it is usually very difficult to attribute ultimate impacts to causes in a public participation process. Moreover, because of the long causal chain, any effects of public participation on environmental quality are typically indirect, mediated by implementation and other intervening events. Exceptions occur when the participants have the power collectively to implement their decisions, as in some watershed partnerships (Leach and Pelkey, 2001; Lubell and Leach, 2005), in negotiated rule-making (Langbein, 2005), and in many instances of collective governance of common-pool environmental resources, such as local fisheries, forests, or irrigation systems (see, e.g., Ostrom, 1990; National Research Council, 2002a). To the extent that the participants in a public participation process have the power to implement their decisions, ultimate impacts and outcomes can provide good metrics for evaluation. To the extent that the process is only advisory, however, the results that can be most readily linked causally to a public participation process are those that can be observed at or shortly after the end of the process.

In most contexts, then, it makes sense to distinguish (1) the effects of public participation on such immediate results of assessments and decisions, (2) the effects of these immediate results on implementation, and (3) the effects of implementation on environmental quality. Immediate results include the *outputs* of public participation, such as completing an assessment, reaching a decision, and making recommendations for action by the responsible agency or others. They also include *immediate outcomes*, such as changes in the attitudes, beliefs, knowledge, skills, and practices of the various participants (including scientists and the convening agency), and changes in relationships or mutual understanding among the participants at the conclusion of a participatory process.

Conclusions about the effects of public participation on subsequent outcomes that depend on implementation of recommendations from the process and on impacts must be built on inferences from information about the causes of success in terms of immediate results (point 1 above), as well as separate evidence about the effects of immediate results on implementation outcomes (point 2) and of implementation on impacts (point 3). Our main focus is on the effects of public participation on immediate results, because unless positive effects on these can be achieved, investments in public participation would not seem worthwhile.

There is evidence that public participation that is successful in terms of immediate results promotes good implementation, and that good implementation in turn promotes positive impacts on environmental and other socially valued endpoints (e.g., Langbein, 2005; Lubell and Leach, 2005). However, most of the research and analysis has appropriately focused on the first critical link in the causal chain, from public participation activities to the results immediately expected or desired from it. Evaluating the entire causal chain, though obviously important, will require a much more substantial investment in research on environmental decision making than has been made to date (National Research Council, 2005a). Some evaluation efforts are being organized under the auspices of the U.S. Institute for Environmental Conflict Resolution (information available at http://www.ecr.gov/multiagency/program_eval.htm).

Other fields of research have long dealt with the problem of inferring the quality of a decision process from the subsequent events it is intended to affect. For example, this is a central theme in discussion of research methods in international relations (e.g., Tetlock and Belkin, 1996; Stern and Druckman, 2000). A key insight from that work is that the effect of a decision on a complex system is most meaningfully assessed in comparison to a counterfactual situation, that is, the conditions that would have resulted if a different decision had been made or a different decision process had been used. Because of the difficulties inherent in specifying such counterfactuals, it is unwise to uncritically take environmental changes, or a lack of such changes, after a decision as evidence of the environmental impact or lack of impact of the decision. In addition, it is important to judge outcomes against reasonable expectations of how much change a decision, even if fully implemented, might make in the relevant impact measures within a given period of time. Finally, outcomes often entail the resolution of uncertainty that was present at the time of the decision and thus reflect additional knowledge that was not available to those making the decision. Inappropriate reliance on outcome knowledge in the evaluation of decision processes and decision makers (i.e., they "should have known it would turn out this way" when in fact they could not have) has been labeled *outcome bias* (Baron and Hershey, 1988; Hershey and Baron, 1992) and is closely related to *hindsight bias* in cognitive psychology (Fischhoff, 1975, 1982). Although it is reasonable to expect the quality of decision processes to be positively associated with the quality of outcomes over the long run (Frisch and Clemen, 1994), it is very difficult (and often improper) to infer the quality of an individual decision process directly from its long-run outcomes.

Given these considerations and the relatively greater amount of evidence concerning immediate results relative to implementation outcomes and impacts in most studies of environmental public participation, it is much more feasible to evaluate most environmental public participation processes on the basis of immediate outputs and outcomes than against implementation or impact criteria. The further down the list of implementation stages, the more difficult data collection and interpretation become. For these reasons, our focus in evaluating the evidence on environmental public participation is mainly on evidence that can be collected at or near the end of the processes studied, that is, on outputs and immediate outcomes. However, we acknowledge that further research on the effects of participation on implementation and impacts is certainly warranted, and some important progress in that direction is being made (O'Leary and Bingham, 2003; Dukes, 2004; Sabatier et al., 2004; Koontz and Thomas, 2006).

What to Evaluate: Types of Results

The published literature on public participation includes numerous typologies of results or evaluation criteria (e.g., Quinn and Rohrbaugh, 1983; Fiorino, 1989, 1990; Laird, 1993; Renn, Webler, and Wiedemann, 1995; Tuler and Webler, 1995; Steelman and Ascher, 1997; Rowe and Frewer, 2000; Webler, Tuler, and Krueger, 2001; Beierle and Cayford, 2002; Renn, 2004, 2008; Rowe et al., 2004; Abels, 2007; Blackstock et al., 2007), as well as many works identifying desired outcomes and potential pitfalls, as discussed above. There is no clear consensus among researchers or practitioners on which results are the most important. However, an examination of the literature suggests convergence on some of the key ones. In this book, we use a classification scheme that emphasizes three main types of results: *quality*, *legitimacy*, and *capacity*. We believe these types cover most of the key results of public participation processes that can be assessed soon after the completion of an assessment or decision process.

Quality of assessments or decisions corresponds closely to the concepts of substantive quality as described by Beierle and Cayford (2002) and competence in works following the tradition of Habermas (e.g., Renn, Webler, and Wiedemann, 1995). It has characteristics identified with good decision making in the field of decision analysis (e.g., Howard, 1966, 1968; von Winterfeldt and Edwards, 1986). A high-quality assessment or decision has these main elements:

- identification of the *values, interests, and concerns* of the agencies, scientists, and other parties that are interested in or might be affected by the environmental process or decision;
- identification of the range of *actions* that might be taken (for decisions);
- identification and systematic consideration of the *effects* that might follow from the environmental processes or actions being considered, including uncertainties about these effects, in terms of the values, interests, and concerns of interested and affected parties;
- outputs consistent with the *best available knowledge and methods* relevant to the above tasks, particularly the third; and
- incorporation of *new information*, methods, and concerns that arise over time.[4]

A number of attributes of outputs and immediate outcomes may be used as indicators of quality; see Box 2-2.

Legitimacy is related to the traditional concept of consent of the governed in U.S. politics. A legitimate decision is one that is fair, competent, and accountable to existing law (Susskind and Cruikshank, 1987; Wondoleck and Yaffee, 2000; Van de Wetering, 2006). The minimal definition of legitimacy is the narrow one of acceptance of the environmental assessment or decision as having conformed to standards of fair and legal process. An assessment or decision can be seen as legitimate in these terms even by someone who disagrees with it. More expansive concepts of legitimacy follow from normative concerns about fairness, for example, about the equitable distribution of the benefits and costs of public decisions or of influence on those decisions among segments of the public (see discussion above). It is difficult to put such concepts of fairness on scales for objective measurement because of differences of opinion about which distributions of cost, benefit, influence, etc., are most equitable or legitimate. However, good proxy indicators can be developed that reflect the extent to which claims of inequity are made after an assessment or decision and the extent to which such claims develop political traction or legal standing. Early

> **BOX 2-2**
> **Three Types of Results of Environmental Decision Processes with Illustrative Indicators**
>
> 1. Quality of Assessments or Decisions
>
> - Concerns expressed by publics were addressed in analysis
> - Information was added; more information was considered in the process
> - Technical analyses were improved
> - Outputs reflected a broad view of the situation that addressed all issues considered important by participants
> - Conclusions were based on and consistent with the best available evidence
> - Innovative ideas were generated for solving problems
>
> 2. Legitimacy of Process and Decisions
>
> Preexisting conflict was reduced or dissent clearly acknowledged and dealt with
>
> - Mistrust among participants, including government agencies, was reduced
> - Participants accepted the assessment or decision process as having conformed to standards of sound analysis and decision making, even if they did not agree with the final assessment or recommendation for action
> - The assessment or decision was widely accepted, even among nonparticipants
> - Participants went outside the process to overturn its results, for example, with legal challenges or attempts to influence legislation (a negative indicator)
>
> 3. Capacity for Future Decisions
>
> - Public participants became better informed about relevant environmental, scientific, social, and other issues
> - Participants and public officials gained a better understanding of each other
> - Public officials gained skill in organizing decision processes
> - Participants gained skill in participatory decision making
> - Scientists gained understanding of public concerns
> - Scientists developed, or committed to develop, new data or methods

claims of inequity are imperfect measures, though, because parties may be mistaken about the impacts a decision will have on them. However, claims of inequity that are widely considered implausible are unlikely to gain traction. Some illustrative legitimacy indicators are listed in Box 2-2.

Improved *capacity* includes having better educated and informed publics, publics more skilled at participating in environmental decisions, more competent and skillful public officials, improved methods for scientific analysis of environmental issues, better communication among interested and affected parties, better relationships among the various participants in making and implementing environmental decisions, improved institutional systems for environmental communication and decision making, and a more widely shared understanding of the nature of environmental issues and decision challenges. Box 2-2 lists some illustrative capacity indicators.

One additional kind of immediate result is an important link between outputs and implementation outcomes and is relevant to assessing public participation in many contexts we have examined. We refer to this result as *support for implementation*. Two examples illustrate the concept. Some participatory processes, including some watershed partnerships (e.g., Leach and Pelkey, 2001; Lubell and Leach, 2005) and many nongovernmental arrangements for managing common-pool resources (see, e.g., National Research Council, 2002a) produce as an output a tacit or formal agreement among the participants to continue to collaborate on implementing management plans in the future. Such agreements arguably predispose to implementation, but they still can be distinguished from implementation itself. Another example is regulatory negotiation (e.g., Langbein, 2005), which normally ends in a recommendation to an agency to adopt a specific regulation. Such participatory processes are reasonably judged more successful if the participants support the regulation they recommended by testifying for it, refraining from lawsuits or other blocking actions, and so forth. In both these examples, the participants bear some of the responsibility for implementation, and it is reasonable to judge the processes in part by how well the participants keep their explicit or implicit commitments for the implementation phase. Support for implementation is an early outcome that may reflect both the legitimacy and the perceived quality of the output of a process. When a public participation process results in such commitments, it is appropriate to judge the process in part by how well the participants keep the commitments.

It is worth emphasizing that although decision quality and legitimacy and changes in decision-making capacity can be analyzed as immediate outcomes of participatory processes, as we do here, implementation can also affect each of these outcomes at later times, sometimes profoundly. The most obvious example arises when policy officials "summarily dismiss a deliberative group's judgment," leaving participants more disenchanted than before and therefore less willing to accept or even participate in future similar processes (a review by Pyle, 2005:62, cites several studies that document this phenomenon; see also Bora and Hausendorf, 2006).

Using Indicators of Results to Evaluate Processes

The primary task of evaluation is to establish causal relationships among aspects of participation processes and aspects of results. Without experimental control, all such causal inferences are problematic. An additional inference problem concerns keeping the measures of processes and of results independent of each other. For example, if the quality of an assessment or decision (a result) is defined in part by the extent to which public concerns are considered in the analysis, processes that are legally or administratively required to respond formally to these concerns will almost automatically score higher on measures of how fully public concerns were addressed. It does not make sense to put much trust in such a statistical association. One would have more faith in an inference that considering concerns improves quality if the quality indicator were derived from scientific or judicial review of the quality of assessments. Thus, it is important for future research to take care in selecting indicators of results to ensure that they do not prejudge research hypotheses in this way.

Another problem arises if the quality of the process is measured by the degree of personal satisfaction of the participants. Although satisfaction is certainly one element to consider and is commonly examined in the literature, it can be influenced by aspects of the experience that are unrelated to quality. An example is cognitive dissonance (Festinger, 1957). Long and tedious processes can lead to higher degrees of satisfaction among participants who, having devoted much time and effort, justify their efforts by a belief that the process was successful. Participants who are not normally consulted may express satisfaction that is rooted in the opportunity to participate.

In contrast, representatives of organized stakeholder groups may judge processes according to the interests to which they are committed (Abelson et al., 2003). Although subjective indicators are problematic, there is little agreement on objective criteria to judge the quality of the process. Multiple criteria have been suggested in the literature (Quinn and Rohrbaugh, 1983; Fiorino, 1989, 1990; Tuler and Webler, 1995; Steelman and Ascher, 1997; Rowe and Frewer, 2000; Webler, Tuler, and Krueger, 2001; Beierle and Cayford, 2002; Renn, 2004; Rowe et al., 2004; Abels, 2007; Blackstock et al., 2007), but given the diversity of goals for participation, not all these criteria are appropriate for every participatory process. However, the three major criteria of quality, legitimacy, and capacity are broad enough to cover most of the important kinds of results and can be made concrete enough to help discriminate between different degrees of performance quality (see Box 2-2).

CONCLUSION

Writers on environmental public participation have generated a wealth of hopes, fears, and other expectations about the effects of public participation on a variety of important social and environmental values. As we note, this literature provides a wealth of hypotheses awaiting tests. The available evidence suffers from the diversity of concepts and the lack of agreed measures, creating a daunting task for anyone seeking clear answers to questions about the effects of public participation and the conditions under which particular results are likely to occur. In our judgment, considering the current state of knowledge, it makes sense to assess the evidence by considering three kinds of results: the quality of assessments or decisions, their legitimacy, and changes in the capacity of public participants, scientists, and agency officials to participate in similar decisions in the future. To draw inferences most confidently, it is important to consider these kinds of results at or soon after the end of the public participation process.

NOTES

[1] These are not the only useful functions public participation can perform. Research on methods for managing the use of common-pool resources identifies some "governance requirements" not listed in Box 2-1 that might be promoted by public involvement. They include dealing with conflict, inducing compliance with rules, and encouraging adaptation and change (Stern, Dietz, and Ostrom, 2002; Dietz, Ostrom, and Stern, 2003).

[2] Gastil (2008) and Parkinson (2006) also posit connections between theoretical framings for public participation and methods for conducting participatory processes. Unlike Renn's taxonomy, their work is not focused on environmental assessment and decision making.

[3] We acknowledge as a problem that assessment immediately after a process may be premature to the extent that the process itself helps shape participants' values and preferences regarding environmental issues (see, e.g., Gregory and McDaniels, 1987; Fischhoff and Furby, 1988; Fischhoff, 1991; Gregory, Lichtenstein, and Slovic, 1993; Dietz and Stern, 1998). In principle, one indicator of success, especially for emergent environmental issues, might be that the process helps shape public values and preferences on emergent issues. However, we see no way to determine in which direction public preferences should change as a result of successful public participation and so do not propose this type of indicator.

[4] These elements elaborate on the injunction offered by the National Research Council (1996) with regard to risk assessments, to get the right science and get the science right. The revised language here partly reflects our concern with decisions as well as assessments.

3

The Effects of Public Participation

As Chapter 2 shows, there are many claims about the positive and adverse effects of public participation in environmental assessment and decision making. Our goal in this chapter is to assess the evidence regarding the degree to which public participation achieves what its proponents hope it will achieve and the degree to which it yields the problematic results its critics expect. We address two specific questions:

- Do processes that are more participatory yield better results in terms of criteria of quality, legitimacy, and capacity than processes that are less participatory?
- Are there trade-offs among results, such that success in terms of one of these criteria compromises success on another?

There is the possibility of a pro-participation bias in the literature because researchers and practitioners predisposed in favor of public participation may be more likely to do research on participation and because reviewers with similar biases may be more critical of studies with negative findings. We have tried to take this possibility into account in reviewing the evidence on the overall effects of public participation. It is important to note, however, that findings on fine-grained issues, such as about which processes produce which kinds of desired results, are much less vulnerable to such biases. The literature shows considerable variation in the degree of success, depending on the context and the process used, so whatever publication biases may exist have not precluded reporting of less than ideal results. The literature on meta-analysis provides guidance on detecting and correcting for publication bias, and these methods can be used in the study of public participation.

DOES PUBLIC PARTICIPATION IMPROVE RESULTS?

We have already noted that the term "public participation" connotes a highly diverse set of activities. As discussed in Chapter 1, processes can be seen as more or less participatory along several dimensions, notably breadth (who is involved), timing (how early and at how many points in the overall decision-making process they are involved), intensity (e.g., the amount of time and effort participants spend and the degree of effort made by conveners to keep them involved), and influence. Although processes can be considered more participatory to the extent that they score more highly on these dimensions, the available research does not always make explicit distinctions among the dimensions. In reviewing the evidence, we comment on the effects of particular dimensions of participation when the evidence allows. We return in Chapter 5 to the issue of whether increases in participation along particular dimensions are associated with better outcomes. As we detail in Chapter 2, public participation processes also vary in their objectives (e.g., to make assessments or inform decisions, to reach consensus, or only to identify options and issues) and in the kinds of decisions they address. In Chapter 7, we examine how these differences may affect results.

This section shows that, on average, public participation is associated with better results, in terms of criteria of quality, legitimacy, and capacity. However, participatory processes can sometimes lead to undesired results that may be worse than what would have resulted from less participatory processes. The considerable variation in results is due largely to variation in the processes used to conduct public participation activities and in the extent to which these processes address the challenges posed by specific aspects of the context of participation. This evidence comes from a convergence of results from several sources.

Experimental and Quasi-Experimental Studies

An important source of evidence comes from experimental and quasi-experimental studies. Only a few experimental studies of participation processes have been conducted using control groups and random assignment to provide internal validity. Even fewer of these address environmental decision processes. Moreover, their practical value is unclear because it is hard to get an adequate sample of processes for statistical comparison, even if only one variable is manipulated. Quasi-experimental studies are a more common source of evidence on environmental decisions. These studies compare more and less participatory processes that occur naturally for similar assessments or decisions. Although such studies cannot ensure the level of internal validity provided by randomized control groups, they permit comparative observation in real-world settings.

An experimental study by Arvai (2003) shows that when people believe that a decision resulted from a public participation process, they are more likely to accept the decision, an indication of legitimacy. Arvai surveyed 378 individuals about a decision by the National Aeronautical and Space Administration to deploy a nuclear generator in space exploration. All individuals received the same information about the risks and benefits involved in using the nuclear generator. However, some were told that mission planning, including the decision to use the generator, was based on expert knowledge and experience, while others were told that decisions about mission planning, objectives, design, and the use of the generator were based equally on active public participation and on expert knowledge and experience. The individuals who were told that the decision incorporated public participation were significantly more supportive of the decision itself, as well as the process by which the decision was reached. They also expressed greater support for similar future missions, even though the two groups ranked risks from nuclear generator use similarly.

A number of studies by Fishkin and collaborators (e.g., Fishkin, 1997; Farrar et al., 2003, 2006; Fishkin and Luskin, 2005; List et al., 2006) used random samples of individuals in carefully planned participatory events, called deliberative polls, addressing a number of public policy issues. Participants were provided with balanced briefings on a policy issue, engaged in informal discussions in their everyday milieus, and participated in professionally facilitated small-group deliberations with opportunities to question experts. They were interviewed before and after the process. In some of these studies, control groups of individuals were interviewed without receiving briefings or participating in organized deliberations. These studies found that participation changed people's opinions on the issues and that people who engaged in deliberative polls were more likely to vote afterward, which we interpret as a positive outcome. Participants gained factual information about the issue at hand, as well as more general political knowledge. People who learned the most changed their opinions the most, and changes in opinion were unrelated to social status. Furthermore, participants were more consistent and predictable in their opinions after their participation. Although the deliberative groups did not come to consensus (after deliberation, they were as likely to become more polarized as less), there was an increase in the extent to which the order of preferred options became more consistent across participants.[1] These findings suggest that participatory processes increase participants' capacity through learning and increased motivation to participate, as well as developing greater consensus on at least some aspects of preference ordering.

A few studies have considered the effects of a deliberative process on responses to a standard willingness-to-pay survey. For example, Dietz, Stern, and Dan (in press) randomly assigned survey participants to express willingness to pay to plant trees to offset carbon emissions, either in a survey-only mode or after a small-group deliberation structured by the nominal group technique. Deliberation increased the number of issues considered in answering the question, had some effect on attention to social costs and benefits, and reduced the effect of personal predispositions on stated willingness to pay, but it did not change mean or median willingness to pay. There was no evidence of within-group convergence indicative of "group think." Overall, the results seem to indicate that even rather minimal deliberation enhanced the quality of decision making: more factors were considered, and there was less influence of personal predispositions.

A form of quasi-experimental evidence specific to environmental decision making is provided by a few studies that compared sets of decision processes in the same organizational context that used systematically different formats, one more participatory than another. Such comparisons lack experimental control in that decisions were not randomly assigned to more or less participatory formats. In one such study, Langbein (2005) compared six traditional and eight (presumably) similar negotiated rule-makings conducted by the U.S. Environmental Protection Agency (EPA). Langbein interviewed and collected survey responses from 152 representative participants in these rule-makings and found consistently better outcomes (see below for criteria) for the negotiated rule-makings, which are more participatory than conventional rule-makings on the dimensions of intensity and influence.

Across cases, negotiated rule-makings received more positive ratings from participants on 13 of the 15 dimensions studied, including quality of the scientific analysis, cost-effectiveness, ability of the rule to survive a legal challenge, and overall assessment of the rule-making process. They received worse ratings only on the cost of the rule for the respondent's organization. The negotiated and conventional processes were judged equal on the ability of the EPA to implement the rule. Participants were found to prefer negotiated "to conventional rulemaking mostly because they believe they get a better rule out of the process, and partly because some aspects of the process, but not all, work well" (Langbein, 2005:20).

Negotiated rule-making resulted in participants' judgments that more issues were settled and that the issues were clearer. Perceived complexity of the issues, however, was greater in the negotiated rule-makings, and complexity was associated with significantly more negative evaluations, although not enough to overwhelm the positive overall effects of the negotiated process. Although participants believed more issues were settled in negotiated rule-makings, there was no difference in the frequency of subsequent litigation. On the basis of participants' subjective judgments, their participation appears to have resulted in equivalent decision quality and legitimacy, possibly increased capacity, and overall greater satisfaction with the process.

Much of the empirical literature on public participation is based on surveys of participants. Although such research is valuable, results should be interpreted cautiously. One reason is that the statistical comparisons are made across individuals rather than across rule-makings. Analyzing individuals' responses increases the ability to find statistically reliable results, but the responses from participants in the same rule-making are not statistically independent (this lack of independence can be overcome in some analyses, such as Langbein's use of robust standard errors). Responses are also subject to "halo effects," in which a participant's judgment on one outcome variable colors judgments on others (see, e.g., Coglianese, 2003a,b,c). If many of the participants in the same process form a common judgment—or have a common halo effect—these judgments can be multiplied to generate a spurious conclusion. And as already noted, the attribution of causation with quasi-experimental designs is never as certain as it is with pure experiments.

A series of studies compared the results of watershed management planning in 20 estuaries operating under the National Estuary Plan (NEP), which sets guidelines and provides technical support for a participatory planning process, with results in 10 non-NEP estuaries (Schneider et al., 2003; Lubell, 2004a,b; Lubell and Leach, 2005). On average, stakeholders in NEP watersheds rated their estuary policies as more effective than did stakeholders in non-NEP watersheds (Lubell, 2004a); reported stronger perceptions of trust, fairness, and conflict resolution (Schneider et al., 2003); and showed a higher level of consensus (Lubell, 2004a). Lubell and Leach (2005) reported that the intensity of the participatory processes, as measured by an indicator of stakeholder teamwork, was positively related to all measures of success used in the study.

Although NEP estuaries did better on a variety of indicators of success based on participants' reports, they did more poorly on a behavioral indicator of cooperation among the participants, apparently an indicator of support in the implementation phase (Lubell and Leach, 2005). Possible explanations of this finding include that it is harder to change behavior than attitudes, that the management problems were more difficult in watersheds that entered the NEP (i.e., a selection effect), and that the NEP process yielded only symbolic progress. As with the Langbein review, the results should be interpreted with caution because the data compared individuals rather than cases of participation.

Multicase Studies

Many hundreds of case reports of environmental public participation can be found in the peer-reviewed literature and in reports that have not been subject to peer review. Because case studies are idiosyncratic in how they assess degrees of public participation and how they define successful outcomes, we do not consider single case reports by themselves as strong evidence in developing an overall evaluation of public participation. More useful are studies that examine multiple cases, some more participatory than others, using internally consistent definitions of the key variables. In addition to the studies already discussed, useful evidence comes from studies that examine the outcomes of multiple public participation processes that are not systematically different because of a specified difference in format, but that vary in degree along dimensions of participation and are coded consistently on such dimensions and on indicators of results.

The most extensive such study was by Beierle and Cayford (2002), who coded 239 cases into five categories from least to most intensively participatory, according to the mechanism used: from public meetings and hearings at the low-intensity end of the spectrum, through advisory committees not seeking consensus to advisory committees seeking consensus, and finally to negotiations and mediations. More intense mechanisms were strongly associated with high ratings on an aggregate success measure: less than one-quarter of the processes featuring public meetings and hearings were rated highly successful, compared with over 90 percent of the negotiations and mediations. Beierle and Cayford (2002:48) noted, however, that the more intensive mechanisms sometimes achieve consensus by "leaving out participants or ignoring issues"—they look more successful from inside the process but may not yield better results when the participation moves out to the broader society. Several other attributes of the participatory process had stronger influences than the mechanism used. These results are discussed in Chapter 5.

A series of studies of watershed partnerships, with participatory processes aimed at collaborative decision making and implementation (Leach and Pelkey, 2001; Lubell and Leach, 2005; Sabatier et al., 2005), supports the conclusion that when the participation was broader and more intensive, indicators of success in watershed management improved. For example, the studies found that participants' perceptions of human and social capital (an indicator of capacity) were more positive when there were more participants, when all critical parties were present, and when deliberation was more intense (in hours per month), although the first two effects were labeled as only "marginally significant." In separate analyses at the level of partnerships rather than of participants, the strength of agreement (an indicator of legitimacy) and the breadth of project implementation were both positively related to the duration of the process (other intensity variables were omitted from these analyses).

Leach (2006) reviewed the results of 25 empirical studies based on one or more cases that drew conclusions about the factors that were keys to success in public participation in the U.S.D.A. Forest Service activities. He coded each stated conclusion in these studies as relating to one of 21 "themes" that represented potentially important attributes of the process, the participants, or the context, and noted the number of studies supporting or detracting from the conclusion that each was a key to success. Three of the keys to success are reasonably interpreted as reflecting the intensity, breadth, or duration of participation. Having a "comprehensive and sustained process" was noted as a key to success in 12 studies, and none concluded that it detracted from success. Having "broad or inclusive participation" was a key to success in 10 studies, though 6 studies detracted from success. Typical of the negative results was the study by Floyd et al. (1996), which reported a negative correlation between number of parties and positive outcomes. Finally, seven studies cited continuity of participation as a key to success, and none saw it as a negative factor. Most of the studies highlighted the need for continued participation of the same Forest Service officials.

Bradbury (2005) reviewed 6 years of observations and surveys of eight sites at which the U.S. Department of Energy (DOE) had organized public participation processes to advise on the cleanup of hazardous materials, mostly associated with nuclear weapons production. She also reviewed a study of four similar sites managed by the Department of Defense (DOD). The review of all 12 studies showed that intensity of the public participation process was strongly associated with success, as shown by subjective indicators of quality, legitimacy, and capacity gathered both from observation and from participant surveys. Specifically, the two DOE sites where advisory boards met for 2 days bimonthly and had both standing executive committees and technical committees were rated highly on all indicators in both 1997 (survey data) and 2002 (observational data). The four sites that had the same committee structure but met for only 3-4 hours per month received ratings slightly above "medium" on all indicators in both time periods, and the two sites that lacked both the intensive committee structure and the intensive meeting times were rated at or near the lowest level on all indicators at both time periods. Outcomes were generally less satisfactory at DOD sites, where participation was less intensive than at DOE sites. (At two DOD sites, participation was little more than provision of information to the public.)

Representation of the parties had a more complex relationship to outcomes. The two DOE sites that were noteworthy for the diversity of viewpoints among the participants consistently scored at the top of the group on subjective measures of success, but the three sites where important parties remained uninvolved (activists in two cases, business and local government in the third) had average outcomes slightly better than the three sites with intermediate breadth of representation. Bradbury (2005) provides a more detailed discussion of the complexities of the representation issue in these cases.

Mitchell et al. (2006) summarized the results of a rich set of studies of "global environmental assessments"—efforts to evaluate the state of knowledge about particular global environmental phenomena, such as climate change and ozone depletion, the likelihood of various scenarios of future change, and the likely benefits and costs of alternative policies. The usual practice for conducting such assessments has not been participatory: many assessments strive to ensure the credibility of the science by involving only scientists and insulating them from interested publics and political actors. However, some global environmental assessments have varied considerably from the experts-only norm, and this variation was a source of some of the major insights of the review. The extent of stakeholder participation was strongly and positively associated with the perceived impacts of the assessments.

It is worth emphasizing that studies of global environmental assessments come from a research tradition quite separate from public participation research: they rarely cite or draw on major works on environmental public participation. Mitchell and colleagues developed their conceptual framework inductively from the insights of assessment practitioners, their own initial studies of a few cases, other case-comparison studies of global assessments that appeared during the life of the project (Andresen et al., 2000; Young, 2002), and previously published analyses from their research group (Jasanoff and Martello, 2004; Farrell and Jäger, 2006). They defined their topic in terms of "the influence of scientific information on policy" (Clark, Mitchell, and Cash, 2006:6); consequently, the main outcome of interest was the degree to which the assessments influenced the thinking of and particularly the actions of various policy-making audiences—an implementation-dependent rather than an immediate outcome, in the terms of this study. These studies did not set out to be studies of public participation.

The results, however, are strikingly consistent with those from research on environmental public participation. Mitchell and colleagues (2006:326) concluded that "participation explains much of the variation in the influence of our assessments." Influence depended not only on the scientific credibility of assessments, which was usually the primary stated objective of the assessments, but on their legitimacy with audiences and on their "salience," that is, on whether they provided results that audiences saw as relevant to their decision making. There were sometimes trade-offs among credibility, legitimacy, and salience, such as when excluding nonscientists sacrificed legitimacy and salience for scientific credibility or when assessments tried to achieve salience by going beyond what the science could support and thus sacrificed scientific credibility. However, most importantly for the present study, the research found that there are ways to avoid such trade-offs: these ways depend on broadening participation to provide access to local knowledge that is essential for a credible assessment and to improve stakeholders' understanding of the foundations of the assessment findings. Mitchell et al. (2006:324) concluded that "the effectiveness of assessment processes depends on a process of coproduction of knowledge between assessment producers and potential assessment user groups" and that "stakeholder participation fosters salience, . . . credibility, . . . [and] legitimacy" (325). Finally, the study emphasized the importance of capacity building, both among scientists and nonscientists, as a way to foster the needed coproduction of knowledge and to increase the influence of environmental assessment processes.

A recent report from the National Research Council (2007a) examined a set of "global change assessments" that overlap with those examined by Mitchell and colleagues and offered several recommendations for action based on judgments about the benefits and costs of public participation in these assessments. The report recommended that "appropriate stakeholders" be identified and engaged "in the assessment design" and noted "the advantages of broad participation," but noted the costs in terms of efficiency and the need to build capacity for "diverse stakeholders and assessment participants" (National Research Council, 2007a:S-7–S-8).

The results of the global assessment studies can thus be fairly summarized in the terms of the present effort as follows: when public participation involves the producers and users of environmental assessments in the coproduction of decision-relevant knowledge, it simultaneously improves quality, legitimacy, and capacity outcomes as indicated by participants' judgments.[2] It may be expected that by so doing, participation increases the likelihood that the assessments will lead to implementation. Capacity building is both an outcome of such participatory processes and a contributor to their success in the policy arena. Moser (2005), who examined the case of the First National Assessment of the Potential Consequences of Climate Variability and Change (a set of processes organized separately for different geographic regions and economic sectors) using the conceptual categories developed for the present study, similarly found that assessments characterized by more intensive involvement of stakeholders were of higher quality, legitimacy, and capacity as judged by participants in the assessment process. These assessment studies thus point to both the breadth and intensity of public participation as important influences on desired outcomes.

Tuler (2003) reviewed the findings of 11 studies (not discussed elsewhere in this section), each of which examined multiple cases (from 7 to 118) in a specific environmental policy area. He classified the factors associated with success in terms of competence, legitimacy, and capacity. Broad representation of interested and affected parties appeared as a contributing factor to each of these outcomes in the multicase reviews. It was found to contribute to the authors' aggregate measure of success in a study of 30 cases of forestry planning (Selin, Schuett, and Carr, 2000). Broad representation was also related to the indicators of decision quality and legitimacy used in Aronoff and Gunter's (1994) review of 7 cases of public participation in managing technological hazards and in the review by Henry S. Cole Associates (1996) of 11 cases of public participation in cleanup of hazardous waste sites. These studies also suggest, however, as did Leach's (2005) study in the forestry context, that different ways of implementing broad participation can have very different results (see Tuler, 2003, for a more detailed discussion). Finally, Williams and Ellefson (1996), in a study of 40 natural resource partnerships, found that representation of all stakeholders led to decreased resistance to the efforts of the partnership, a measure of support for implementation.

Practitioners' Experiences

Practitioners' judgments about the overall effects of participation are based on their assessments of the results of processes that they consider to be either more or less participatory. These judgments are reflected in numerous handbooks for practice (e.g., Carpenter and Kennedy, 1988; Pritzker and Dalton, 1990; Creighton, 1992, 1999, 2005; Canadian Round Tables, 1993; Doyle and Straus, 1993; World Bank, 1996; Presidential/Congressional Commission on Risk Analysis and Risk Management, 1997a,b; SPIDR, 1997; Policy Consensus Initiative, 1999; Susskind et al., 1999; Bleiker and Bleiker, 2000; U.S. Environmental Protection Agency, 2000a,b; Institute for Environmental Negotiation, 2001; Adler and Birkhoff, 2002; Bingham, 2003; Chambers, 2003; International Association for Public Participation, 2003; McKeown, Hopkins, and Chrystalbridge, 2003; Gastil and Levine, 2005; Leighninger, 2006; Pomeroy and Rivera-Guieb, 2006). Although their definitions of degree of participation are qualitative and not always explicit, and they are therefore almost certain to vary across practitioners, every one of these practitioner handbooks strongly supports two fundamental conclusions about environmental public participation from the practitioner's point of view: that making environmental decisions more participatory can yield improved results, and that such results are contingent on a variety of process and contextual factors. We discuss these factors more thoroughly in the following chapters.

Summary

These lines of evidence demonstrate that there is great variation in the immediate outcome of public participation. They show that under many conditions, including some that are likely to apply often in environmental contexts, processes that are more participatory along the dimensions of breadth, timing, intensity, and influence lead to improved overall outcomes. The evidence also strongly suggests that public participation processes can lead to undesired results that may be worse than what would have resulted from less participatory processes. The evidence on both counts consists largely of associations between aspects of process and aspects of results, as discussed in Chapters 4-6.

The strength of this evidence could be bolstered by complementing the existing literature with more experimental studies and carefully structured quasi-experiments that provide strong internal validity regarding cause-and-effect relationships, even if at some cost in external validity. Nevertheless, the available evidence converges on fairly consistent results across several methods of measuring participation and outcomes and across a wide variety of environmental assessment and decision contexts. Evidence from such varied sources and from varied families of cases that are quite similar on variables other than participation provide confidence that the observed associations are due in substantial degree to causal links between participation and outcomes. However, the very limited evidence going beyond immediate effects is mixed. On one hand, regulatory negotiations lead to greater participant satisfaction, but no fewer cases of litigation, than conventional regulations. On the other hand, studies of groups of environmental assessments seem particularly consistent in indicating that broad participation increases not only the scientific credibility of assessments, but also their effects on policy.

Of course, there are many cases in which participation processes have not managed to enhance quality, legitimacy, and capacity (see Coglianese, 1997; Carr, Selin, and Schuett, 1998; Imperial, 1998, 2005; Weber, 2003, among many others). Indeed, one of the bases of our analyses in the rest of the report is studies that show variation in the outcomes of participation processes and identify the reasons for that variation. So while our overall conclusion is that participation enhances environmental assessment and decision making, it remains critical to understand the causes of the considerable observed variation among outcomes. The issue for practice is to specify the conditions that favor successful outcomes and to seek them out or create them. In Chapters 4 through 8, we review the evidence on how conditions in the practice and context of public participation affect its outcomes.

ASSOCIATIONS AMONG RESULTS: CAN YOU HAVE IT ALL?

The evidence discussed in this section shows that the desired immediate results of public participation are positively correlated: one generally finds similar levels of success in terms of quality, legitimacy, and capacity. Available evidence supports with high confidence a conclusion that trade-offs among these types of results are not inevitable. Across a wide variety of environmental assessment and decision contexts, there are practices that can simultaneously promote all three positive results.

Government officials and critics of public participation sometimes express the concern that although intense public involvement may increase the legitimacy of decisions, it is likely to reduce their quality (see Chapter 2). The available evidence does not support either the hypothesis that such trade-offs are inevitable or the hypothesis that participatory processes that promote the legitimacy of an assessment or decision necessarily detract from its quality. Rather, the data strongly indicate that there are positive relationships among the various desired results, such that processes that perform well on one outcome measure are likely to perform well on other measures also.

Case-Based Evidence

The broadest relevant database comes from the work of Beierle and Cayford (2002) in their ratings of 239 public participation cases covering a great diversity of environmental decision contexts and processes on five outcome variables, which they called social goals: improving the substantive quality of decisions, incorporating public values into decisions, resolving conflict among competing interests, building trust in institutions, and educating and informing the public. In our terms, the first is a measure of quality, the second may relate to both quality and legitimacy, the third and fourth are legitimacy indicators, and the last is a measure of capacity. The study found that each type of result was positively correlated with each of the others at a statistically significant level, with a median correlation coefficient of 0.46 (range = 0.16–0.57). Beierle and Cayford inferred that "aggregate success" was a meaningful concept; their aggregate measure had correlations with the individual outcome measures that ranged from 0.56 to 0.73, although these part-whole correlations are somewhat inflated because the whole was created by averaging the parts. A reanalysis of that dataset for this study (Dietz and Stern, 2005) indicated that the five outcome variables could be represented as measures of a single underlying success factor based on standard factor analysis and scaling criteria.

The evidence from the case-comparison studies conducted for this study is consistent with the conclusion that different types of desirable results are positively correlated. These studies compared various outcome indicators across multiple public participation processes dealing with the same type of environmental problem. One example is the work, already mentioned, of Lubell and Leach (2005), who explored the effectiveness of environmental decisions in collaborative watershed partnerships involving 20 estuaries under the National Estuary Program (and 10 nonpartnership estuaries) and over 70 partnerships under the Watershed Partnership Project (WPP), combining and reanalyzing data from previous studies. In the NEP cases, indicators of legitimacy, such as perceived fairness, were among the strongest predictors of indicators of benefits to the watershed, project implementation, and degree of consensus and policy agreement, with various contextual factors held constant statistically. For the WPP cases, similar indicators were among the strongest predictors of level of agreement (an indicator of support for implementation) and of benefits to human and social capital (an element of capacity). Moreover, agreement was in turn a strong predictor of perceived effect of the program on the watershed. Bradbury's (2005) review of environmental restoration programs at eight DOE and four DOD facilities found similarly that sites that were rated positively or negatively on one type of result, whether by participants or researchers, tended to be rated similarly on others.

The review by Tuler (2003) conducted for the panel covered 17 case-comparison studies, most of them not summarized above and almost all of them essentially qualitative in the methods used. He found that several specific attributes of the participatory process that were conducive to one type of outcome, such as legitimacy, were also found to be conducive to other types of outcome, such as improved quality and expanded capacity, using the indicators employed in the studies. For example, broad representation of stakeholders was found to contribute to indicators of both legitimacy and capacity in participation about decisions regarding technological hazards (Aronoff and Gunter, 1994), Superfund sites (Henry S. Cole Associates, 1996), and forest planning (e.g., Selin, Schuett, and Carr, 2000). High-quality communication and quality of decision-relevant information were also associated with increased legitimacy, capacity, and decision quality in some of the families of cases. There was no case family in which a process attribute was positively related to one of the major types of results and negatively related to another. Tuler (2003) did conclude, however, that the intensity of the mechanism used may be negatively associated with legitimacy if intensity is achieved by methods that limit representation. Intensity was positively associated with competence, a measure of quality in our framework.[3] Thus, there may be conditions under which promoting success on one dimension can interfere with another dimension of success; however, these may result from specific process choices rather than an inherent incompatibility of outcomes.

In sum, the best available case-based evidence indicates that positive results in terms of any one process outcome of environmental public participation are usually associated with positive results on others and that negative associations are rare. The evidence strongly suggests that there are ways of implementing public participation that can promote multiple desired results together and, moreover, that when practitioners are successful, they are frequently successful on multiple dimensions at once. These correlations do not necessarily mean that the different types of results are causally interrelated so that efforts focused on producing one of them will necessarily produce the others. However, it has been suggested that some positive causal relationships exist.

For example, Lubell and Leach (2005) provide evidence consistent with one causal chain: capacity promotes legitimacy, which in turn contributes to decision quality. But their data do not provide strong evidence that building consensus causes better decisions for two reasons. Decision quality was measured by participants' judgments, which could have been influenced by feelings about the process as well as by characteristics of the decision themselves (the unit of analysis in the relevant regression analyses was the participant, not the case). Also, the analyses were based on causal modeling of data at one time rather than repeated observations over time. Thus, the data are consistent with Lubell and Leach's causal model but, as in other areas of research on public participation over time, experimental and quasi-experimental evidence would enhance understanding of causal processes.

Practitioners' Experience

Unlike researchers who collect data only at the end of a process, practitioners of environmental public participation observe these processes over time and often have observed dozens of such processes. Their observations provide a particularly useful window on the associations and possible causal connections among quality, legitimacy, and capacity.

There are three somewhat distinct traditions of practice in environmental public participation. One has roots in the practices of conflict resolution, negotiation, mediation, and game theory (e.g., Bacow and Wheeler, 1984; Goldberg, Green, and Sander, 1985; Bingham, 1986; Moore, 1986; Carpenter and Kennedy, 1988; Crowfoot and Wondolleck, 1990; Fisher, Ury, and Patton, 1991; Raiffa, 1994; Dukes, 1996; Saunders, 1999; Daniels and Walker, 2001). A second emerges from planning and organizational development (e.g., Arnstein, 1969; Avery et al., 1981; Mansbridge, 1983; Susskind and Cruikshank, 1987; Doyle and Straus, 1993; Gastil, 1993; Chrislip, 1994; Schwartz, 2002; Creighton, 2005; Gastil and Levine, 2005; Schuman, 2005; Kaner, 2007). The third has its origins in risk assessment, ecological assessment, impact assessment, risk communication, and other kinds of environmental analysis (National Research Council, 1983, 1989, 1994, 1996, 2007b; Dietz, 1987; Fiorino, 1990; Chess, Tamuz, and Greenberg, 1995; Presidential/Congressional Commission on Risk Assessment and Risk Management, 1997a,b; Wondolleck and Yaffee, 2000).

Public participation practitioners in the dispute resolution and planning traditions, in which consensus is often the most salient objective, devote much attention in their handbooks to broader involvement and improved communication. They emphasize understanding and communicating the interests that underlie diverse views, sharing information and developing creative options, using objective criteria, building trust, including the public early in the process, keeping the process open and flexible, maintaining transparency, and monitoring outcomes.

This emphasis suggests that practitioners in this tradition believe that processes focused closely on improving legitimacy yield benefits in terms of the full range of desired outcomes, including decision quality and capacity. Experienced practitioners of environmental public participation know that legitimacy does not guarantee quality—that it is a mistake to ignore aspects of the process that directly promote the quality of assessments and decisions. In other words, the popularity of or support for a particular assessment or decision does not ensure the technical quality of the assessment or decision. Practitioners' advice for consensus building in the context of contested science recognizes the importance of making assumptions transparent, coproduction of information and analyses, getting the right information, clarifying the relevance of the information to decisions, and addressing problems that participants may have in understanding and communicating about scientific information (e.g., Ozawa, 1991; Adler and Birkhoff, 2002; Bingham, 2003).

It is worth noting that at least some participants in environmental decision processes are also well aware of the dangers of poorly informed decisions. Consequently, they are unlikely to accept as legitimate any process that does not pay specific attention to elements of decision quality, such as gathering all the relevant information, subjecting it to criticism from a variety of sources, and ensuring that decisions are consistent with the best available information. Thus, it seems likely that at least certain practices that are conducive to quality are also conducive to legitimacy.

Practitioners in the environmental assessment tradition tend to focus primarily on decision quality: ensuring that public policy decisions are based on accurate scientific analysis. In early writings in this tradition, the values and judgments of nonscientists—especially public officials, but also other parties—were to be insulated from the scientific analysis (National Research Council, 1983, is frequently referenced in this regard). For some writers in this tradition, too much concern with legitimacy in the eyes of nonscientists was considered a potential threat to good science (e.g., Starr, 1969).

Over time, practitioners in this tradition have been coming to many of the same conclusions as practitioners in the dispute resolution and planning traditions. Writings about public participation in the environmental assessment tradition have been relatively scarce until recently, in part because this tradition is primarily concerned with the practices of analysts. An important statement about public participation practice in this tradition was the 1996 National Research Council (NRC) report, *Understanding Risk*, which drew heavily on practical experience in characterizing risks for nonscientists. The recommendations offered for practice in that report were justified primarily in terms of their contribution to the quality of environmental decisions, although legitimacy was also considered. The report specified four guidelines for deliberative practice (National Research Council, 1996:4-5):

- seek broad participation, especially in the early "problem formulation" phase of the process, and with respect to interested and affected parties that "are particularly at risk and may have critical information about the risk situation" and lack the capacity to participate effectively, "it is worthwhile for responsible organizations to arrange for technical assistance to be provided to them from sources that they trust."
- "clearly and explicitly inform the participants at the outset about . . . external constraints likely to affect the extent of deliberation possible or how the input from deliberation will be used."
- "strive for fairness in selecting participants and in providing, as appropriate, access to expertise, information, and other resources for parties that normally lack these resources."
- build flexibility into deliberative processes.

This set of guidelines is clearly aimed at achieving a full and informed deliberation that can provide the best available information relevant to the decision at hand. Nevertheless, it includes elements that are familiar from the writings of practitioners in the dispute resolution tradition, whose initial driving concerns were presumably legitimacy and agreement. Like the NRC's *Understanding Risk*, the more recent work of the Global Environmental Assessment Project (Mitchell et al., 2006) and the National Research Council (2007a) review of global change assessments also begin with a primary interest in decision quality and end with practical recommendations that strongly emphasize broad public participation and considerations of legitimacy as well. Similarly, the Presidential/Congressional Commission on Risk Assessment and Risk Management (1997b:41), drawing its conclusions from practical experience, recommended that "public stakeholders" be placed in "prominent roles" in risk assessment and management processes and concluded (1997b:48) that public participation in the comparative risk process improves understanding of competing priorities, provides an appreciation of the complexity of decision-making, and can stimulate new insights into solutions. As a result of increased communication among institutions and interest groups, new avenues of cooperation might be established. Adversarial relationships among interest groups and jurisdictional conflicts among agencies might not disappear, and could even be intensified, but [public participation in] comparative risk projects have revealed unexpected agreement among parties and enhanced understanding of differences in perspectives and values in some cases. Most important, experience has shown that the process itself can help to build coalitions that favor priority setting and shifting resources to the identified priorities. Broader public support for a common agenda might allow agencies, state legislatures, and Congress to move money and staff into priority problems with less litigation and less controversy.

We see this convergence as an important trend: Practitioners of public participation from these three, somewhat distinct traditions of dispute resolution, planning, and environmental assessment have increasingly converged on similar ideas of best practice. Over time and with increased experience regarding difficult environmental decisions, each group of practitioners has come to recognize the critical importance of the central concerns of the other, with writings on risk assessment increasingly recognizing the importance of broad representation and of deliberation about judgments and values, and writings on environmental dispute resolution increasingly emphasizing the need to incorporate relevant scientific knowledge. This observation suggests that in most areas of environmental assessment and decision making, it is both imperative and possible to develop practices that promote decision quality and legitimacy at the same time. It also suggests that practices recommended in the various practitioner traditions, such as representation of the full range of interested and affected parties, early involvement of those parties, and flexible decision processes capable of taking new information into account, in fact contribute to both quality and legitimacy. Technical assistance to important parties that are not able to participate fully may contribute to all three objectives: capacity, legitimacy, and decision quality. In this light, the empirical finding should not be surprising that participatory processes that are successful on one of these dimensions are also likely to be successful on others.

CONCLUSION

The evidence from the best available empirical studies of environmental public participation processes and from the experiences of practitioners in the dispute resolution, planning, and environmental assessment traditions converges on the conclusion that best practices in public participation can advance decision quality, legitimacy, and capacity simultaneously. This does not imply that a standard set of guidelines for practice will work equally well and achieve all three objectives in all situations. In some situations, particular issues require special attention. For example, when the interested and affected parties to a decision seriously mistrust each other or the responsible public authority, special attention to building legitimacy may be necessary. When the relevant science is known to be in dispute, special attention to issues of scientific quality may be necessary. When certain critical parties lack sufficient scientific understanding to participate effectively, technical assistance to these parties may be essential to any desirable outcome.

It is also likely that certain attributes of public participation practice or its context strongly affect particular desired results but are unrelated to others. For example, some public participation practices may be particularly valuable for enhancing legitimacy, while others may be particularly valuable for educating the public. We consider such possibilities in the following chapters, in discussing how various practice and context factors may affect particular kinds of outcomes. It is also possible that in certain situations that are not well represented in the available data, there may be trade-offs among desirable outcomes.

NOTES

[1] In technical terms, participants approached greater "ordinal single-peakedness" in their opinions. That is, following deliberation of an issue with more than two possible decision options, more people ordered their preferred options in a way that was logically consistent with others' ordering of preferred options. Whether or not they agreed on which specific options they preferred, they did agree more on the relationships among the options. Ordinal single-peakedness is a desired characteristic of preferences in theoretical work on social choice paradoxes (Condorcet, 1785; Black, 1948; Arrow, 1951; Niemi, 1969; Miller, 1992; Knight and Johnson, 1994; List, 2001; Dryzek and List, 2003; Gehrlein, 2004). In theory, it increases the ability to avoid, with democratic majorities, the instability and manipulability of "cycling," which can occur when equal proportions of a population prefer each of three or more options, but voting takes place between only two options at a time. Single-peakedness helps to meaningfully aggregate individual preferences into social choices by arraying options along a continuum so that the median preference can prevail.

[2] Certain negotiated rule-making efforts, such as EPA's federal advisory committee that produced the Disinfectants and Disinfection By-Products

Rule, also have attributes of coproduction of information (National Research Council, 1996).

[3]Practitioners also express concerns that unexamined assumptions about the relative priorities among desired characteristics of public participation may affect process choices implicitly or explicitly, creating trade-offs, for example, between the desire to make decision making more efficient and the time and attention needed to achieve joint gains or improved quality in the agreements reached.

4

Public Participation Practice: Management Practices

The way a public participation process is conducted can have more influence on overall success than the type of issue, the level of government involved, or even the quality of preexisting relationships among the parties. Thus, those variables over which the convening agency has the greatest control turn out to be key to achieving the desired results.

In this chapter and the two that follow we review the evidence with regard to the practice of participation. We find that good outcomes can be obtained even in difficult circumstances. The evidence discussed in this chapter supports generic principles of program management that apply to a broad range of programs managed by government agencies and other organizations and specifically to public participation programs.

Many of the principles stated in these three chapters strongly echo those offered in other studies (e.g., Presidential/Congressional Commission on Risk Assessment and Risk Management, 1997a; Office of Management and Budget and President's Council on Environmental Quality, 2005) and in multiple works by practitioners and scholars of public participation practice (e.g., Dukes, 1996; Daniels and Walker, 1997; Susskind, Thomas-Larmer, and Levy, 1999; Wondolleck and Yaffie, 2000; Dukes and Firehock, 2001; Institute for Environmental Negotiation, 2001; U.S. Environmental Protection Agency, 2001; Creighton, 2005). We reiterate such previously stated principles for two reasons. First, it is important to recognize principles that are supported by a convergence of evidence—not only by practitioners' experience, but also by the kinds of careful case-study research, case-comparison studies, and basic social science knowledge examined in these chapters. Second, some often-stated principles bear repeating because they are so often violated in practice.

The principles presented in these three chapters overlap to some degree, as is often the case in describing effective practices. We begin by summarizing the main finding from our review of the evidence on management and then discuss more specific points along with the supporting evidence.

Basic principles of program management apply to environmental public participation. When government agencies engage in public participation processes with clarity of purpose, commitment, adequate resources, appropriate timing, an implementation focus, and a commitment to learning, they increase the likelihood of good results. When they fail to do these things or lack adequate organizational capacity, the results are likely to fall short of the potential of public participation.

Public participation activities share a number of features with other programs that government agencies and other organizations run: they require planning, resources, coordination, implementation, and the like. It should therefore not be surprising to see that much of the advice on how to run these programs echoes basic principles of program management such as can be found in the research literature on organizational management and on management of relationships between organizations and outside constituencies (Blundel, 2004). This section discusses aspects of the practice of public participation that are matters of basic program management and reviews what is known about good management practice, drawing from both the general management literature and from experience with environmental public participation.

CLARITY OF PURPOSE

When responsible agency develops a clear set of objectives for a participatory process, integrated with a plan for how the outcomes of the process will be used and with serious efforts to share that understanding with the participants, it increases the likelihood of acceptance of agency decisions and of public willingness to engage in future participation efforts. By doing these things, government agencies fulfill widespread expectations that they will play a leading role in setting the agenda for policy discussions and making public purposes clear (Hibbing and Theiss-Moore, 2001). Public participation processes tend to yield better results when the clear purpose reflects an agreement about goals among the convening organization and the participants and when it takes account of the objectives of all parties involved, the scope of legally possible actions, and the constraints on the process.

Several lines of evidence support the proposition that clarity of purpose is conducive to success in public participation. This proposition is, first of all, supported by the general literature on organizational management (Blundel, 2004). Clear purposes reduce certain kinds of uncertainty for participants and thus reduce an impediment to high-quality thinking (e.g., Janis and Mann, 1977; Klein, 1996; Covello et al., 2001; Van den Bos and Lind, 2002). This research also shows, however, that uncertainty makes people more vigilant about evaluating the credibility of information sources (Halfacre, Matherey, and Rosenbaum, 2000; Brashers, 2001; Van den Bos, 2001) and leads their judgments about a process to be influenced more strongly by procedural fairness (Van den Bos, 2001; Van den Bos and Lind, 2002). In our judgment, the uncertainty of information for environmental decisions is almost always sufficient to trigger such effects, even when purposes are clearly stated. A process with clear purposes and procedural fairness is consistent with reducing these uncertainty-related cognitive effects.

In the context of environmental public participation, Wondolleck and Ryan (1999) have argued that agencies can engage in public participation processes as leader, partner, or stakeholder, and that when it is not clear which role an agency is playing, the process can suffer. This argument is borne out by empirical work on participation processes. For example, Bradbury (2005) found, in an examination of public participation in the cleanup of multiple Superfund sites, that clarity about agenda setting and prioritization of issues are important factors influencing the perceived competence, legitimacy, and capacity of public participation processes.

Leach's (2005) review of 25 empirical studies of public participation in U.S.D.A. Forest Service decisions beginning in 1960 found strong evidence supporting the importance of focused scope and realistic objectives. Some of these studies highlighted the importance of clear purpose, goals, and objectives (Schuett, Selin, and Carr, 2001), along with measurable, quantifiable, or tangible goals (Doppelt, Shinn, and John, 2002). Others focused on the importance of defining results in terms of action rather than talk (U.S.D.A. Forest Service, 2000) and of focusing on attainable goals to build momentum, confidence, and reputation (Wondolleck and Yaffee, 1997). Findings highlighted the importance of addressing a manageable number of projects with a reasonable level of complexity (Daniels and Walker, 1997) and the importance of recognizing milestones throughout the process by setting and acknowledging short-term and long-term goals (U.S.D.A. Forest Service, 2000). Clarity in objectives of the process has also been found helpful for keeping decision processes focused on negotiable disputes rather than on discussion of values (Walters et al., 2003). The National Research Council (2007a) analysis of global change assessments, which examined a very different environmental context, concluded that a clear audience for an assessment product is essential to success, which also implies the importance of the processes having clear goals.

Conflicts about the scope of public participation efforts have often derailed them. For example, the failure of community members and ex officio members to reach agreement about the goals of the Site Specific Advisory Boards of the U.S. Department of Energy (DOE) was cited as one of the factors that led to the inability of four boards to deal with substantive issues (Branch and Bradbury, 2007). In the advisory boards of the U.S. Department of Defense (DOD), when community interests did not reach the table, in part due to failure to reach an understanding on the scope of the discussion, this lack of clarity about purpose led many discussions to devolve to procedural issues (Branch and Bradbury, 2006).

The ways in which agency decision makers intend to use the output of a process may or may not be clear to participants at the outset. In most cases, public participation is an informal element of the decision-making process, and thus the agency needs to clarify how the results of the participatory process will be incorporated into the decision process. When the participatory process is a formal part of the decision process, agency rules sometimes clearly specify the role of the public process in informing or making the decision. However, the role and influence of public participation may not be clearly specified in advance, as when a participation process is initiated at one level of an agency while the ultimate decision is made at another, higher level. Participants who take their charge seriously and who devote considerable effort to reaching a consensus may grow to assume that what the process recommends will be implemented.

Explicit, honest agency statements about what it wants from the process and how the results will be used ensure realistic assessment by the other parties of the reasons for them to participate and reasonable expectations about results. A strong commitment to act on the results of a participatory process obviously increases incentives for parties to participate. Lack of clarity about how decision makers intend to use the results encourages skepticism. Because agencies are not monolithic, it is not always easy for participants to gain such clarity. While one unit may state its interest in incorporating suggestions, another may be less invested in doing so (O'Leary and Summers, 2001; Bradbury, 2005). Public participation processes can be caught in the middle, reducing the usefulness of and trust in them.

It is common at the outset of a participatory process for assumptions regarding the nature of the environmental problem being addressed and the possible paths to a solution—the "frames" for the issue—to differ across participants (Snow et al., 1986; Bradbury, 1989; U.S. Environmental Protection Agency, 1992; Kroll-Smith and Couch, 1993; Thompson and Gonzalez, 1997; Pellow, 1999). Yet few external participants will understand in any detail the concerns of and constraints on the agency. Thus, developing clarity of purpose involves the emergence of mutual understanding of the alternative frames.

Differing frames can contribute to undesired results. When individuals believe that decision processes have not adequately taken into account important values, the results can include a loss of trust, exacerbated conflict, and prolonged negative affective reactions (e.g., Fisher, 1991; Rich et al., 1995; Shah, Domke, and Wackman, 1996; Susskind and Field, 1996; Baron and Spranca, 1997; Thompson and Gonzalez, 1997). As we discuss in Chapter 7, discordant framing is also a source of conflict. When agencies have sufficient flexibility to allow problems to be reframed through deliberation so as to incorporate participants' definitions, and they allow this to happen, these results may be avoided (National Research Council, 1996; Renn, 2004; Lemos and Morehouse, 2005). Clarity may include explicit recognition that the goals of the process may evolve as it is codesigned with citizens who may have somewhat different goals and expectations from the agency's initial ones. A convergence of purposes has the potential to reduce conflict and enable cooperation. However, agency constraints sometimes limit flexibility in this regard. In our judgment, clarity about such real constraints is preferable in the long run to a lack of clarity that allows participants to become engaged in a process they may later conclude was organized under false pretenses.

AGENCY COMMITMENT

Public participation processes are more likely to be successful when the agency responsible for the relevant environmental decisions is committed to supporting the process and taking seriously the results. This is in part because the more committed a decision-making agency is to act on the results of a public participation process the more likely the parties are to engage seriously. Commitment involves support of both agency leadership and staff at all levels for the objectives of the process, stated at the outset and updated periodically as the participation process and the context evolve. It implies clarifying how and by whom the outputs will be used and a commitment to open-minded consideration of those outputs.

Basic understanding of group processes and decision making suggests the importance of clear agency commitment. Ambiguity about how information will be used increases uncertainty, which, as already noted, makes high-quality thinking less likely. The research literature suggests, however, that if the convening agency is committed to a high-quality process, rather than to a particular kind of decision outcome, participants are more likely to engage in evenhanded and effortful consideration of the available options, rather than defensive justification of their preferred alternative (Simonson and Staw, 1992) and arrive at higher quality judgments (Siegel-Jacobs and Yates, 1996; for more detailed reviews, see Lerner and Tetlock, 1999, 2003). This implies that a public participation process is likely to go better if the responsible agency can honestly signal to the participants that it has not made a decision and does not have a strong predisposition for one course of action over another but is sincerely looking for input.

Studies of environmental public participation reinforce the importance of agency commitment (e.g., Bingham, 1986; Wondolleck and Yaffie, 2000; Schuett, Selin, and Carr, 2001). Leach (2005:8), reviewing several studies of public participation in the U.S.D.A. Forest Service, stated that "support from line officers and agency-wide Forest Service Policy is the dominant contextual factor in the reviewed studies." But such support must be stable. Lubell's (2004b) analysis of estuary partnerships demonstrated that changes in agency plans at high levels degraded the quality of the participatory planning processes. Changes may be interpreted as signals from the agency that the process is not important in its decision making: this may lead participants to opt out and seek alternative mechanisms for being heard in the political process, which results in less careful and thorough consideration of the issues.

A series of research studies suggests that the DOD advisory boards were less successful than similar ones convened by DOE because DOD failed to convey the importance of public participation (Branch and Bradbury, 2006). In contrast, the relative success of the DOE advisory boards in the late 1990s was attributed to "clarity and commitment to public participation in both policy and implementation" (Branch and Bradbury, 2006:746).

Similarly, a study of alternative dispute resolution by the U.S. Environmental Protection Agency (EPA) in the 1990s (O'Leary and Summers, 2001) found that despite a stated institutional commitment nationally, implementation varied regionally and with the enthusiasm and skill of staff, rather than on the basis of any consistent method of institutional assessment of when alternative dispute resolution would be most valuable. The resistance of mid-level managers was seen as particularly problematic. In addition, EPA often signaled a lack of commitment to the process by sending representatives who lacked the authority to make decisions.

We have emphasized how too little agency commitment can hamper a public participation process, as when agency officials are not available to provide information about the issues or the decision context or to build relationships with the participants. Too much engagement can also be a problem, as when agency officials so completely dominate the process that participants cannot take an active role in shaping the questions for discussion or deciding how the process is organized (e.g., Delli Priscoli, 1983; Stewart, Dennis, and Ely, 1984; Plumlee, Starling, and Kramer, 1985). Whatever an agency's level of commitment, it is best for the agency to make clear to the participants how it intends to use the results of the participatory process.

ADEQUATE CAPACITY AND RESOURCES

Public participation processes are more likely to be successful when agencies have adequate capacity and resources and deploy them appropriately according to the scale, complexity, and difficulty of the issues involved. The commitment of resources is both a practical matter and a signal from the agency that the participatory process is important.

The difficulty of conducting a process without adequate resources is obvious, and perhaps for this reason it has not been the subject of much empirical research. However, the need for adequate funding and other resources for achieving goals of a participatory process is among the most frequently mentioned lessons from practitioners' experience (e.g., Creighton, 1999; Leach and Pelkey, 2001; U.S. Environmental Protection Agency, 2001).

Lack of resources is always a challenge. Budgetary constraints on agencies or decisions within an agency can cause problems in the process, such as "stop and go" funding or uneven funding among different entities in the process (Moser, 2005). In the Forest Service context, one study (Wondolleck and Yaffee, 1994) pointed to the need for startup costs to be considered; especially when skilled facilitators are needed or public outreach needs to be undertaken. Frentz et al. (2000) recommended that convening organizations consider setting aside dedicated funding so that staff can consistently be present during and participate in the processes. A U.S.D.A. Forest Service (2000) internal study suggested that for longer processes that involve multiple parties, participants should contribute staff and financial resources. Bradbury (2005) found that the problems of two DOE advisory boards were reduced by an infusion of funds to facilitate access to information and provide for neutral facilitation, technical assistance, and support for a sufficient number of meetings.

Organizational resources include more than money. Several studies have found that organizational capacity in the form of skilled and enthusiastic staff is vital to program success (e.g., Henry S. Cole Associates, 1996; O'Leary and Summers, 2001), providing an invaluable reservoir of experience (Henry S. Cole Associates, 1996). An assessment of EPA's alternative dispute resolution program stated bluntly that the future of the program depended on the agency's ability to find trained mediators (O'Leary and Summers, 2001). Other studies have found an association between staff expertise and the extent to which communication activities were two-way and "symmetrical" (see Grunig and Grunig, 1992). Organizations that employed "technicians" to develop communication materials were more likely to engage in one-way processes of message transmission in which the organization attempted to control the process. When senior communications managers were part of the organization's decision-making structure, it was more likely to engage in two-way communication processes that featured listening to outsiders.

Continuity of agency personnel has also been found to benefit participatory processes. The set of empirical studies regarding Forest Service public participation points to the importance of staff continuity (Leach, 2005). A policy in the Forest Service to prevent conflicts of interest among agency personnel by rotating them between forests had the unexpected consequence of placing stress on long-range participation processes (Clarke and McCool, 1985). New personnel faced the challenges of assuring participants of the agency's continued commitment to the process (Wondolleck and Yaffee, 1997; Tuler and Webler, 1999). Similar findings have been reported with public participation at hazardous waste sites (Henry S. Cole Associates, 1996).

Limited resources may not be only an external constraint. They may also reflect a lack of agency commitment, as reported in the examination of DOE and DOD public advisory boards by Bradbury (2005). Creating expectations that cannot be met can be a bigger problem than lack of resources per se. To a large extent, public participation processes can be scaled to the resources available. To do so, however, requires careful planning. Some practical planning guides have been developed by environmental agencies that address resource issues in planning and developing participatory processes (e.g., U.S. Environmental Protection Agency, 2001). Practical experience suggests that diagnosis and process design efforts can determine the amount of time and resources available for the process and that it is important for the convening organization to make resource constraints clear to the participants, so that a realistic set of objectives for the public involvement process can be set. Being clear about resource constraints can also help an agency allocate resources so as to invest in meeting the most important challenges or obstacles that have been identified.

It is often useful to be creative in looking for additional resources, including from participants, the public at large, and the nonprofit sector (U.S.D.A. Forest Service, 2000; Delli Carpini, Cook, and Jacobs, 2004:316). The practitioner literature reinforces this point: agencies are increasingly emphasizing partnerships, with stakeholders and with one another, through which resources as well as perspectives are brought together to accomplish environmental and other public policy objectives in a participatory or collaborative manner. For example, a watershed management effort in New Jersey was able to include extensive public participation in part due to in-kind contributions and financial resources provided by a variety of organizations, including the local water purveryor, nonprofit watershed organizations, and municipalities (http://www.raritanbasin.org/). A substantial grant from EPA for the larger watershed management effort and limited funding awarded to municipalities by the state department of environmental protection were also critical. (Numerous other case examples can be found at http://cooperativeconservation.gov/stories/index.html.)

TIMELINESS IN RELATION TO DECISIONS

Public participation processes are more likely to have good results when planned so that they can be informed by emerging analysis and so that their outputs are timely with regard to the decision process. That is, participatory processes need to be designed so that closure is achievable and outcomes are available to decision makers in a timely manner.

Timing presents a "Goldilocks problem" with regard to both scientific analysis and decision processes (National Research Council, 2007a:3-11). If a public participation process is started too soon, key information may not yet be available. If the process is started too late, there may not be adequate time to develop trust and understanding and to process scientific and technical information. Furthermore, if the outputs from the participatory process come too late to influence decisions, it becomes impossible for an agency to fulfill promises to take the process seriously. And yet, if the process does not have sufficient duration, it may not be possible to develop the mutual understanding that underpins a successful participatory process.

Most of the literature on environmental public participation emphasizes the importance of starting the process early enough. Including the public as early as possible is one of the most frequently mentioned lessons learned by public participation practitioners in the dispute resolution tradition (e.g., National Park Service Division of Park Planning and Special Studies, 1997; Cestero, 1999; Bleiker and Bleiker, 2000; McKeown, Hopkins, and Chrystalbridge, 2002). Similarly, the Consensus Building Institute (1999:14) stated that "mediation should be used when it is started at an early stage of conflict, before going to public hearings." This lesson is also often cited by expert groups in the risk analysis tradition (e.g., National Research Council, 1996; Presidential/Congressional Commission on Risk Assessment and Risk Management, 1997a,b), suggesting that in practitioners' judgment, early involvement contributes to both the legitimacy and the quality of decisions. The U.S. Environmental Protection Agency (1992) concluded that many weaknesses in ecological risk assessments emerge at the problem formulation stage, and thus public participation can be especially helpful at exactly the point at which such assessments go awry. Germain, Floyd, and Stehman (2001) found that participants' satisfaction was higher when they were involved early in scoping activities. Bradbury, Branch, and Malone (2003) found that the involvement of stakeholders in "scoping and framing of issues during the initial stages of a decision-making process" was associated with higher quality decisions, and Duram and Brown (1998) found that public participation was perceived to be most helpful in the planning stages involving outreach and identifying and prioritizing issues. Peelle et al. (1996) identified early involvement in a long list of factors that influence success. Early involvement may not preclude and can enhance access to and generation of high-quality scientific information and analyses, particularly when time and resources are available for technical representatives of stakeholder groups to be involved in collaborative information generation processes (Susskind, Thomas-Larmer, and Levy, 1999).

While early involvement may aid success in many contexts, it appears to be neither necessary nor sufficient. Ashford and Rest (1999), who studied a series of relatively successful public participation processes, reported that this success occurred even though the public became involved fairly late in the process. Mitchell, Clark, and Cash (2006:314) concluded that it is also possible to be too early: "Information must be . . . timely, coming before—but not too long before—relevant decisions get made." Multicase comparative studies of DOE and DOD advisory boards (Bradbury, 2005) and of the National Assessment of Climate Change (Moser, 2005) examined cases in which the parties appear to have been involved very early. Nevertheless, the DOD and DOE projects had mixed success, and the climate change assessments received mixed evaluations from participants, suggesting that early involvement is not sufficient for success.

We suggest that some, though not all, of the benefits that typically accrue from early involvement may be a result of having enough duration in the participation process to overcome obstacles raised by the context. For example, in analyzing watershed partnership programs, Lubell and Leach (2005) found that having a long enough duration for a planning process had a significant positive effect on both the scope of the policy agreements achieved and on project implementation and ultimately on watershed conditions. Sufficient duration allows mutual understanding and a degree of trust to develop. And without sufficient duration, it is difficult for the participants to develop familiarity with the science or to have input into the scientific analysis (issues to which we return in Chapters 5 and 6). Nor is it easy to allow participants to influence the design of the participatory process if the process starts relatively late and is of short duration. Yet, as discussed below, such collaborative design is important to successful participation.

In sum, timing presents significant challenges for the conveners of public participation. Sometimes the best way to meet these challenges is to adjust the intensity of participation and the scope of issues to be covered so that the timetable is realistic. However, rushing the process and compressing its scope have downsides: a hurried timetable may not allow enough time for a process appropriate to the challenges of the context, and a reduced scope will sometimes be seen as having the effect of putting aspects of the decision out of bounds for discussion.

A FOCUS ON IMPLEMENTATION

Participation processes tend to be more successful when designed to relate in clear ways to policy decision making and implementation. The responsible agencies need to be clear from the outset about what they can and cannot implement, a point already raised. Especially for public participation processes intended to inform decisions, it is important for implementation to be part of the purview.

Many case studies have noted the importance of considering implementation issues in defining the scope of a participatory process. For example, Bradbury (2005:13) noted in one decision-making context that being "able to identify and prioritize the issues on which to focus and to prevent issues that were not considered part of their designated scope" from being on the agenda was one of the biggest challenges faced by the advisory boards she studied. Mitchell, Clark, and Cash (2006), in their analysis of global assessments, concluded that the salience of an analysis to potential users is a key factor in determining how much impact the assessments have.

Experienced practitioners often advise that it is useful to identify in advance roles and responsibilities of various groups following the formal public involvement process and to be sure to involve those who are needed for the implementation of decisions that result from the participation process. Implementation considerations include possible partnerships for implementation, monitoring and oversight mechanisms, and incentives and disincentives to implementation. Many practitioners believe that anticipating difficulties in implementation from all perspectives and discussing contingencies makes public participation processes better informed and increases the chance that they will produce results that participants consider useful. Implementation raises the issue of limitations on an agency's scope of authority. Researchers and practitioners often advise that goals match what can be implemented and that the scope of a public participation process be defined accordingly. For example, Wilbanks (2006), considering the experience of local "smart growth" decision processes, advised convening organizations to "deliver on promises. . . . It is better to indicate a positive intent but to limit one's promises than to take a chance that resource limitations or political complexities will lead to disillusionment." But limiting scope can be frustrating to members of the public who have broader interests in the issue. Thus, an early understanding of what an agency can and cannot do will enhance the chances of an effective participatory process. This understanding does not, of course, preclude participants from raising issues outside the agency's purview or pursuing those issues in contexts beyond the participatory process.

COMMITMENT TO LEARNING

Public participation processes, as well as the larger assessment and decision processes in which they are embedded, benefit from engaging in self-assessment and design correction as they proceed. The design of participatory processes can benefit from opportunities for participants and sponsors to assess the process both as it is under way and at the end. Designs that allow for midcourse adjustments and that are evaluated to generate lessons for future public participation efforts are most conducive to learning.

Learning can be greatly advanced by independent evaluations of public participation efforts. Evaluation studies repeatedly demonstrate that careful research can reveal knowledge that does not emerge from intuitive judgments of what works and what does not (Rowe, Marsh, and Freaer, 2004; Blackstock, Kelly, and Horsey, 2007). Although some systematic retrospective studies of public participation now exist, the state of knowledge would be much advanced if organizers of participation supported careful evaluation studies, particularly including prospective studies comparing different modes of participation, which can provide evidence about the causes of participation outcomes. Systematic evaluation is the most trustworthy way to gain understanding of the effects of participation practices and thus to ensure institutional learning and improvement in practice. Even when resources are limited, expenditures on systematic evaluation can add a great deal of value (Rohrmann, 1992). However, the scope and resources needed for the evaluation need to be appropriate to the public participation effort. A small, short-term public participation effort may not need as detailed evaluation as a more extensive effort, or one that may become a model for future participatory processes.

Evaluation is not merely a report card that agencies get (or give themselves) at the end of a project. So-called formative evaluation is aimed at improving programs in progress and provides managers with feedback during program development and implementation (Posavac, 1991). Multiyear assessments of DOE's Site Specific Advisory Boards (Bradbury, 2005) are an example of an effort to improve the participation process over time and use evaluation data as a basis for making programmatic decisions. Because the boards were a major new agency initiative, it invested in multiyear, qualitative, and quantitative evaluation. A more limited effort would merit a more limited formative evaluation, perhaps brief surveys after each meeting or routine debriefings with participants.

Other agencies have included formative evaluation as a critical component of their public participation efforts. For example, before the New Jersey Department of Environmental Protection conducted a trial release of a genetically modified rabies vaccine, it interviewed key opinion leaders to develop participatory processes that met local needs, and it modified its plan to release the vaccine on the basis of this feedback. According to staff, both this participatory process and the formative evaluation that kept it on track accounted for the marked difference between the programmatic success of the effort and the failure of other states to test rabies vaccines due to public opposition (Chess, 2001). The New Jersey agency's formative evaluations of the rabies effort were initiated by its public participation staff and involved program staff in implementation. Retrospective studies, even though they cannot employ contemporaneous measures to track change, can also serve to improve future agency programs (Rosener, 1981).

Evaluations conducted by professional evaluators external to the convening agency can bring objectivity and insight that may not be available from internal evaluations. Several useful efforts have been made to develop evaluation measures for environmental public participation processes (e.g., Lauber and Knuth, 1999; Rowe and Frewer, 2000, 2004). The evaluation research community has also pioneered participatory evaluation, which involves stakeholders in designing the evaluation process (Fetterman, 1994, 1996). In this approach, participants clarify goals and expectations, as well as processes for ascertaining whether these goals are being met. Some evaluators feel that those who participate in evaluation design are more likely to use the results (e.g., Guba and Lincoln, 1989; Syme and Sadler, 1994). An example of participatory evaluation is the evaluation of DOE's Site Specific Advisory Boards, which involved local and headquarters agency staff, as well as participants in the boards, in developing the goals, criteria, and instruments for evaluation (Bradbury and Branch, 1999).

Evaluation is only one step in improving agency practice. Learning from the results and institutionalizing them are equally important. Theorists suggest that organizational learning goes beyond the learning of individuals, so that agencies develop an institutional memory. According to one often-cited definition, organizational learning is "encoding inferences from history into routines that guide behavior" (Levitt and March, 1988:320). This definition implies that organizational learning is reflected in changes in policies, procedures, and systems. Without such institutionalization, learning about public participation may not extend beyond the personnel involved in public participation. This insight is consistent with a history of policy studies research that has long emphasized the need to treat policies as experiments and the value to policy effectiveness of instituting evaluation strategies that deploy both external reviewers and review by the participants and the sponsors (Campbell, 1969).

Learning also involves questioning assumptions and operating systems. This can be particularly important when agencies are in the midst of controversy. According to one widely accepted model (Argyris, 1982), agencies must learn how to learn. This goes beyond making strategic changes in specific programs, so-called single loop learning. "Double loop" learning includes questioning the larger systems in which a process is embedded. It may involve changes in those systems, in the rules and methods for deciding, or in other organizational routines (Scott, 1992). Thus, an evaluation may yield feedback that can be used to change more than the specific participatory process being evaluated. Such feedback might lead to internal dialogue about the goals of the overall public participation program, the systems that support it, and the institutional memory needed for ongoing improvement (Chess and Johnson, 2006).

In sum, accumulated experience and research on program management support the conclusion that successful practice is more likely to be found in agencies that develop a culture and set of procedures that allow them to learn not only from past experience as organized in the research literature and practitioners' knowledge, but also from recent and ongoing experience in their own agencies and in other organizations convening public participation in similar contexts. The notion of ongoing learning from participatory processes is congruent with the idea of integrated, repeated analysis and deliberation endorsed in *Understanding Risk* (National Research Council, 1996). Advice from public participation practitioners is also consistent with this view. Zarger (2003), reviewing a set of practitioner handbooks, listed "commitment to iterative, resilient, responsive processes and monitoring" as one of the 10 most frequently mentioned lessons learned, citing Creighton (1999), Leach and Pelkey (2001), U.S. Environmental Protection Agency (2001), National Environmental Justice Advisory Council (1996), and McKeown, Hopkins, and Chrystalbridge (2002), and the work of Pierce Colfer (2005) as sources for the lesson. Another of the top 10 lessons learned from watershed partnerships was openness and flexibility to respond and to change course if necessary to get to the end goal (Leach and Pelkey, 2001). Both of these lessons from practice speak to the importance to success of a commitment to learning in the responsible agencies.

Learning is important not only in the agencies, but also among the parts of the public who have limited experience with participation and limited resources to devote to participation. As we have noted, many scholars have concluded (Bowles and Gintis, 1986; Delli Carpini, Cook, and Jacobs, 2004:322) that the apathy and alienation found in much of the public is substantially a consequence of limited opportunities for meaningful participation, so that, over the long term, increased public involvement depends on learning and enhanced capabilities among the public as well as in agencies. Leach (2005) found the Forest Service's use of adaptive management a sound example of an approach to policy that built capacity for participation.

CONCLUSION

The evidence indicates that public participation processes have better results when they follow basic principles of program management: clarity of purpose, commitment, adequate resources, appropriate timing, an implementation focus, and a commitment to learning. However, it is not always easy to follow these principles. Difficulties can arise from a variety of factors, including internal differences of purpose within the responsible agency, shortages of money or skilled personnel, timing of participation, and various other contestable factors outside the agency. When such difficulties exist, success depends on how well the process is organized to avoid or overcome the problems they present. Chapters 7 and 8 discuss this issue.

5

Practice: Organizing Participation

This chapter reviews the evidence on how different ways of organizing the participation itself—the interactions among the agency and the various participants—affects the results of public participation. As in the previous chapter, we begin by stating this main finding from our review and then discuss more specific points and the supporting evidence.

A few attributes of the ways environmental public participation is organized are associated with the likelihood of successful outcomes and can be treated as principles of good practice. Successful outcomes are more likely when a process includes the full spectrum of parties who are interested in or will be affected by a decision and encourage their voluntary commitment to it; involves the parties in formulating the problem for assessment or a decision and in designing the participatory process; is transparent to participants and observers; and is structured to encourage the parties to communicate in good faith.

There is no single best format, set of procedures, or level of intensity for implementing these principles or for achieving good outcomes in all situations. The best results follow when the participation is organized so as to be responsive to context-specific challenges, such as those discussed in Chapters 7 and 8.

PUBLIC PARTICIPATION FORMATS AND PRACTICES

Public participation processes can be organized in many different ways. So it is reasonable to ask if certain formats work better generally, or in some circumstances, than others. However, the evidence does not support such conclusions. Various public participation formats have been successful in achieving the goals of high-quality and widely acceptable assessments and decisions, and each format has also failed at times in achieving these goals.

Many terms have come into use over time to describe ways of organizing participatory processes. Numerous typologies of them can be found in the published literature (e.g., Creighton, 2005). Some terms refer to broad "formats." Examples include public hearings, scoping meetings, focus groups, workshops, open houses, charrettes, listening sessions, advisory committees, blue-ribbon commissions, summits, policy dialogues, negotiated rule-making, task forces, town meetings, citizen juries, study circles, future search conferences, online deliberation, and deliberative polling. Other terms refer to more specific practices, tools, or techniques that can be used together with particular formats. These include working groups, panels, debates, field trips, web sites, listservs, voting, consensus-building exercises, professional facilitation, process steering committees, visioning exercises, decision analysis exercises, scenario-building exercises, participatory budgeting, media campaigns, surveys, various educational or outreach activities, and so forth. The International Association for Public Participation, for example, offers a "toolbox" of dozens of such tools and techniques, classified by the purposes for which they are commonly used (see http://www.iap2.org/associations/4748/files/06Dec_Toolbox.pdf).

Frequently, different formats share practices in common or a single format is flexible enough that it can, under the right circumstances, integrate practices that are usually associated with a another format. As a result, processes called by the same name can look quite different in use, and processes with different names can have many specific components in common. For example, an expert panel can be assembled and integrated into virtually all the formats above, as can many other specific practices. Table 5-1 identifies and distinguishes three broad classes of public participation formats for purposes of reference.

Even a process that is tightly controlled by an agency may include a limited participatory role for the public. Information dissemination, not listed in the table, is sometimes considered to be a kind of public participation, albeit a very passive one (Creighton, 1999; Zarger, 2003). Conventional rule-making procedures often require that agencies publish a proposed regulation and allow a period of time for public comment before finalizing and implementing the rule (see Chapter 2). Such processes are open to the public, and they may or may not influence decision outcomes (e.g., Creighton, 1999; Beierle and Cayford, 2002; Zarger, 2003; Langbein, 2005). Information exchange, as noted in the table, allows for somewhat more interaction but still leaves little space for public influence.

Advisory committees and similar activities encourage a more active role

TABLE 5-1 Classes of Participation Formats Often Used by Government Agencies

Format Type	Breadth of Public Participation
Information Exchange (used both to inform and consult) Includes public hearings, comment periods, scoping meetings, focus groups, workshops, open houses, and listening sessions	Open access; often oriented toward individual citizens, but often includes interest group representatives
Involvement Includes citizen panels, deliberative polling, charettes, some advisory committees, citizen juries, study groups, town meetings, future search conferences, and online deliberation	Predefined group selected to represent diverse perspectives; may include individual citizens or group representatives
Engagement (in both decision making and collaborative action) Includes joint fact-finding, policy dialogues, negotiated rulemaking, blue-ribbon commissions, summits, community partnerships, and comanagement of projects or programs	Predefined to represent interested groups, sometimes geographically defined in the cases of partnerships or comanagement of projects to include stakeholders with local ecological knowledge

NOTE: We use descriptive terms to describe generic approaches that are distinct enough to be readily distinguished by a nonspecialist. Some of the terms in the first column of the table are sometimes used to refer to very carefully defined procedures. We do not mean to imply that all the formats in the same row of the table are alike, but rather that they have more in common with each other than they do with formats described in other rows.
SOURCES: Compiled from Renn et al. (1995); Beierle and Cayford (2002); International Association of Public Participation (2003); Zarger (2003); Leach (2005).

for some people. Committee members are chosen to represent a range of interests, and they may be asked to produce recommendations or other deliverables (Lynn, 1990; Lynn and Kartez, 1995; Renn, Webler, and Wiedemann, 1995; Bradbury and Branch, 1999; Zarger, 2003). Their meetings may be structured to encourage intracommittee interaction or participation by other citizens and groups. Agency personnel might or might not play a substantive role. Variations from traditional advisory committee structures include citizen juries, policy dialogues, citizen panels, study groups, and consensus conferences (Stewart, Kendall, and Coote, 1994; Dienel and Renn, 1995; Renn, Webler, and Wiedemann, 1995; Beierle and Cayford, 2002).

More collaborative formats may include a commitment to shared decision making among agencies and citizen groups, usually extending over a relatively long time period. They may incorporate interagency and intergroup relationships, and they may evolve over time (Pinkerton, 1994;

Zarger, 2003; Lubell and Leach, 2005; Leach, 2006). Some collaborative decision-making activities encourage interactions between agency scientists and citizens with local ecological knowledge (Murphree, 1991; Pinkerton, 1994; Berkes et al., 2001; Kemmis, 2002). Many require that agencies and public participants invest in building capacity and trust (Pinkerton, 1994; Cestero, 1999; Zarger, 2003; Lubell and Leach, 2005).

Consensus-building formats tend to promote binding agreements between an agency and a relatively small number of participants, selected to represent a range of stakeholders (Zarger, 2003; Birkhoff and Bingham, 2004; Langbein, 2005). These formats include negotiated rule-making, dispute resolution, and other mediated processes. They require some level of facilitation, whether by an involved agency or stakeholder or by an uninvolved professional. Outcomes might be constituted as a proposed rule, a memorandum of understanding, a statement of principles, a legal settlement, or a less formalized agreement.[1]

Public participation processes are commonly tailored to the specific circumstances of an assessment or decision, drawing on elements or practices from the various available formats to suit the context, explicitly addressing potential obstacles to success that are diagnosed at the planning stage and incorporating different participatory modes at different project stages (Creighton, 1999; Bleiker and Bleiker, 2000; Bradbury, Branch, and Malone, 2003; Zarger, 2003). Each decision process might schedule a number of participatory events, each tailored for a different procedural purpose and a different mix of participants (U.S. Environmental Protection Agency, 1998; Cestero, 1999; Creighton, 1999; Lawrence and Deagen, 2001; Bingham, 2003).

As already noted, most types of formats can incorporate specific mechanisms, tools, or practices. Most of these formats also can be more or less formal with regard to organization, protocol, and overall tone of communications. Most can be pursued with small or large budgets.

Few studies have rigorously and empirically compared participation formats, incorporating multiple cases with two or more formats. Such work would in fact be difficult to do because most of the formats are not very rigorously defined. The few comparative studies have examined relatively well-defined and distinct formats (e.g., conventional and negotiated regulation; e.g., Langbein, 2005) or have coded processes into categories based on reports about them (e.g., Beierle and Cayford, 2002). Moreover, almost all the studies are *ex post* assessments and have all the problems of this research form in attributing effects to causes.

Moreover, as we note above, it may not be the format itself that matters, but practices carried out within the format, on which researchers may or may not have data. In short, the ability to draw conclusions about format is limited by varying conceptualizations of what the formats are and inadequate theory about how they produce their effects.

A discussion of formats would not be complete without considering the potential role of the Internet in public participation. Though not a format in itself, it is attractive because it might provide a way to deal with some of the practical difficulties of conducting participatory processes when the participants are geographically dispersed or have limited available time and cannot be available for face-to-face contact. The individual and institutional costs of information transfer and acquisition can be lowered by Internet technologies, while convenience can rise. However, concerns persist about the "digital divide" in which access to and facility with online communications are not equitably distributed, with the possible results that use of the Internet will diminish the participation of some groups (Mossberger, Tolbert, and Stansbury, 2003; Chakraborty and Bosman, 2005; Martin and Robinson, 2007). Moreover, one cannot assume that interactions via the internet will have the same effects as face-to-face processes.

The limited evidence on using the Internet for public participation suggests that successful participatory processes can be conducted online, but the conditions for success are not yet established (Beierle, 2002). Studies suggest that online participation yields many of the benefits of face-to-face participation, but that in some cases it can also increase polarization (Capella, Price, and Nir, 2002; Price and Capella, 2002; Price, Nir, and Cappella, 2002; Inyengar, Luskin, and Fishkin, 2003; Price et al., 2003). A randomized comparison of face-to-face and electronic participation in deliberative polls found the two to perform similarly, with the benefits of electronic participation being slightly weaker (Iyengar, Luskin, and Fishkin, 2005).

In our judgment, electronic participation processes are most likely to be valuable when the assessment or decision problem will affect people who are not in geographic proximity to each other and thus are hard to bring together. However, we believe electronic processes should be tried only when adequate representation of the interested and affected parties can be obtained from among the subpopulation that has access to, and is comfortable using, available technology for online participation. Clearly, however, this is an area for continued experimentation.

DIMENSIONS OF PARTICIPATORY PROCESS

This section focuses on attributes of participatory process that cut across different formats. Evidence suggests that attributes are more relevant than formats to developing principles of practice. As noted in Chapter 3, one process can be more participatory than another in several distinct ways. The dimension most commonly used to organize typologies of public participation in the practitioner literature is variation in the nature of the public's role (termed *influence* below and often described by practitioners as the objective of the process), with some process formats being designed to inform the public, some to elicit public perspectives, others to involve the public in consensus decision making or recommendations, and still others to engage in collaborative action. As Table 5-1 indicates, formats can be roughly classified in terms of such objectives.

We have identified four dimensions of participatory process:

1. **breadth:** the number and variety of participants involved;
2. **openness of design:** the degree to which participants are involved at early stages of the process, the number of points in the process at which they are involved, and their influence on the design of the process;
3. **intensity:** the amount of time and effort participants put into the process and the amount of interaction that takes place among them, as well as between public participants and the agency officials and scientists who would otherwise be involved; and
4. **influence:** the extent to which the process allows for or provides mechanisms by which the public participants can affect how the convening agency defines, considers, and acts on the issue.

In much of the literature on public participation, the dimensions of breadth, openness of design, intensity, and influence are treated as highly correlated, in the sense that advocates often favor more participation on all these dimensions at once, and researchers do not always clearly distinguish them so that their effects on results can be assessed independently of each other. The largest multicase comparison study of public participation (Beierle and Cayford, 2002) found that the dimensions are strongly correlated in practice. Reflecting this correlation, Beierle and Cayford's classification of formats closely corresponds with the intensity of the deliberative process they required. Comparisons of negotiated and traditional regulations (e.g., Langbein, 2005) similarly link a difference in formats with differences in intensity.

In addition to the above dimensions, formats differ in what we call "boundedness." Some processes are bounded in that they identify and target specific parties and stakeholders or individuals representing those interests. Other processes are unbounded, in the sense that they are open to all parties and constrained only by the extent to which individuals and organizations have sufficient interest and resources to participate. Unbounded processes often tend to attract well-organized interests. Agencies sometimes use bounded processes to make sure that important interests or perspectives that might not find representation in an unbounded process get a place at the table—that is, to increase the breadth in comparison with what would happen in an unbounded process. However, some parties may still be left out.

Generally, processes designed to involve stakeholders in consensus building, such as advisory committees, summits, or commissions, are usually quite bounded even when meetings are open to the public, in that specific individuals are named as members of the group. In contrast, processes designed to inform or consult the public, such as scoping meetings, listening sessions, and online deliberations, often are more open. In some circumstances, process formats are combined so that both characteristics can be found in a single public participation effort.

Some observers suggest that less bounded formats may be appropriate early in a process for the purpose of problem formulation, when organized interest groups have not yet formed, or when there may be affected groups that are unorganized. As interests become more organized and the needed information and expertise clarified, it may be useful for the process to become more formalized and less open to new participants. Hypotheses like these are attractive because they specify intervening variables that might explain how formats affect outcomes through mechanisms of social interaction—but they are very hard to evaluate given the current state of knowledge.

Although dimensions such as intensity and boundedness are often correlated, decisions about how to organize public participation also can affect these dimensions independently. Agencies commonly face separate process choices about whether to invite or include some participants (breadth), whether to include public participants in particular discussions, including discussions about the process itself (openness of design), how many public meetings to hold (intensity), and how much influence to allow ideas and suggestions from public participants to have in making decisions (influence). We thus divide most of our discussion about the effects of the way participation is organized according to these four dimensions.

Before discussing the evidence on the effects of various ways of organizing the participation, it is worth characterizing this evidence. The best evidence of cause-effect relationships comes from controlled case-comparison studies in which an aspect of a participatory process is systematically varied and differences in outcome indicators are observed. Although such studies could be carried out to investigate the effects of various aspects of participation practice, very few have been done. A notable exception is the work of Fishkin (Farrar et al., 2003, 2006; Ackerman and Fishkin, 2004; Fishkin and Luskin, 2005) on the deliberative polling method, which has addressed energy policy options (but not other environmental topics). The lack of experimental research probably reflects some combination of practical difficulties, tight budgets for public participation, the absence of a culture of evaluation in many of the relevant agencies, and perhaps a lack of expertise in social research design in those agencies. Whatever the reasons, the shortage of studies with true experimental or quasi-experimental designs considerably weakens the confidence that one can have in inferences from experience. Experimental studies alone, however, will never be sufficient for the study of public participation. Although experiments greatly enhance the ability to understand causation, it is usually at the cost of creating a somewhat artificial situation and that may lead to different behaviors than would be found in real-world participation. Understanding of public participation will require "triangulation" across multiple methods. While more evidence accumulates, our conclusions should be assessed as provisional and requiring stronger evidence for certainty. In Chapter 9, we return to the issue of improving the quality of research on environmental public participation.

BREADTH

Public participation processes are more successful when they include the full spectrum of parties who are interested in or will be affected by a decision.

The argument to include all interested and affected parties goes back at least to Dewey (1923), who uses the idea of affected parties to define "the public." The idea is foundational to democracy. A major rationale for public participation processes is that, without such processes, many who are interested in or who will be affected by a decision may not have a chance to influence that decision, a situation widely judged inappropriate in a democracy. However, this normative justification does not necessarily entail that public influence is best organized through the kinds of direct participation in agency activities that are the focus of this volume. There are other avenues for public influence in environmental decisions. Moreover, governmental agencies, in exercising their duty to act in the public interest, sometimes see broad and direct public involvement as an impediment to their work. The issue for the present discussion is an empirical one: whether public participation improves decision quality, legitimacy, and the capacity of agencies and participants.

A number of analyses of environmental assessments and decision-making processes argue that inclusiveness is important for achieving legitimacy with the public (reviewed in National Research Council, 2007a). Mitchell et al. (2006) noted that global change assessments may lack credibility with key audiences if local expertise has not been included, and often participatory mechanisms are the only effective way to engage such local knowledge. In her study advisory boards at the U.S. Department of Defense (DOD) and the U.S. Department of Energy (DOE), Bradbury (2005) emphasized the importance to widespread acceptance of active outreach on the part of boards to make sure diverse perspectives were represented. Leach's (2005) assessment of Forest Service participation also emphasized the importance of inclusiveness.

However, a number of studies cite practical justifications for some limits to inclusiveness. Some of the justifiable restrictions that Leach extracted from the literature include excluding journalists, avoiding having too many parties involved, and being selective in including those who have the time and knowledge to effectively engage and are able to accurately reflect the views of their organizations (Leach, 2005; see also Wondolleck and Yaffee, 1994). For example, Floyd et al. (1996) found a negative correlation between the number of parties and perceived efficiency and equity of outcomes. A number of studies have suggested that participation should be restricted to those who can commit for the duration of the process and who have some expertise or can speak authentically for interested and affected groups (Selin and Myers, 1995; Shindler and Neburka, 1997; Yaffee, Wondolleck, and Lippman, 1997). Selin and Chavez (1994) argue that a proper mix of participants, including those with collaborative personalities and diversity of skills and resources, is helpful.

Some of these suggestions about whom to include or not to include are related to concerns with decision quality and some to legitimacy. Such suggestions for restricting participation have to be balanced with the importance of building capacity among communities of interested and affected parties, one of the objectives of public participation. Concerns about legitimacy tend to generate recommendations to involve parties that are already organized or that might become organized if they object to a decision—either by open invitation or through representatives. Concerns about quality tend to generate recommendations for breadth in terms of getting all significant viewpoints and sources of knowledge represented. There may be categories of individuals that it makes sense to exclude because of their professional relationship to the policy process. For example, attorneys who might be invovled in litigation about decisions or journalists who are covering the issue may not be appropriate participants.

A problem deserving special attention in considering breadth is the balance between national and local interests. In some cases, such as many toxic contamination problems, local issues logically predominate because there are few effects distant from the site. But for many natural resource management problems, there is a strong national interest in both economic development and environmental protection. The U.S.D.A. Forest Service (2000) has noted that local and national perspectives have to be balanced in participatory processes, although Shindler and Neburka (1997) argue that the inclusion of national interest groups may lead to their strong influence on and even domination of a local process. This is a problem deserving further research and careful thought when designing a process. As noted above, the Internet may be helpful in dealing with extra-local participation.

Global interests and affected parties are also relevant in some environmental decisions, such as those affecting climate or biodiversity. As discussed in Chapter 7, the spatial scale of the environmental issue affects who the parties are, and particular constellations of parties can present specific challenges to effective public participation.

Some of the issues regarding inclusiveness are pragmatic ones. Processes that are too large for real communication are not likely to be effective. The large body of research on collective action includes strong theoretical arguments and evidence that the ability to develop and realize a common agenda is inversely related, among other things, to the size of the group involved (e.g., Olson, 1965; Wade, 1994; Baland and Platteau, 1996). However, there is also contrary evidence (e.g., Marwell and Oliver, 1993). It seems most likely that the effect of group size is mediated by other factors. The ones most likely to be relevant to a public policy process are the levels of heterogeneity and interdependence among those in the group (Agrawal, 2002). In most environmental policy contexts, heterogeneity is fairly high and interdependence relatively low, suggesting that group size is likely to be an impediment to consensus. However, several process formats are specifically designed to accommodate large numbers of participants, including "world cafes," study circles, online dialogues, and large "town meetings" using electronic polling technology. Having large numbers of interested stakeholders is a contextual variable discussed in more detail in Chapter 8. The larger the number of participants, the greater the transaction costs of engaging with them (National Research Council, 2007a). Some of the process formats for accommodating large numbers can be expensive. An agency has to balance these costs against the value to be gained by engaging all the interested and affected parties and by helping individuals and groups to build capacity.

Many practices are available to identify and include all stakeholders. For public hearings and workshops, agencies often provide public notice of the event through advertisements in local newspapers and through press releases or, at the federal level, the *Federal Register*. Agencies may also send electronic messages to individuals who have attended past workshops or meetings and ask stakeholder organizations to forward meeting notices through their membership communication channels. Public participation professionals often call key stakeholders as part of the initial assessment and process design phase to be proactive about learning who might be interested in participating. In some circumstances, such as under the Negotiated Rulemaking Act and the Federal Advisory Committee Act at the federal level (see Chapter 2), agencies are required to publish a notice of the intent to form an advisory committee and to solicit comments on whether the proposed participant list represents all affected interests.

In sum, the literature indicates that it is very important to have representation for the spectrum of interested and affected parties and that, especially at the outset, it is important to identify all such parties in order to engage them. The research and practitioner literature indicates that it is better to err on the side of too much rather than too little inclusiveness. However, some of the advice emerging from recent studies appears to suggest that the notion of a "spectrum of interested and affected parties" (National Research Council, 1996:30) may be reasonably interpreted differently in relation to the needs of different parts of an assessment or decision process. For example, problem formulation requires enough of a range of parties to get all the plausible problem formulations considered. This may be a somewhat different range of parties from the ones needed to design the process, consider decision options, or interpret scientific information. Because a convening organization cannot always know in advance who will be needed to fulfill these roles, the admonition to err on the side of too much inclusiveness makes sense.

Public participation processes are more likely to be successful if they are structured to encourage voluntary commitment to them. Achieving inclusiveness requires that the interested public actually participate. Thus, it is important to understand what would motivate people, whether organized into interest groups or not, to engage seriously in a participatory process. Successful designs maximize existing incentives to participate, create new ones at times, and minimize disincentives and obstacles.

Effectively engaging those who might otherwise be absent from or not effective in the participatory process sometimes requires extensive outreach efforts and special ways of engaging those with limited time and skills. Issues that may seem minor to agency staffers, such as scheduling meetings at times and places that are accessible to those who should participate, can be consequential in terms of including certain parties (Tuler and Webler, 1999). Particular care may be needed to engage representatives of organized economic interest groups that may not see the advantage of engaging in a public forum (National Research Council, 2007a).

Special efforts to increase attractiveness make a difference because, as noted in several studies of families of public participation cases, it can be difficult to get a broad spectrum of the public to participate and easy for key individuals to drop out of the process as it goes on, either because they feel it is not of value or simply because of the press of other priorities (Bradbury, 2005; Leach, 2005; Moser, 2005). A key feature that makes a participation process attractive is the likelihood that the results of the process will influence agency decision making. Collaborative design of the process can also help make a process attractive to those who should be participating. We address these issues below.

OPENNESS OF DESIGN

The degree to which the participants have influence over the process itself is a critical element in the design of a process. This is closely related to the timing of the process—if there is little time for the process, there won't be time for the participants to influence the design. If the process is concluded too early, the latest information may not be available to inform decisions; if it is too late, a decision may be required before the process is completed. These concerns lead to the commonly stated dictum to involve the public "early and often," which could be understood as a proposition that the broadest possible participation is advisable in each phase of an assessment or decision process. For example, the Presidential/Congressional Commission on Risk Assessment and Risk Management (1997b:122), in the context of facility siting decisions by the U.S. Environmental Protection Agency (EPA), concluded that "inclusion of affected communities from the start as partners in the investigation and remedy selection processes can improve the likelihood that the choice of remedy will reflect reasonably anticipated uses of the site and the wishes of the community. Involving community members should also reduce the dissonance and long delays that often occur when EPA proposes solutions before discussing goals and costs with stakeholders."

As the discussion of breadth suggests, although public involvement from the start may be good advice as a general rule of thumb, the evidence suggests that it may be useful to differentiate the kinds of public input that are needed in different phases of the process. It may be wise to refine the "early and often" dictum to take into account the possibilities that somewhat different kinds of input are needed in different phases, and that the importance of maximally broad public participation may be greater at some phases of an assessment or decision process than at others.

Process designs developed collaboratively by those convening the processes and those participating in them yield benefits, particularly in terms of legitimacy. When participants co-invent and co-govern a process, they have a direct way to communicate information about what would motivate them to participate actively and to express their views about how to organize the process in a manner that is likely to engage effectively their capabilities and promote their acceptance of the process. Collaborative process design is likely, therefore, to increase legitimacy; a reasonable hypothesis is that it is important to include all those parties whose acceptance of the process is important. Inclusiveness may also affect the quality of the outcome if a component of the interested public that might be left out has expertise or interests relevant to the formulation of decision options. Furthermore, stakeholder groups that do not participate are not likely to receive the benefits of capacity building, nor is the agency or community likely to improve its capacity for engaging excluded publics.

Co-invention requires that the potential participants be identified and brought into the planning process as early as possible and that they participate to the extent possible in the choice of formats and decision rules and be able to seek assistance as needed from sources (e.g., scientists, impartial facilitators, mediators) selected by and accountable to all the parties. Participation specialists can make recommendations and advise all parties on what is likely to work best in the given context, but engagement and legitimacy depend on final decisions about the process design being made in a collaborative effort with the main parties involved.

The importance of participation in process design is consistent with the findings from cognitive science, already noted, that under conditions of uncertainty, people use the fairness of the process as an indicator or proxy for the trustworthiness of its outputs. This implies that a process that participants accept as fair from the outset is more likely to be legitimate and to be judged as of high quality.

The value of collaborative process design has long been articulated in the literature on public participation as well (Renn et al., 1993), and the empirical and practitioner literature is replete with studies showing the value of collaborative design of the participatory process. Leach (2005) notes the many analyses in the literature on Forest Service planning processes that emphasize the importance of collaborative planning of the process. Bradbury (2005:17, 20) found that DOE Superfund sites she studied, because of organizational tradition, were more flexible in the design of the participatory processes than the DOD sites. She argues that this may be one reason that the DOE processes were generally more successful. As she notes, "The public played an important role in helping DOE interpret what public participation meant" (Bradbury, 2005:20).

Public participation processes benefit in terms of both quality and legitimacy if the spectrum of interested and affected parties is involved in formulating the problem for assessment or decision. This finding reinforces one of the main conclusions of *Understanding Risk* (National Research Council, 1996), and the case-based evidence since then remains consistent with it.

Extensive social science research helps explain why broad participation is important during problem formulation. People use frames of reference to impose order on and give structure to complex environmental situations, and these frames thus sensitize them to specific aspects of problems (e.g., Couch and Kroll-Smith, 1994; Rich et al., 1995; Kamenstein, 1996; Pellow, 1999) and determine what scientific evidence will be used and will be judged as relevant and acceptable (National Research Council, 1996; Pellow, 1999; Halfacre, Matheny, and Rosenbaum, 2000). The research shows that incompatible framings and differing mental constructions of issues are two of the most important factors underlying conflict about risk and environmental management (e.g., Miller, 1989; Fisher, 1991; Carnevale and Pruitt, 1992; Kunreuther and Slovic, 1996; Fischhoff, 1996a; Pellow, 1999; Bazerman et al., 2000; Lewicki et al., 2002). Many disputes about technological and environmental risks appear to involve fundamental disagreements about the definition and nature of the problem to be addressed (Snow et al., 1986; Bradbury, 1989; U.S. Environmental Protection Agency, 1992; Couch and Kroll-Smith, 1994; Thompson and Gonzalez, 1997; Pellow, 1999).

There is evidence that some differences in frames are associated with membership in particular social, cultural, and economic groups. Cross-national comparisons, as well as studies of different social groups in the United States, consistently have found significant ethnic differences in risk responses, perceptions, and preferences (e.g., Kleinhessenlink and Rosa, 1991; Slovic, Kraus, Lappe, and Major, 1991; Vaughan and Nordenstam, 1991; Vaughan and Seifert, 1992; Flynn, Slovic, and Mertz, 1994; Bord and O'Connor, 1997; Weber and Hsee, 1998; Bechtel, Verdugo, and Pinheiro, 1999; Langford et al., 2000). Recent research has begun to elaborate on these findings by examining decision making under chronic loss (Rivers, 2006; Rivers and Arvai, 2007). In addition, culture influences the affective significance of a conflict (Kruglanski, Bar-Tal, and Klar, 1993) and judgments about what are acceptable ways to resolve disputes (Bazerman et al., 2000). In short, the culturally derived value priorities of individuals shape their frames of reference for environmental decisions (e.g., Kroll-Smith and Couch, 1991; Rich et al., 1995; Kamenstein, 1996).

Group differences in the subjective meaning of decision problems may underlie the commonly observed difficulty of engaging minority and economically disadvantaged communities (e.g., Pellow, 1999), as well as contributing to difficulties in arriving at analyses of environmental problems that are accepted in these communities as legitimate. For example, environmental risk issues are especially likely to be framed as moral or value dilemmas in ethnic minority and lower income communities whose structuring of risk problems frequently reflects a justice and equity frame (e.g., Bullard, 1990; Mesquita and Frijda, 1992; Vaughan and Seifert, 1992; Pellow, 1999). African Americans in particular are more likely than others to evaluate and structure risk problems in terms of fairness, equity, and justice (e.g., Vaughan and Seifert, 1992), and such a framework contributes to more intense and durable emotional responses (e.g., Mikula, Scherer, and Athenstaedt, 1998), including attributions of blame and the expression of such emotions as anger, disgust, and fear (e.g., Mikula, Scherer, and Athenstaedt, 1998). Even if groups are similar in framing issues in terms of fairness, their notions of what is fair in a situation may vary (Scherer, 1997;

Langford et al., 2000), resulting in different predominant intensities and types of emotions and appraisals (Scherer, 1997; Bohm and Pfister, 2000).

The degree of difficulty of conflict resolution is influenced by several factors, including the framing of a conflict and the level of congruity among participants' cognitive and affective representations of decision problems (Bazerman et al., 2000). During successful negotiations, frames of reference often evolve, and parties develop shared or compatible perspectives on the nature of the problem to be solved and beliefs about whether goals are compatible (Kruglanski, Bar-Tal, and Klar, 1993). By contrast, when individuals believe that decision processes have not adequately taken into account important values, trust in the process tends to be undermined, conflict exacerbated, and negative affective reactions prolonged (e.g., Fisher, 1991; Rich et al., 1995; Shah, Domke, and Wackman, 1996; Susskind and Field, 1996; Baron and Spranca, 1997; Thompson and Gonzalez, 1997). This evidence suggests that efforts to formulate problems for assessment and analysis in ways that fit with participants' frames of reference are more likely to be accepted. Similar conclusions have been drawn from analyses of large-scale environmental assessments related to global environmental change (e.g., Lemos and Morehouse, 2005; National Research Council, 2007b).

Finding common ground may require framing decision problems in more than one way, so that those analyzing the problem and considering the choice options have the opportunity to do so from multiple perspectives. Evidence from research on group process and deliberative decision making suggests, however, that it makes a difference how process conveners go about eliciting multiple perspectives. (We discuss this evidence in more detail below, in discussing the intensity and formats of participation.)

Broadly based problem formulation can ensure that the agency's initial frame is not the only frame applied. Collaboration in problem formulation at the start can also ensure that the process is attentive to the most important values of interested and affected parties. It can also help elucidate cultural differences in risk perceptions, values, beliefs, expectations, and decision-making styles, any of which can be quite substantial and can potentially undermine the success of a participation process.

Because a key purpose of problem formulation is to develop a set of questions on which participants need good information, a breadth of perspectives is essential. Broadening the perspectives considered thus has a positive effect on both legitimacy and decision quality. Scientific analysis that addresses the problem from multiple perspectives is typically more robust than a more narrow formulation.

Broad participation is also important at the point of interpreting scientific information—what *Understanding Risk* calls the synthesis phase—because individuals or groups that apply different frames to a problem are likely to interpret scientific information through different lenses. Public participation processes may well benefit by using elements of the standard scientific process of independent peer review. This is especially true if practical constraints require that a process engage with people who are presumed to represent interested and affected parties rather than with all who are interested. In such situations, peer-review-style processes can ensure that the representatives are well calibrated with those they are intended to represent. It also may be useful to submit any interpretations of information resulting from the process to a structured peer review process (National Research Council, 2007a). This process is likely to increase the quality of the product, as well as its legitimacy, and it allows the outputs to be refined as a result of comments received. There is some evidence that holding a group's outputs "accountable" through a peer review process increases objectivity and reduces bias (Delli Carpini, Cook, and Jacobs, 2004:328).

INTENSITY

Most studies in the public participation literature find a positive association between the intensity of deliberation—such variables as the number and length of face-to-face interactions and the amount of time participants spend in the process—and desired results (e.g., Beierle and Cayford, 2002). In our judgment, this overall association between intensity and outcomes reflects the great importance of intensity in many situations in which a major controversy or mistrust demands intense interaction to reach a resolution. Intensity may not be as important in other situations. The key point is to have a process for which the intensity is appropriate to the context. Of particular importance is the structure of the face-to-face interactions that are the heart of a participation process. Results can be highly sensitive to the extent to which the participatory process is organized so as to ensure that the advantages of group deliberation are enhanced and the potential adverse effects are minimized.

The intensity of the deliberative process does not have a simple or universal relationship to results. The best results follow from processes whose intensity is dictated by responding to context-specific challenges (see Chapters 7 and 8) with appropriate participation strategies. Contexts that present challenges that require intensive interactions, such as those involving serious potential for conflict, can benefit more from high-intensity processes than contexts that do not present such challenges. However, when the context calls for intense interactions, results are highly dependent on how those interactions are organized.

Intense deliberative processes create significant potential to promote desired results from participation, but at some costs. They can increase opportunities to improve mutual understanding among those who participate, to modify the process as it proceeds, and for all participants to develop a solid understanding of both the relevant scientific information and the perspectives of various communities used in understanding the issues. However, there can be a trade-off between the intensity and the breadth of participation. Beierle and Cayford (2002) found that in their large sample of cases, more intensive processes tended to be less inclusive and less representative. Furthermore, intense processes may create distance between those participating and other interested and affected parties not involved, thus reducing transparency. They can also lead to consensus on novel solutions among those participating that may meet resistance among the constituencies that the participants are expected to represent. And, of course, more intense processes can be financially costly and so may not be feasible in some circumstances.

The research literature does not give a clear overall message regarding intensity, perhaps because the effects of intensity are so dependent on context. Tuler (2003) reviewed 15 multicase studies from different policy arenas, with somewhat mixed results for the intensity of deliberation. For example, Duram and Brown (1998) found no significant relationship between meeting frequency and perceived effectiveness, whereas Henry S. Cole Associates (1996) found that ongoing committees or panels were more effective than forms of participation with less continuity. Although Tuler (2003:18) concluded that "more extensive roles for participants within a process can improve the competence of decisions," much of the evidence cited seems more closely related to the early involvement of various parties and the incorporation of community knowledge than to the intensity of interactions. Early involvement may imply a longer engagement, and hence greater intensity, but the issue is which variable influences the results. Tuler reported mixed findings about the effect of duration of the process. For example, among the studies examined by Leach, Pelkey, and Sabatier (2002) found that older watershed partnerships had higher perceived impacts, but Gericke and Sullivan (1994), in a study of 61 Forest Service forest land management plans, found that total time spent on public participation did not reduce the number of appeals of forest plans or the time spent to resolve them. However, this same study found that processes involving two-way communications between the Forest Service and stakeholders had a lower probability of high numbers of legal appeals compared with processes in which the Forest Service only gathered information from or provided information to stakeholders.

Ashford and Rest (1999), in reviewing successful participation processes at seven hazardous waste cleanup sites, indicated that both longer ongoing procedures and shorter, more intense procedures (e.g., summits) can be successful. Since their study did not include unsuccessful cases, it does not provide evidence of whether or not success is more likely with intensive processes but provides an important existence proof that less intensive processes can yield success. Another such indication comes from a comparison of responses to a willingness-to-pay survey with and without a brief structured group discussion before answering that demonstrated some significant positive effects of even minimal deliberation (Dietz, Stern, and Dan, in press).

A few studies of negotiated regulatory rule-making provide evidence based on comparisons between processes that are quite similar in many respects but that differ systematically in the intensity of participation. For example, negotiated rules require much more intense public participation than the conventional, notice-and-comment style of rule-making. A study by Coglianese (1997) reviewed archival sources to compare conventional and negotiated rule-making processes at EPA and found that negotiation neither lowered the probability of subsequent lawsuits nor significantly shortened the amount of time required for rule-making. The study's methods and conclusions, however, have been strongly disputed (Harter, 2000).

Kerwin and Langbein (1995; Langbein, 2005) examined six conventional and eight negotiated rule-makings at EPA and found no difference in litigation rates, but they drew several more nuanced conclusions. Compared with participants in conventional rule-making (who only submitted written comments on proposed rules), participants in regulatory negotiation had more favorable overall assessments of the processes, both in terms of their assessments of the final rule and of some aspects of the process. This was true even though participants in regulatory negotiation perceived the regulatory issues at hand to be more complex than those involved in cases of conventional rule-making and perceived negotiated rules to address more difficult issues of regulatory compliance and implementation (Langbein, 2005). These differences persisted even after controlling statistically for participants' perceptions of the final rule's benefit to their own organizations and of economic efficiency for society. In addition to being more satisfied with final rules compared with participants in conventional rule-making, participants in negotiated rule-making varied less in their levels of satisfaction, indicating higher levels of interparticipant agreement or consensus (Langbein, 2005).

This study also found differences in evolving relationships between participants and the convening agency and also among participants. Participants in regulatory negotiation were significantly more likely to perceive EPA to have encouraged their participation, compared with participants in conventional rule-making. Although participants in the two formats were equally likely to perceive disproportionate influence by some parties, participants in negotiated rule-making tended to see this disproportionate influence as affecting only aspects of the process, not of the rule. Participants in conventional rule-making more frequently perceived undue influence to manifest in the rule itself, and more frequently named EPA as the interest group with disproportionate influence. In addition, informal negotiations among the participants with the goal of seeking greater unanimity or consensus were more frequent in negotiated rule-making, whereas in conventional rule-making, external informal negotiations were more likely to be aimed at negotiating a rule that would work for EPA (Langbein, 2005). Finally, parties in negotiated rule-making indicated that they learned more than did participants in conventional rules, suggesting an association of deliberation intensity with increased capacity among participants.

Some additional studies have examined how intensity influenced the perceptions of the participants. Lubell and Leach (2005) reported that intensiveness of deliberation was positively associated with perceived partnership effectiveness for the National Estuary Program (NEP) and the Watershed Partnership Project (WPP) but noted that causation was uncertain: individuals' perceptions that a process is effective might cause them to participate more intensively. They also noted that continued participation could lead individuals to make more positive judgments as a way to reduce cognitive dissonance—that is, to justify their investment of time and effort to themselves. Intensiveness in this study was measured by an indicator of teamwork among stakeholders for the NEP and as hours spent on activities for the WPP. In multiple regressions that controlled for other variables, positive relationships were found between intensiveness and the following dependent variables: perceived effectiveness (an index of 12 ecological and social variables), cooperation, and consensus for the NEP, and perceived effects on watersheds and on human and social capital for the WPP. For the WPP, longer project durations were also positively associated with policy agreement and project implementation. Regression analyses predicting policy agreement and project implementation were conducted using the watershed (i.e., the case) as the unit of analysis, whereas all other regressions used the individual respondent as the unit of analysis.

Moser's (2005:63-67) study of participants in the various regional assessments under the National Assessment of Climate Change found similar results. Individuals with more intense involvement judged the assessment as more legitimate and believed it was "more likely to build lasting and trusting relationships or spawn new collaborations" and to produce broader environmental and social outcomes. Although the regional assessments varied greatly in the intensity of the processes used, the small number of respondents per regional assessment made it impossible to analyze the relationships between intensity and the quality of the output by comparing across assessments.

Advisory committees are a particularly intense form of public participation. Although some studies have associated advisory committees with successful results (Ashford and Rest, 1999; Bradbury, Branch, and Malone, 2003; Lubell and Leach, 2005), participants from low-income and racial minority communities have expressed frustration with advisory committees that they felt diluted their voices (Ashford and Rest, 1999) and a preference for public meetings, in which they saw their concerns as less likely to be muffled by other interested and affected parties, such as businesses. This concern is consistent with our judgment that it is not intensity in itself that matters, but a design that is appropriate to the context, including the nature of the parties and their relationships. Moser (2005) noted another potential negative effect of intensity: that participants in regional climate change assessments who were more intensely involved might be more disappointed when the assessments were not completed.

Taken together, these studies present evidence of a relatively strong positive association between the intensity of deliberation and the success of public participation efforts. However, it is not clear that this association is evidence for a direct causal effect of intensity on success. Rather, the effects of intensity may be spurious, in the sense that more intensive participatory processes tend to be more successful because they tend to include certain kinds of activities or social interactions that are actually responsible for positive effects. It is important to be clear about whether simply increasing intensity (e.g., adding meetings) can be expected to yield better results, or whether those results instead derive from specific actions taken during the meetings that do occur. In our judgment, the latter is more likely the case.

This judgment is difficult to test against evidence. Although researchers on public participation have for some time argued that the rules for and structure of the face-to-face interactions characteristic of public participation settings can be very consequential (Dietz and Pfund, 1988; Renn et al., 1993; Renn, Webler, and Wiedemann, 1995), empirical research on environmental public participation has not given much systematic attention to the details of how the interaction among participants is structured. It has instead focused primarily on the contextual issues reviewed in Chapters 7 and 8 and some of the issues of process management, such as breadth and intensity of participation addressed earlier in this chapter. However, a substantial research literature on small-group interaction and deliberative process, much of it conducted by experimental methods that allow for fairly strong inferences about cause and effect, gives ample evidence for the importance of the interaction process and provides useful insight into what happens in relatively intense interactions that affect the quality and legitimacy of their outputs.

First, research from a variety of experimental settings demonstrates that face-to-face communication enhances the probability of cooperation (Dawes et al., 1990; Bornstein, 1992; Sally, 1995; Ostrom, 1998; Delli Carpini, Cook, and Jacobs, 2004). It is well known that individuals are more likely to support decisions made by a group if they feel that the decision-making process was fair. Fairness often implies that all views are given reasonable hearing, whether they are held by a majority or a minority of participants.

A number of studies have shown that there can be a tendency for groups to converge on majority views (Moscovici and Zavalloni, 1969; Myers and Lamm, 1976; Schkade, Sunstein, and Kahneman, 2000), a result that is likely to bring with it suboptimal-quality decisions, particularly when important information is held by a minority (see, e.g., Gigone and Hastie, 1993, 1997; Wittenbaum, Hubbell, and Zuckerman, 1999; Delli Carpini, Cook, and Jacobs, 2004). However, under well-structured group processes, minority views, rather than being ignored, can cause majority groups to "consider new alternatives and perspectives," "seek out new information," and "empathize with the minority's viewpoints" (Delli Carpini, Cook, and Jacobs, 2004:327). Well-structured interactions can also reduce bias and increase agreement (Gaertner et al., 1999). In their study of willingness to pay with and without small group deliberation, Dietz, Stern, and Dan (in press) found no evidence of convergence on the majority opinion but did find that group participation led respondents to take more aspects of the choice into account in assessing their willingness to pay and to be less influenced by their personal perspectives.

However, strong communication within groups and cooperation can also degrade communication and cooperation across different interacting groups (Bornstein, 1992; Insko et al., 1993; Mendelberg, 2002). This research suggests that although enhanced communication and cooperation within a group engaged in deliberation is a benefit of the process, it holds the potential for alienating those actively participating from the larger groups they represent, with a resulting loss of transparency and trust in the process by the larger community.

A number of studies emphasize the importance of different "languages" that are deployed by various groups in deliberation (see Delli Carpini, Cook, and Jacobs, 2004, and Mendelberg, 2002, for discussions of this research). Public participation in environmental assessment and decision making inevitably involves multiple such languages, including those of technical experts, such as scientists and lawyers; experts on what is politically and organizationally feasible, such as agency officials and politicians; and interested and affected parties, who bring knowledge of both local context and the concerns of their communities (e.g., Dietz, 1987). There are likely to be substantial differences in languages within each of these three types of participants as well. It is thus crucial to have a process of interaction structured so that language barriers to mutual understanding can be overcome. In particular, it is important that domains of expertise be acknowledged but also delimited, so that expertise in one area (e.g., scientific expertise about the dynamics of an ecosystem) does not become conflated with expertise in another (e.g., expertise on what is important to members of the local community) (Mendelberg, 2002).

A recent review of the literature on public deliberation (Delli Carpini, Cook, and Jacobs, 2004:336) concludes that "although the research summarized in this essay demonstrates numerous positive benefits of deliberation, it also suggests that deliberation, under less optimal circumstances, can be ineffective at best and counterproductive at worst." We echo those findings. A number of specific procedures for conducting face-to-face deliberations have been proposed, which may provide useful guidance. However, because of the great variety of settings and the potential for surprises in dynamic social processes, we think it is more important to identify and focus on principles of good process, such as those enumerated at the beginning of this chapter, than to closely follow any set procedure. We emphasize that the processes and methods used in the face-to-face interaction at the heart of a typical participation process must be tailored to the context, as discussed in Chapters 7 and 8.

INFLUENCE

As discussed in Chapter 2, the power to make public policy decisions about the environment is typically vested by legislatures in government agencies. Thus, the formal power and influence of the public on these decisions lies in the ability to elect legislators and chief executives, to contest administrative decisions in courts or in established appeal procedures, and to influence decisions in administrative procedures established by law. However, agencies also have considerable discretion to go beyond the letter of the law and open themselves to input and influence from interested and affected parties in many ways, both in assessing environmental conditions and in dealing with them. From the perspective of public administration, this is the import of the kinds of public participation mechanisms examined in this volume. Such mechanisms can vary considerably in the degree to which they provide for public influence. This section examines evidence on how the extent to which public participation processes invite or allow such influence affects their results.

Public participation processes yield better results when they are transparent to those involved and to those observing them. Transparency includes an understanding of the purpose and objectives of the assessment or decision process and of the legal and administrative authorities, requirements, and constraints governing it and public participation in it. Transparent processes typically include ongoing communication about the process and public access to information about the process and about informational and other inputs to it.

Transparency is necessary for the possibility of public influence, and most research indicates that it is essential to effective participatory processes. Transparency does not necessarilty require that every meeting and activity be open to the public. There are valid reasons for having some nonpublic activities within a broadly transparent process (see Chambers, 2005). However, when an agency is perceived as having a history of secrecy and decision making "behind closed doors," it can be particularly difficult for the public to accept a process that is not fully open (Bradbury, 2005:9). In such instances, a process that provides open access to information in a way that is appropriate for the participants can help in overcoming mistrust (Bradbury, 2005). The benefits of transparency have been documented from the earliest through the final stages of public participation processes. Several studies found that transparency in the process of selecting participants partly determines the credibility and legitimacy of the overall program (National Research Council, 1996, 2007a; Watson, Bulkeley, and Hudson, 2004). At least one study noted that it can be helpful to ask participants to publicly state their biases at the outset of the process (National Research Council, 2007a). In the later stages of a decision process, some studies have suggested that transparent and extensive reviews can help to increase credibility, partly by expanding the number of involved stakeholders (Edwards and Schneider, 2001; Watson, Bulkeley, and Hudson, 2004; Goldschmidt and Renn, 2006; National Research Council, 2007a).

A substantial literature shows that given an open process and access to scientific information, the public can develop sound understandings of many aspects of enviornmental science and thereby increase their capacity to participate in future decisions (Brown and Mikkelsen, 1990; Kinney et al., 2002; McCormick et al., 2004; McCormick, 2006, 2007a,b). For example, Brown and colleagues (Brown et al., 2004, 2006; McCormick, Brown, and Zavestoski, 2004; McCormick et al., 2004; Zavestoski, McCormick, and Brown, 2004) examined health movements and found that these movements, which are usually disease specific, are quite effective at forming liaisons with health researchers and identifying inadequacies in research on the diseases that are their focus. In particular, these movements have emphasized the need to examine environmental causes of disease.

McCormick (2006, 2007b), in a case study of popular resistance to dam construction in Brazil, shows how citizen activists, working with scientists, produced new scientific conceptualizations of the issues to be addressed in considering the dams. In at least one North American case, stakeholder mediation over plans to dam a river generated a broader slate of options, including not only flood control, but also more general economic, land use, and conservation planning for the affected area (Cormick, 1980). Another stakeholder group, advising the U.S. Department of Energy on remediation of a nuclear weapons facility was credited by the agency for a cost savings of more than $2 billion (Applegate and Dycus, 1998; Beierle and Cayford,

2002). The ability of citizens to produce such pertinent and effective recommendations requires that they acquire pragmatic understandings of agency goals, capabilities, and constraints.

One practice that has been advocated on grounds of transparency is that of establishing fixed rules of process and decision making. The evidence on the effects of this approach is mixed: Leach (2005) found that seven studies of public participation in the Forest Service concluded that clear fixed rules contributed to success, while two studies favored flexible or informal processes. We note, however, that it is possible to achieve transparency and to have a process open to public influence with process rules that are relatively stable or that change substantially over time. The keys are the nature and extent of public influence in setting or changing the process rules and the openness of information about rule-setting and changing.

Some types of participatory process cannot be open to all who are interested, simply for practical reasons. When participation must be limited to representatives of various groups and points of view, transparency regarding who is selected and how they are selected is especially important (National Research Council, 2007a). And as noted above, a peer review process may be useful in ensuring that representatives are in fact representing the views of the groups from which they were selected.

Participatory processes have better results when all parties commit to act in good faith and maintain communications with each other and those they represent. Good-faith communication is partly a matter of attitudes, but it also involves behavior (e.g., responding appropriately to input from others) and creating and maintaining mechanisms for communication to and from decision makers, scientists, interested and affected parties, and the public. It involves, for the responsible agency and the interested and affected parties, keeping the other parties informed of progress and reporting on actions taken or other changes that may affect the process and the reasons why such changes occurrred. A negotiated consensus on rules for communication, deliberation, and decision making, before the substantive issues are discussed, creates a basis for good-faith communication because with agreed-on procedural rules, it is easier for the moderator to enforce these rules and to ensure fair play among all participants. Especially when an agency has problems of public trust, demonstrations of good-faith communication through action and the maintenance of mechanisms provides an avenue for public influence and an indication to interested and affected parties that influence is possible.

Good-faith communication on the part of all parties is important for an open and transparent process and, obviously, for maintaining trust among parties. Research on group processes suggests that perception of outcomes can be influenced by how participants treat each other (e.g., Leventhal, 1980; Tyler and Lynd, 1992). Bradbury (2005) reported several specific examples in which DOE took special steps to ensure open, clear communication among parties that contributed substantially to the success of its processes. She also noted that the agency's clear and ongoing demonstration of a commitment to take public inputs seriously was helpful in overcoming mistrust. Leach (2005) found that both professional facilitation and the training of participants in communications skills enhanced the chances of success. Although facilitation and training do not preclude disingenous communications, they can reduce misunderstandings.

CONCLUSION

The evidence supports four principles of good practice for organizing public participation processes: inclusiveness, collaborative problem formulation and process design, transparency of process, and good-faith communication. As with the management principles described in Chapter 4, it is not always easy to implement these principles, and some contexts can make it especially difficult to implement certain principles. In difficult situations, success depends on identifying the likely difficulties and finding ways to address them. We return to these issues in Chapters 7 and 8.

NOTE

[1]This summary of public participation formats is limited to those in which government agencies might choose to become directly involved. As discussed in Chapter 1, other modes of public participation may be initiated outside the auspices of government agencies. They include voting in elections, citizen ballot initiatives, citizen referendums, New England–style town meetings, lobbying of legislatures and executive offices, formal and informal debates and deliberations, public information campaigns, public demonstrations, civil disobedience, lawsuits, and other activities. These extra-agency activities may proceed on timelines that either support or interfere with agency-led forms of public participation.

6

Practice: Integrating Science

Because of the substantial scientific content of environmental policy issues, efforts to engage the public must address the issue of integrating science and public participation. This chapter elaborates on the need for integration, particularly in terms of achieving the objective of quality. We note a series of challenges posed by this need to integrate, assess available knowledge relevant to meeting them, and identify several norms and procedures that promote the successful integration of science and public participation in environmental policy. In Chapter 4, we note that many of the practices required for effective public participation are also sound management practices. As this chapter shows, the practices required for shaping processes to integrate science and participation are also sound practices for organizing science to inform public policy. This chapter concludes with a discussion of the issue of implementing the principles of good practice set forth in this chapter and the previous two.

The evidence reviewed in this chapter shows that integrating science and public participation through processes that iterate between analysis and broadly based deliberation—as recommended in *Understanding Risk* (National Research Council, 1996) and subsequent National Research Council (e.g., 1999a, 2005a) reports—promotes the quality, accountability, and legitimacy of environmental assessments and decisions. In contrast, processes that treat analysis and deliberation in isolation from each other impede both analysis and deliberation.

Evidence from various sources suggests that efforts to integrate science and public participation are more likely to produce satisfactory results if they follow five specific principles:

1. *Availability of decision-relevant information:* The processes ensure that decision-relevant information is accessible and interpretible to all participants and that decision-relevant analyses are available in open sources and presented in enough detail to allow for independent review.

2. *Explicit attention to both facts and values:* Efforts are made to identify the values at stake, to consider different formulations of the problem to be analyzed that may embody different values or concerns (especially in the initial design phase of a public participation process), and to analyze how the available choice options affect various values.

3. *Explicit description of analytic assumptions and uncertainties:* The analysis and deliberation include the implications of different assumptions and different possible actualizations of uncertain factors.

4. *Independent review:* Official analyses are reviewed by other competent analysts who are credible to the parties.

5. *Iteration:* Past conclusions are reconsidered on the basis of new information and analysis.

INTEGRATION

In Chapter 2, we define quality in environmental assessments and decisions in terms of five elements:

1. identification of the *values, interests, and concerns* of the agencies, scientists, and other parties that are interested in, or might be affected by, the environmental process or decision;

2. identification of the range of *actions* that might be taken (for decisions);

3. identification and systematic consideration of the *effects* that might follow from the environmental processes or actions being considered, including uncertainties about these effects, and consideration of kinds of impacts that deserve consideration given the values, interests, and concerns of those affected;

4. outputs consistent with the *best available knowledge and methods* relevant to the above tasks, particularly the third; and

5. incorporation of *new information*, methods, analyses, and concerns that arise over time.

A good decision has been defined as one that is logically consistent with what is known (e.g., information, including uncertainties), what the decision maker (or the constituencies that he or she represents) wants (i.e., values and preferences about the possible effects), and what the decision can do (management alternatives or actions) (Howard, 1966, 1968; Raiffa, 1968). Approaching decision making in this way seems like common sense

(North, 1968), but in practice it is difficult to do, especially with complex decisions affecting the environment. Furthermore, environmental decisions also involve considerations of fairness and learning from experience that are not explicit parts of most formal decision frameworks (Dietz, 2003).

All of these criteria for good decisions can be met only if scientific analysis is used effectively. Uncertainties in the *best available knowledge* (element 4 above), regardless of whether this knowledge comes from data collected using scientific methods, from the judgments of scientific experts, or from observations made without the use of formal methodologies, must be considered as a part of scientific analysis, using appropriate qualitative and quantitative methods. Of course, the interested and affected parties to a decision are generally the best judges of what they want and of their values—but without scientific analysis, they may not know when or how environmental decisions affect those values. For example, as science showed that climate change will affect not only average temperature, but also the frequency and intensity of coastal storms, floods, droughts, wildfires, and so forth, some people who had considered themselves at little risk reconsidered their positions.

Scientists are usually in the best position to identify and systematically consider the effects of environmental processes and actions. However, good scientific analysis often requires information about local context that is most likely to come from people with close experience with local conditions. In a well-known example, British authorities advised sheep farmers after the Chernobyl nuclear accident that they could avoid radioactive contamination of their flocks by simply keeping the lambs out of the valleys. But the farmers knew that the fields were unfenced, so the solution was not practical and the risk was greater than the government scientists thought (Wynne, 1989).

These examples are among many that could be cited to show that integrating scientific analysis and public input requires more than a handoff of tasks from one group to another. The public cannot make good value judgments without good science, and scientists cannot do good decision-oriented analysis without public input. Recognizing the latter point, many policy reviews have advocated integration of public input into environmental assessment processes that have traditionally been dominated by science (e.g., National Research Council, 1996, 1999a, 1999b, 2005a, 2007a; Presidential/Congressional Commission on Risk Assessment and Risk Management, 1997a,b). They recognize that past nonintegrated assessment efforts have suffered in terms of both quality and legitimacy because they did not fully incorporate information (including appropriate consideration of uncertainties) and concerns coming from various affected parties.

These studies represent an important departure from previous thinking about how to conduct environmental assessments for informing practical decisions. They show that public input holds the potential to avoid the repetition of past failures, although they provide few specifics on how best to integrate public input to achieve its theoretical benefits. Most importantly, they do not address in much detail several well-known difficulties that present significant challenges to successful integration of public input and scientific analysis.

CHALLENGES OF INTEGRATION

Integration is challenging for several reasons, including attributes of the science involved, characteristics of the public, and the difficulties of communication.

Challenges Related to Science

One set of scientific challenges arises from lack of data, the complexity of environmental processes, and the uncertainty of scientific knowledge about environmental processes. In the absence of precise knowledge, a traditional decision rule is to be conservative: either choose options that include a margin of safety that is adequate to avoid bad effects or outcomes, or choose analytic procedures to avoid underestimating the probability of the bad effects or outcomes. Another approach is to analyze how each bad outcome might occur, with a major effort to understand the sequence of events or the characteristics of situations in which bad outcomes or effects may occur. Guidelines for good practice in such analyses may be a useful way to deal with many decision situations without a need for repeated, expensive, and time-consuming analyses. But it must be recognized that such guidelines typically include important value judgments on how conservative the analysis or decision process should be.

Especially in cases in which scientific knowledge is evolving, it may be appropriate to have both guidelines and a procedure to depart from the guidelines (see Box 6-1). Departure from the guidelines may involve detailed analysis including formal probabilistic methods to characterize uncertainties affecting what can go wrong and lead to bad outcomes. Such methods make the value judgments about how to deal with uncertainty or how to make trade-offs among different bad outcomes explicit rather than being left implicit in guidelines and therefore not open for review or discussion. We emphasize that such trade-offs, although not part of science, are a critical input into the decision-making process.

Another challenge comes from the possibility that informing environmental decisions may require some reconsideration of standard approaches to scientific epistemology (Funtowitz and Ravetz, 1993; Rosa, 1998). For example, standard scientific practice places the burden of statistical proof on the data that show associations or causal relationships. The default

> **BOX 6-1**
> **Guidelines for Analysis Under Uncertainty and Departures from Guidelines**
>
> A 1983 National Research Council (NRC) report developed the idea of default assumptions as a means to bridge across uncertainties surrounding the risk posed by chemicals that may cause cancer in humans. In the context of a great many decisions on regulating a multitude of such chemicals in the environment, it was judged appropriate to have a standard set of assumptions, called "inference options" (National Research Council, 1983) and later "defaults" (e.g., National Research Council, 1994), so that cancer risks could be estimated for many chemicals in a consistent and standardized way, rather than using different procedures for different chemicals. The default assumptions were to be chosen conservatively, so that human cancer risk is more likely to be overestimated than underestimated. *Guidelines for Carcinogen Risk Assessment of the U.S. Environmental Protection Agency* (EPA) (1986:21) described the estimate as a "plausible upper limit to the risk that is consistent with some proposed mechanisms of carcinogenesis. Such an estimate, however, does not necessarily give a realistic prediction of the risk. The true value of the risk is unknown, and may be as low as zero." Following the 1983 NRC report, federal and EPA cancer risk guidelines were established so that agency procedures for calculating risk from human exposure to potentially carcinogenic chemicals became more predictable for all of the interested and affected parties.
>
> Guideline-based procedures for cancer risk assessment have been criticized as slow, resistant to change in response to new scientific information, opaque to nonscientists, and, perhaps most important, as making value judgments invisible and less open to discussion. Although both the NRC reports endorsed the use of defaults, both also stressed the need for iterative processes in which standard procedures using the guidelines could be used for screening, priority setting, and more routine decision making, while exceptions would be permitted when warranted by the importance of the decision situation and by new scientific information. The Presidential/Congressional Commission on Risk Assessment and Risk Management (1997a,b) also recommended an iterative process that includes both risk analysis and public deliberation. Although EPA's cancer risk guidelines have recently been modified (U.S. Environmental Protection Agency, 2005) to have more flexibility, only a few examples of departures from defaults have occurred in EPA's cancer risk assessments for specific chemicals.

presumption or "null hypothesis" is that there are no effects; to reject that presumption, the data must be so inconsistent with it as to render it highly unlikely (i.e., a likelihood of less than 5 percent or 1 percent). This conservative approach comes at the cost of treating associations and causal relationships that do not meet that high standard of proof as if they are not present. In risk management contexts, this practice followed naively could lead to ignoring consequential risks and costs—risks and costs that might be of great concern to the public and its well-being. Thus, standard conservative scientific practices for making knowledge claims may lead to misunderstandings in risk management contexts.

Alternative practices are available, such as eliciting scientists' judgment in the form of probability estimates. Characterizing uncertainty in terms of probabilities based on judgment has been widely advocated (e.g., Raiffa, 1968; Morgan and Henrion, 1990; Intergovernmental Panel on Climate Change, 2001; National Research Council, 2002b), and some applications have been carried out (e.g., Howard et al., 1972; Morgan and Keith, 1995; Moss and Schneider, 2000). However, many scientists remain skeptical, and this approach is not yet standard practice for environmental assessments.

A third challenge is for scientists to gain sufficient understanding of what the parties need to know to direct their efforts toward providing decision-relevant information. The possibility that the available science may not be seen as useful by the intended users has often been noted as an impediment to public use of and trust in scientific information that is meant to be useful for decision making (e.g., National Research Council, 1989, 1999a, 2007b). Thus, it is important to ensure that the analyses being conducted make sense to the parties involved.

Challenges Related to the Public

Many interested and affected parties lack sufficient technical and scientific background to understand the scientific issues as scientists present them. It is impractical to educate all participants, so this challenge requires that someone perform a translational role in linking publics to the relevant science.

Another challenge is that most people, under most conditions, do not carefully consider all information relevant for analyzing complex issues. Instead, they apply cognitive shortcuts, called heuristics, that do not follow rules of logical reasoning and that affect their understanding of environmental, health, and safety risks, as noted in Chapter 2. But as noted in Chapter 5, given the time, resources, and motivation, nonscientists can become quite adept at critically understanding complex scientific analyses. At the same time, scientists are also imperfect analysts, also subject to heuristic processing as well as disciplinary blinders, and subject to other factors that predispose them to deviations from normative ideals, although scientific communities have developed various norms and procedures that provide some safeguards against individual biases and overconfidence (see below).

Another challenge is that what people want can appear unstable or inconsistent. The values that people consider in expressing preferences can be influenced by the way choices are framed (e.g., Tversky and Kahneman,

1981; Tversky, Slovic, and Kahneman, 1990; Payne, Bettman, and Johnson, 1992; Gregory, Litchtenstein, and Slovic, 1993; Slovic, 1995; Payne, Bettman, and Slovic, 1999). There are strategies for eliciting people's values that show promise for addressing this limitation (e.g., Saaty, 1990; Keeney, 1992; Hammond, Keeney, and Raiffa, 1999).

Additional challenges arise from the diversity of participants' values, interests, and concerns. People can be expected to attend to different aspects of an environmental issue, to draw different conclusions from the same information, and otherwise to engage in modes of thinking and analysis that may be mutually unintelligible or even mutually provocative. There are analytic techniques, such as benefit-cost analysis and risk analysis, that can help organize information and thinking about choice options and values. However, the use of these techniques, especially as formulas for decision making, have been questioned on methodological grounds because of the concern that they, like default assumptions, make value judgments opaque and inaccessible for political debate (see., e.g., Presidential/Congressional Commission on Risk Assessment and Risk Management, 1997a,b). When value judgments are embedded in analytical techniques that nonscientists do not understand, risks are being taken with the legitimacy of an assessment process.

These challenges are outlined in greater detail in a background paper prepared for this study by DeKay and Vaughan (2005). They underscore the need to find ways to organize thinking at the collective level that can overcome the limitations of individual cognition. This hope motivates many calls for public deliberation, but it too faces challenges, in the form of some well-established pathologies of group discussion and decision making such as the possibility of increased polarization or inappropriately strong influence from high-status individuals (discussed in Chapter 5; see also Levine and Moreland, 1998; Mendelberg, 2002; Stern, 2005b). One recent review concluded that "left to their own devices, groups tend to use information that is already shared, downplaying unique information held by specific individuals that arguably could improve the situation" (Delli Carpini, Cook, and Jacobs, 2004:328). Research on group process also suggests, however, that under appropriate conditions, group discussion can improve decisions by increasing the use of information that is not commonly shared (e.g., Winquist and Larson, 1998; Kelly and Karau, 1999).

Challenges of Communication

Given the differences among participants in funds of knowledge, habits of thinking, analytic languages and methods, and values and concerns, communication is likely to be problematic. One challenge is presented by the fact that to understand environmental systems and their complex relations to human activity, scientists often use mathematical models and statistical and probabilistic methods of analysis that are difficult for nonspecialists to understand. Moreover, the extent of uncertainty or disagreement among scientists on a complex environmental issue may also be hard for nonscientists to understand. The challenges of making science understood have been described in extensive bodies of research on risk communication and the use of science to support environmental decisions (for some reviews, see Fischhoff, 1989; National Research Council, 1989, 1999a, 2007b).

There are also differences in how knowledge is validated between the scientific community and others. On one hand, scientists have learned to trust the norms of their community as a control on the honesty and quality of scientific work, while other participants may not share these norms or trust that community as an arbiter (National Research Council, 2007a). On the other hand, the valuable, locally grounded knowledge that nonscientists can bring to the analysis of environmental problems usually is not developed via scientific inquiry, so melding it with traditional science and vetting it through review processes that scientists accept can be challenging. The Millennium Assessment (Reid et al., 2005), the U.S. National Assessment on Climate Change and Variability (U.S. National Assessment Synthesis Team, 2000), and the Arctic Climate Impact Assessment (Arctic Climate Impact Assessment, 2004) all have made special efforts to include locally grounded knowledge in environmental assessments, but effective processes for doing so are only beginning to be explored and represent a special challenge for the future.

MEETING THE CHALLENGES

This section reviews and draws conclusions from four sources of knowledge and insight about how to effectively integrate science and public input: decision analysis, research on environmental assessments and decisions, the practice of science at the frontiers of knowledge, and experience dealing with uncertain and disputed knowledge in various social arenas.

Decision Analysis

Decision analysis is a paradigm for supporting decisions that has emerged over the past half century, building on ideas from economics, systems analysis, and many other areas of science, that is now widely applied in business and governmental decision making (Howard, 1966, 1968; North, 1968; Raiffa, 1968; Fishburn, 1981; Behn and Vaupel, 1982; von Winterfeldt and Edwards, 1986; Clemen, 1996; National Research Council, 1996; Hammond, Keeney, and Raiffa, 1999). Essentially, it uses logical methods to break a problem into elements; describe what is known (including what is known about uncertainties), what is desired (goals, objectives, values), and what is possible (the management actions available), and apply an analytical structure to evaluating the alternative actions. North and Renn (2005) provide a more detailed discussion in relation to the present context.

The key aspect of decision analysis is the use of logic, especially mathematics and probability theory, to represent relationships between environmental management actions and subsequent effects. Such use of logic is basic practice in most fields of science and engineering, as well as in such common activities as making a household budget or preparing a tax return. However, in both scientific practice and everyday life, systematic errors in decision making are common, so decision analysis identifies logics and procedures to overcome these common and often subtle mistakes. Decision analysts have developed useful insights about how to organize complex, incomplete, and uncertain scientific information in ways that are both logically consistent and intelligible to nonscientists. These tools are not unique to decision analysis but are widely used in science, government, and business mangement. A decision analysis approach has been used to inform a great variety of environmental decisions, including weather modification applied to hurricanes (Howard, Matheson, and North, 1972), control of sulfur oxide emissions for coal plants (North and Merkhofer, 1976), the probability of contaminating Mars from Viking landings, and applications to sanitary and phytosanitary protection standards (North, 1995; National Research Council, 2000), acid rain (North and Balson, 1985), U.S. government commercialization of synthetic fuels (Tani, 1978), and drinking water contamination by arsenic (North, Selker, and Guardino, 2002).

Five principles for organizing and presenting scientific information flow from research and practice in decision analysis: they cover accuracy, uncertainty, the use of models, the use of sensitivity analysis, and the use of disagreements.

Ensure accurate calculations. Quantitative analysis of how management options affect outcomes can be very complex. Calculations must be done correctly, and inputs and assumptions in scientific analyses must be explicit and available for critical review. The process needs to be transparent and subject to review for correctness by outside parties.

Characterize uncertainty in the form of probabilities. An important analytical tool in decision analysis is the use of probabilities to characterize uncertainty (Savage, 1954; Raiffa, 1968). Indeed, probability theory is the only logically consistent way to reason about uncertainty (Cox, 1961; Jaynes, 2003). Probabilities are usually calculated from statistics on past events. However, probability is also useful for characterizing uncertainty in the absence of statistical data (e.g., uncertainty about the outcomes of possible management choices; see, e.g., National Research Council, 1996,

2002b, 2007b). In one tradition, various experts are asked to express their best professional judgment as a probability. It could be stated as willingness to place a bet on an uncertain outcome for which the probabilities are known (e.g., flipping a coin, rolling a die, or spinning a ball onto a roulette wheel; see, e.g., Savage, 1954; Raiffa, 1968; Morgan and Henrion, 1990) or by using ordinary qualitative expressions, such as "very unlikely," that have been keyed to probability numbers (Moss and Schneider, 2000). Another tradition in science (Cox, 1961; Jeffreys, 1961; Jaynes, 2003) holds that probability theory is a logic for inference, and that probabilities should reflect the available evidence relevant to the uncertain event or variable. Those carrying out or interpreting such probability assessments need to understand the subtleties of human judgment about uncertainty (Spetzler and Staël von Holstein, 1975; Kahneman, Slovic, and Tverskey, 1982; Wallsten and Budescu, 1983) and to recognize that such judgments are imprecise. Important uncertainties may warrant careful assessment by multiple experts under carefully designed protocols.

Use models carefully to represent complex realities. Describing how management actions will affect the environment and human health usually involves a large number of relationships, such as growth processes, interaction of various species within an ecosystem, transport and transformation of pollutants in the environment, etc. These relationships are often too complex to describe and comprehend in nonscientific language, so scientists often describe them quantitatively in the form of mathematical models that are implemented as computer programs. Modern computers enable very large numbers of relationships to be specified, so that effects following environmental management decisions can be calculated through a large number of steps from assumptions and data input to the model. The use of such models is widespread in scientific disciplines related to the environment and in federal agencies (National Research Council, 2007c) Such models are often very difficult to understand, especially for nonscientists, and yet such understanding is critical for effective participation in decisions that use them.

Although a model may be presented as representing science, participants in an analytic-deliberative process need to understand that a model is not scientific reality. A model is an abstraction from reality that is based on many assumptions and judgments about how nature works, which elements are most important to represent, and which elements may be left out without compromising accuracy. Thus, a model is always an embodiment of scientific judgment that was developed with a purpose in mind. As Levins (1966) noted, any model must make a trade-off among precision, realism, and generality. Although experienced modelers may understand how a particular model deals with these trade-offs, these tradeoffs are not obvious to the public or even to scientists from other fields. Models can be very useful for describing how environmental systems work and how management actions may affect the environment. But models and results calculated by models can be wrong or misleading, especially when a model was developed in a different context, the assumptions are not appropriate, or the input data are incorrect. Scientific peer review and comparison with results calculated by other models that describe the same environmental system are very important for interpreting results from a model. Some models in science can provide the basis for very accurate predictions, but in the area of environmental assessment and decision making, the accuracy of model predictions may not be high. In many cases it is useful to examine how uncertainties on inputs (data, parameters, model assumptions) may translate into uncertainty on the effects or outcomes of interest for environmental systems.

Use sensitivity analysis to find out which elements are important and which are not. Environmental decisions typically turn on projections of the impacts on an environmental or human system over time of each of the options being considered. Thus, projections, whether from models or from the judgment of experts, depend on assumptions about how the system works and on the data being used to make the projections. Sometimes the projected impacts differ greatly depending on particular assumptions. It is useful to carry out a systematic analysis of sensitivity to data inputs and assumptions. Sensitivity analysis is done by listing each input or model assumption and asking how the projected impact would change if this factor were different, within a range judged to be reasonable.

For example, in a study of water management in a river basin, one would want to evaluate management policies not only under an assumption of average precipitation, but also under scenarios of a series of wet years or dry years. If the evaluation of the management alternatives changes as the precipitation levels vary from wetter to drier, one might choose options that yield good results across scenarios or undertake further study of ways to manage the river to reduce the sensitivity to precipitation. If variation in precipitation is found not to be important, further study of this topic and associated refinement of the model is not appropriate. An important value of sensitivity analysis is in identifying factors that do *not* have a strong effect on the impacts of interest. Then discussion and debate then move away from those factors and concentrate on the ones that appear to be more critical to the decision.

Often models have a great many assumptions and input parameters. Outside review of the model by experts in the relevant technical disciplines may be useful in identifying assumptions and inputs that might be sensitive. When the input is uncertain and model results are sensitive to these inputs, it may be useful to represent the uncertainty explicitly for each input factor and the resulting overall uncertainty in the results calculated using the model. For example Patil and Frey (2004) have suggested that food safety models should be designed to facilitate sensitivity analysis, and that sensitivity analysis methods are a valuable tool in supporting food safety regulation. Such conclusions also seem appropriate for other areas of environmental assessment and decision making. A recent National Research Council report evaluating a health risk assessment prepared by a federal agency cited lack of sensitivity analysis as a major failing (National Research Council, 2007d).

Use disagreements to focus analysis and promote learning. Achieving a logically consistent integration of what is known, what society wants, and what society can do may require that many disagreements must be resolved: about policy goals, about the state of knowledge, and so forth. Good decision analysis helps make the nature of the disagreements clearer, often allowing some of them to be addressed through further data collection and analysis. Thus, getting high-quality information may require multiple rounds of interaction, in which the parties learn from each other.

Research on Environmental Assessment and Decision Processes

Strong traditions of research on integrating science and public participation have emerged in the overlapping literatures on risk (National Research Council, 1996; Jaeger et al., 2001; Rosa, Renn, and McCright, 2007), common-pool resource management (Dietz, Ostrom, and Stern, 2003), ecosystem and natural resource management (Shannon, 1987, 1991; Dietz and Stern, 1998) and impact assessment (Cramer, Dietz, and Johnston, 1980; Dietz, 1987, 1988).

Stated briefly, the core of this literature acknowledges that both the public (interested and affected parties, in our terminology) and the scientific community have substantial expertise, but expertise of different kinds and on different matters (Dietz, 1987). On one hand, as noted in Chapter 2, the public often has detailed knowledge of the local context and everyday practices that is not readily available to the scientific analyst. And of course the public, rather than the scientific community, is the legitimate source of information about public values and preferences. On the other hand, scientific analysis is essential for understanding the dynamics of complex systems and assessing the uncertainty in how such systems evolve over time with different management actions. It is also valuable in systematically eliciting public views about values and preferences (for a discussion of systematic techniques for value elicitation, see Gregory and McDaniels, 2005).

The literature also shows that differences in perspective sometimes arise that can be so fundamental as to call into question the methods, or even the use, of decision analysis. Perhaps the most prominent example arises with choice options that have a nonzero probability of resulting in catastrophes such as the extinction of species, the elimination of major ecosystems, or the development of biological weapons of mass destruction. Faced with such possibilities, some segments of the public advocate precautionary approaches that rule out such options absolutely. They may reject any decision analyses that might be used to justify trading these risks against potential benefits of the choice options or advocate decision-analytic methods that emphasize the possibility and highlight the importance of worst-case possibilities (e.g., Marshall and Picou, 2008).

Because of the complementary nature of scientific and other kinds of knowledge and the potential for conflict about how best to use knowledge to inform decisions, researchers have often concluded that high-quality environmental assessment and decision making require a dialogue between scientfic analysis and public deliberation in which science both informs and is informed by the public. Because public concerns in large part determine which scientific questions are decision relevant and because public knowledge of local contexts and practices must inform scientific analysis, successful linking of analysis and deliberation is also considered critical to the legitimacy of the analysis and of decisions that use it. In addition, this literature gives strong reason to believe, though still very little hard evidence, that by participating in iterated processes of analysis and deliberation, public participants will become more sophisticated about the scientific analysis over time and scientists and agency officials more sophisticated about the public's needs for understanding. Such changes increase capacity in all those involved.

Many empirical studies support the value of linking science and public deliberation for obtaining good outcomes. For example, Mitchell et al. (2006:320), in their synthesis of work on environmental assessments, noted the importance of users understanding the science well enough that "credibility by proxy" is replaced with "credibility through understanding." Bradbury (2005:15) concludes in her analysis of Superfund sites that "access to timely and accurate information is a prerequisite for community members' ability to participate effectively." Leach (2005) identified nine studies that show the importance of access to adequate scientific and technical information in Forest Service planning processes. Bingham (2003), drawing on multiple sources of practitioner experience, has noted the special importance of being explicit about the character of science and data when the interested and affected parties (sometimes including responsible government agencies) define the problem differently, when decision makers' objectives are not clearly defined, when the conceptual framework for the issue is shifting, when the parties disagree about the methods for data collection and/or analysis, or when arguments over science are masking an underlying conflict about something else. The need for iterating between analysis and deliberation has been repeatedly emphasized in assessments of environmental decision making (e.g., National Research Council, 1999a, 2003, 2005a; Renn, 2005) and is reflected in much existing agency guidance.

Such studies do not point to any *a priori* optimal format or set of rules for integration. Rather, they suggest that each process must be designed around the problems and opportunities of a specific context. The literature on formats for public participation, discussed above, does not provide systematic knowledge for reasons already discussed.

An analytical and empirical literature is beginning to emerge that identifies and examines possible processes and techniques for integration (Burgess et al., 2007; Chilvers, 2007, 2008; Bayley and French, 2008; Webler and Tuler, 2008). This is a promising area for future research.

Scientific Practice at the Frontiers of Knowledge

Uncertainty and disagreement are always present at the frontiers of science, so scientific communities have developed methods, norms, and procedures that help them test knowledge claims so that shared knowledge can advance despite the limitations of individuals. Some of these methods (e.g., nonhuman instrumentation, the use of mathematics and logic) aim to reduce or circumscribe the role of human judgment, but others actually rely on "the subjective judgments of fallible human beings and social institutions to detect and correct errors made by other fallible humans and institutions" (Stern, 2005a:976).

Judgment is especially critical under conditions that are common in environmental policy arenas: when there are multidimensional and inequitable impacts, scientific uncertainty and ignorance, value uncertainty and conflict, an urgent need for scientific input, and mistrust among the interested parties (Dietz and Stern, 1998). Under such conditions, scientists often disagree about which scientific problems must be solved to provide needed information, which assumptions are reasonable when knowledge is incomplete, how to interpret uncertain or conflicting information, and so forth.

Scientific communities have developed several norms and practices that help them advance knowledge despite uncertainty, disagreement, and human frailty and that seem capable of adaptation to practical problems of environmental assessment and decision making. The key norms and practices seem to be:

- Define concepts concretely and operationally
- Make assumptions explicit
- Test sensitivity of conclusions to different assumptions
- Make analytic methods transparent

- Make data available for reanalysis
- Apply logical reasoning in drawing conclusions
- Restrict arguments to matters of substance, method, and logic and avoid *ad hominem* arguments
- Test conclusions for consistency with other data
- Publish results and conclusions in open sources
- Subject scientific analyses and reports to independent (peer) review
- Sanction fraud and misrepresentation

These norms and practices seem to have considerable generality and are understandable to nonscientists who may not understand scientific data, assessments, or inference techniques. They create a climate of scientific openness that contributes to the legitimacy of deliberations; that moves the deliberation toward clarity on matters of fact; and that tends to clarify the ways in which disputes turn on matters of evidence, judgment, and values.

Experience with Uncertain and Disputed Knowledge

Practitioners of environmental dispute resolution are among the most knowledgeable about procedures for integrating science and public input in ways that advance understanding. A summary of insights from this experience (Bingham, 2003) identifies five principles for making better public choices in the face of contested science: clarify the questions jointly before gathering more data; focus on decision-relevant information; let science be science, and do not confuse it with policy; learn together; and remember that science is not necessarily the underlying cause of disputes and draw on other basic consensus-building principles and tools. Bingham calls on process conveners to consult widely about the scientific questions that need to be addressed, to talk explicitly about trust, uncertainty, and the role of information, and to identify and address disagreements over information through a process that builds trust. We return to this approach in more detail in Chapter 7.

Citizens in daily life have also found ways to make good use of the expertise of physicians, accountants, attorneys, architects, and other professionals, even though they are personally not equipped to evaluate that expertise. An important strategy relies on seeking independent assessments from others who have relevant knowledge and different interests or perspectives from the original expert. Principles such as getting second opinions and independent checking of claims are widely familiar and understandable from practices in medicine, the adversarial system of trial by jury, news reporting, and other social institutions. This observation suggests that it should be possible to devise systems of independent review for use in environmental public participation that are credible to most, if not all, participants. Doing this would seem to require efforts at the start to seek agreement among all parties on how to identify, share, evaluate, and apply decision-relevant information of various kinds (scientific, cultural, technical, etc.). To integrate the science well, it is important for relevant information to be accessible to all participants and to make special efforts to ensure that the participants can understand this information. It is also important to agree on processes for joint fact-finding or other strategies for shared learning that respect that any participant may have special expertise, that acknowledge the different scopes and domains of such expertise, and that provide ways of checking knowledge claims that are credible to participants who lack expertise in the specific area.

CONCLUSION

Integrating science and public participation through processes that iterate between analysis and broadly based deliberation promotes the quality, accountability, and legitimacy of environmental assessments and decisions. Such processes are more likely to produce satisfactory results if they are transparent regarding decision-relevant information and analysis, are attentive to both facts and values, are explicit about assumptions and uncertainties, and provide for independent review and iteration to allow for reconsideration of past conclusions.

Understanding Risk (National Research Council, 1996:3) concluded:

> [S]uccess depends critically on systematic analysis that is appropriate to the problem, responds to the needs of the interested and affected parties, and treats uncertainties of importance to the decision problem in a comprehensible way. Success also depends on deliberations that formulate the decision problem, guide analysis to improve decision participants' understanding, seek the meaning of analytic findings and uncertainties, and improve the ability of interested and affected parties to participate effectively in the risk decision process. The process must have an appropriately diverse representation of the spectrum of interested and affected parties, and of specialists in risk analysis, at each step.

This set of conclusions is generally supported by evidence available since that report's publication and applies to environmental assessments and decisions more generally. The Presidential/Congressional Commission on Risk Assessment and Risk Management (1997a,b) reached similar conclusions and recommended an analytic-deliberative process to improve federal decision making in the management of environmental risks (see Box 6-2). The evidence presented in this chapter allows for some further specification of the advice offered in these earlier reports.

The evidence supports the conclusion that there is no readily available alternative to including the public in decision-relevant environmental assessments that integrate science. Furthermore, there is no readily available alternative to deliberation as a means of resolving disagreement among scientists, and it must be recognized that valuable information often comes from nonscientists. Dialogue among multiple perspectives is necessary for quality in assembling and assessing the relevant information. Respectful evaluation is needed by all parties, including the scientists, and independent review is essential to the credibility of scientific and technical conclusions. Although there are significant challenges to integration, both for scientists and for the public, the challenges can be addressed.

The most promising approach is to extend the norms used in science and in the best of public policy to encourage balanced and substantively focused discussion that advances the quality and legitimacy of analysis and contributes to participants' capacity for future deliberation. There is little careful empirical research on how to do this, but the studies that do exist converge with insights from decision analysis and long-standing practices in science, public policy, and nonspecialist use of expert knowledge on the principles identified here—transparency of decision-relevant information and analysis, explicit attention to both facts and values, explicitness about assumptions and uncertainties, independent review, and iteration—as a shorthand description of a set of norms and practices for integrating science that are conducive to good results.

We note that many federal agencies that invest considerable effort in doing environmental and risk analysis have not institutionalized these norms and practices in their assessment and decision processes. It is more typical to use linear or sequential processes in which the agency assumes responsibility for problem formulation, has its scientific staff and contractors gather information and conduct the analysis, and then submits the analysis to a notice-and-comment process. In such a process, the analysis may receive some level of peer review prior to finalization, but the issues to be addressed in the analysis, the information to be considered, and the underlying assumptions do not get reviewed early enough to shape the analysis. We emphasize the conclusion of the Presidential/Congressional Commission on Risk Assessment and Risk Management (1997a:5; see Box 6-2) after extensive hearings, that "many risk management failures can be traced to not including stakeholders in decision making at the earliest possible time and not considering risks in their broader contexts." Although some federal agencies have had extensive experience with public participation, the processes proposed in this and preceding National Research Council reports are not standard practice across federal agencies. Most federal agencies with responsibility for environmental assessment and decision making do not commonly integrate science and public participation through processes that iterate between analysis and broadly based deliberation.

> **BOX 6-2**
> **The Presidential/Congressional Commission on Risk Assessment and Risk Management**
>
> The Presidential/Congressional Commission on Risk Assessment and Risk Management (hereafter the commission) was established by Congress in 1990 legislation amending the Clean Air Act (Presidential/Congressional Commission on Risk Assessment and Risk Management, 1997a,b). The 10 commission members, appointed by the president and leaders of both parties in Congress, included leading scientists with expertise in biological sciences applicable to public health and environmental problems. The commission was chartered to "make a full investigation of the policy implications and appropriate uses of risk assessment and risk management in regulatory programs under various Federal laws to prevent cancer and other chronic health effects which may result from exposure to hazardous substances" (Presidential/Congressional Commission on Risk Assessment and Risk Management, 1997a:i).
> This purview overlapped considerably with that of the present study, in that federal agencies, such as the Environmental Protection Agency, the Food and Drug Administration, and the Occupational Safety and Health Administration, spend much of their budgets dealing with hazardous substances in the environment.
> The commission conducted an extensive set of hearings with stakeholder groups in a variety of locations across the United States. It developed a six-step risk management framework as the basis for its recommendations for reform of federal agency practices (Charnley, 2003; North, 2003; Omenn, 2003; Presidential/Congressional Commission on Risk Assessment and Risk Management, 1997a,b; see Figure 6-1) and produced a set of recommendations very similar to those made in *Understanding Risk*.

SUMMARY: THE PRACTICE OF PARTICIPATION

Based on an assessment of multiple sources of evidence, Chapters 4, 5, and 6 identify sets of empirically supported principles of good public participation practice—for project management, for organizing the participation, and for integrating the science. We summarize them in Box 6-3. These principles echo those that can be found in sources of guidance derived mainly from practitioners' experiences and, in that sense, the principles are not new. Our findings do, however, reinforce at least certain aspects of the collected experiential knowledge with other sources of support.

The main challenge facing practitioners is to find practical ways to implement the principles of good public participation practice. As we note throughout this volume, practitioners have developed numerous formats,

> [M]any risk management failures can be traced to not including stakeholders in decision making at the earliest possible time and not considering risks in their broader contexts. In contrast, the Commission's Risk Management Framework is intended to:
>
> - Provide an integrated, holistic approach to solving public health and environmental problems in context
> - Ensure that decisions about the use of risk assessment and economic analysis rely on the best scientific evidence and are made in the context of risk management alternatives
> - Emphasize the importance of collaboration, communication, and negotiation among stakeholders so that public values can influence risk management strategies
> - Produce risk management decisions that are more likely to be successful than decisions made without adequate and early stakeholder involvement
> - Accommodate critical new information that may emerge at any stage of the process
>
> The commission's final report is a strong call for a shift to an altered approach. However, it does not prescribe detailed methods for accomplishing such goals as "adequate and early stakeholder involvement." Also, despite the fact that most commission members are scientists, the report did not provide much detail on how scientific information should be assembled or evaluated for the iterative analytic-deliberative approach it proposed. For example, the commission's report does not clarify the phrase "best scientific evidence" or address the role of judgment in determining what is best, particularly when the science available for predicting the consequences of policy alternatives involves pervasive uncertainty.

techniques, and practices for implementing the principles, and many of them can be helpful. As we also note, numerous guidebooks are available that describe the formats and practices and offer advice on how and when to use different ones.

Practical experience makes clear, however, that implementing the principles can be much more difficult in some contexts than others, and that different contexts present different challenges for public participation. In Chapters 7 and 8 we review available evidence on which aspects of context matter, how they matter, and how it is possible to examine the context of public participation to diagnose the situation, that is, to identify and anticipate specific difficulties that are likely to arise in the context at hand, when trying to implement principles of good practice. This kind of diagnostic process has prescriptive value in that it can help practitioners and

> **BOX 6-3**
> **Empirically Supported Principles of Practice for Environmental Public Participation**
>
> MANAGEMENT PRACTICES (Chapter 4)
> Clarity of purpose
> Commitment to use the process to inform decisions
> Adequate resources
> Appropriate timing
> Implementation focus
> Commitment to learning
>
> ORGANIZING PARTICIPATION (Chapter 5)
> Inclusiveness of participation
> Collaborative problem formulation and process design
> Transparency of process
> Good-faith communication
>
> INTEGRATING SCIENCE (Chapter 6)
> Iteration between analysis and broadly based deliberation with:
> - availability of decision-relevant information
> - explicit attention to both facts and values
> - explicitness about analytic assumptions and uncertainties
> - independent review
> - reconsideration of past conclusions

participants select from among the great variety of available formats and practices those that may help address the particular difficulties they can expect to encounter.

However, there are limits in offering prescriptions. As the following chapters show, it is neither possible nor advisable to identify any single "best practice" for conducting public participation or even for overcoming particular difficulties that certain contexts present. Rather, the best that can be done after identifying the likely difficulties is to select practices collaboratively to try to address them and then to monitor the process to see whether the practices are accomplishing the desired results, keeping open the possibility of changing practices or formats when they are unsuccessful.

7

Context: The Issue

This chapter and the next review evidence on how the context in which an environmental decision is made influences public participation. By context we mean factors that are outside the control of those who convene a public participation process, at least in the short term. These factors may explain the variation in results associated with particular modes of practice in public participation, and they must be taken into account in planning and implementing public participation. Unfortunately there is no established theory that specifies which contextual factors matter, and how they matter, to the results of public participation.

We consider contextual factors under five broad categories of attributes that cover much of the variation: the purpose of the process (assessment or decision making); the environmental issue under consideration; the state of the relevant science, including scientific uncertainties and disagreements; the responsible agency and the laws and external organizations that affect the assessment or decision; and the interested and affected parties. We examine the first three, issue-related aspects of context in this chapter; we examine the last two, the people-related aspects, in Chapter 8. In both chapters, we draw conclusions from the range of available evidence about whether these contextual factors make a difference in terms of the likelihood of a successful process or in what a convening organization needs to do to make the process successful.

The available evidence indicates that some contextual factors make little difference to the effects of public participation. Others can make an important difference, although they seldom present insurmountable barriers to successful public participation or determine the results in and of themselves. Rather, contextual factors can make the principles of good public participation described in Chapters 4-6 harder to implement. In this chapter and the next, we describe some of these relationships and provide examples of tools that practitioners have used to attempt to overcome contextual difficulties, although evidence on the efficacy of these tools is weak. For this reason and others elaborated in Chapter 8, we do not recommend any of the tools discussed as "best practice." In Chapter 9, we recommend a process for identifying effective ways to address the various difficulties that can arise in the many contexts of public participation.

The evidence reviewed in this chapter and the next shows that achieving quality and legitimacy and building capacity in public participation depend very much on how well a participatory process is tailored to the challenges or potential difficulties presented in any specific context. Addressing certain key questions can aid substantially in diagnosing them. Such diagnosis, in turn, enables more explicit consideration of processes and approaches that can help overcome potential problems or make accommodations for them.

Our review of available knowledge and experience enables us to describe a set of diagnostic questions that can be useful for identifying those aspects of a situation that are likely to make a difference in the outcome of a public participation process and the ways in which these contextual factors may affect the process. Such diagnosis can form the basis for tailoring participation processes for more successful results.

PURPOSE OF THE PROCESS: ASSESSMENT OR DECISION MAKING

The evidence indicates that the determinants of successful public participation are largely the same for processes focused on assessment and those focused on decision making.

Direct comparisons of the two purposes are not possible because of a lack of studies of multiple public participation cases that include both assessment and decision-making objectives. However, it is possible to consider whether success is easier to achieve or whether different factors are conducive to success when the objectives are different (Stirling, 2006).

Public participation in environmental assessments involves a shift away from an approach in which only scientists participated in gathering and synthesizing information, and reflects increasing acceptance of the idea that nonscientists possess knowledge and expertise that complements the expertise of the scientific community and can help improve environmental understanding, particularly when it is applied to practical problems. Experience is accumulating as the conveners of assessments respond to calls for public involvement in risk assessment (National Research Council, 1989,

1994, 1996; President/Congressional Commission on Risk Assessment and Risk Management, 1997a,b) and global environmental change assessment (National Research Council, 2007a). Government science agencies at various levels and in different countries have increasingly engaged publics in environmental assessments (Kasemir et al., 2003). Examples include the U.S. National Assessment of the Potential Consequences of Climate Variability and Change (http://www.usgcrp.gov/usgcrp/nacc/default.htm), the Millennium Ecosystem Assessment (http://www.maweb.org/en/index.aspx), and the Arctic Climate Impact Assessment (http://www.acia.uaf.edu/) (see National Research Council, 2007a).

The Presidential/Congressional Commission on Risk Assessment and Risk Management (1997b:75-76) offered, in the report subsection on "Identifying Highly Exposed Populations," a compelling example of how public participation in environmental assessment can contribute to "getting the science right" by gathering important information for analysis that is not otherwise available.

> Some population groups are at increased risk for toxic effects of chemical exposures because their exposures are greater than those of other population groups. Cultural practices, occupational exposures, behavior patterns, eating habits, and effects of related chemicals can be responsible. The high-risk subpopulations might be of special concern when risk assessments are conducted and risk management decisions are made. Risk assessors often have not sought information from knowledgeable citizens and consequently have not explicitly considered specific exposure conditions that might be present in minority group communities, certain occupational settings, or areas of particular socioeconomic status.

The commission recommended broad participation and further acknowledged the possibility that public engagement might even enhance the quality of risk management decision making (Presidential/Congressional Commission on Risk Assessment and Risk Management, 1997b:76-77):

> Affected parties should be consulted in the early stages of an assessment to obtain information about all known sources of exposure to a particular chemical and related chemicals and to characterize exposure factors peculiar to particular subpopulations. . . . Specific information gathered from the community and stakeholders could reduce the need for default assumptions and improve the quality of risk assessments. . . . Community assistance in characterizing exposure factors peculiar to particular segments of the population can focus a risk assessment and broaden risk management options.

There are relatively few careful analyses comparing different degrees of public participation in comparable environmental assessments. As we note in Chapter 3, a study of a set of global environmental assessments (Mitchell et al., 2006) found that the extent of stakeholder involvement was strongly and positively associated with the perceived impacts of the assessments and that those impacts were dependent on the scientific credibility of the assessments, their legitimacy, and on whether their results were perceived as decision relevant. Participation, the study concluded, fosters all three results. The National Research Council (2007a) study of global change assessments similarly noted the advantages of broad participation but also noted costs in terms of efficiency. A review of several European environmental assessments found that stakeholder participation has increased the knowledge base of modelers and added credibility to both the assessment process and the concerns of stakeholders (Welp et al., 2007).

Moser (2005) and Morgan et al. (2005) provide detailed analyses of the U.S. National Assessment of the Potential Consequences of Climate Variability and Change. The national assessment was perhaps the most ambitious effort yet undertaken in the United States to engage scientists and citizens in a deliberative process intended to produce an assessment rather than policy recommendations. The results of these analyses are complex. Participants varied considerably in their views of whether or not the national assessment was successful in producing sound information about climate change and its impacts, giving it an average grade of a high "C" or low "B." There seemed to be a general sense among participants that the idea behind the assessment was sound and the process useful, but that the effort did not have sufficient resources, an important issue for participation in decisions as well (discussed in Chapter 4). The meaning of these data is unclear because of the absence of comparable cases, which would make it possible to determine whether respondents' lukewarm evaluations reflected the character of climate change as an environmental problem, the nature of public participation in an assessment rather than a policy process, the shortage of resources, or other issues.

A fairly extensive body of practice-based knowledge exists for evaluating the effects of public participation in environmental assessment and determining which factors affect them, much of it examined in reviews of the practice of risk assessment (e.g., National Research Council, 1996; Presidential/Congressional Commission on Risk Assessment and Risk Management, 1997a,b) and in the recent study of global change assessments (National Research Council, 2007a; Welp et al., 2007). As is discussed in Chapter 3, there is strong convergence between the practical lessons that come out of this experience and the ones drawn by practitioners of regulatory negotiation, environmental dispute resolution, and other decision-focused public participation processes. We see no evidence to support organizing public participation differently for supporting assessments and for decisions.

NATURE OF THE ENVIRONMENTAL ISSUE

The evidence indicates that subject matter of an environmental assessment or decision has little direct effect on the ability of the public participation process to produce good results. Certain environmental issues, because of specific characteristics, often create particular difficulties in participatory processes that, if left unaddressed, can affect the likelihood of success. However, the nature of the issue by itself is not the determining factor in achieving successful results. Much more important is the design of the processes to address potential difficulties.

Subject Matter

Subject matter has little direct effect on the results of participation. Public participation is used in environmental assessment and decision-making processes related to highly diverse substantive issues: air quality standards, biotechnology policy, brownfields remediation, climate change, dam relicensing, forest planning, habitat restoration, highway construction, oceans policy, water allocation, wetlands protection, and many other topics. Subject matter is also diverse at an abstract level: one can distinguish decision processes that focus on collective goods (resources) or collective bads (pollution); involve human health effects or nonhealth effects; do or do not raise environmental justice issues; do or do not concern harm to innocents; and so forth. These differences suggest to some that the ways in which public participation efforts unfold and, in particular, the factors leading to better or worse outcomes, may differ substantially on the basis of the subject. Agencies or divisions of agencies are usually specialized around one or a few substantive areas. Specialists in a substantive area sometimes believe that that area is unique, so that experience in other areas is not relevant. Many are skeptical, for example, that there is any useful transfer of knowledge from forest planning to remediating a contaminated site or to engaging the public in assessments of climate change impacts.

This argument seems to make intuitive sense. Public concerns are different in different substantive domains. They can vary widely. For example, concern in one setting may focus on health risks to a community, especially children, and on the costs of cleanup of toxic contamination. In another setting, concern may focus on the revenues to be generated by timber harvesting and the reduction of ancient forest habitat. In other settings, concern may focus on the competition for water for municipal, agricultural, and recreational uses or any of dozens of other issues. Yet despite these differences, the proposition that successful outcomes are more likely to be achieved for some environmental issues than others finds little support in research on public participation. Studies that have compared public participation in different problem contexts generally fail to find that the type of environmental problem is related to the likelihood of successful results (e.g., Bingham, 1986; Consensus Building Institute, 1999; Beierle and Cayford, 2002; Mitchell et al., 2006).

These studies also suggest that the factors that lead to better or worse outcomes are essentially the same across types of environmental issues. For example, the geographic scale of the issues or the number of agencies with jurisdiction over aspects of the decision to be made are important attributes to consider in the design of a participation process, regardless of whether the environmental issue is toxic waste management, ecosystem planning, or climate change assessment. Beierle and Cayford (2002:40-41) conclude from their extensive comparative study that "differences among environmental issues, preexisting relationships, and institutional contexts appear to play surprisingly small roles in determining whether public participation is successful. . . . [They] play a role in how participatory processes play out, but they do not appear to predetermine outcomes."

Potential Difficulties

Environmental issues may have certain characteristics that predispose to particular difficulties in participatory processes. The available evidence suggests that these characteristics affect public participation by making it easier or harder to implement the principles of good practice described in Chapters 4-6, such as clarity of purpose, inclusiveness of representation, and availability of decision-relevant information. It is the way such difficulties are addressed, more than the environmental issue, that affects the prospects for success.

For example, temporal and spatial scales associated with an issue may affect the number of interested and affected parties and who among them is able to participate. Such characteristics as complexity of the issues, qualitative characteristics of the hazards, and collective action and common-pool attributes can also affect the complexities of the relationships among the parties and the processes of organizing and deliberating necessary for an effective public participation process. We explore each of these separately.

Temporal Scale

Issues of the time scale appropriate for environmental decision making are at the heart of many debates about sustainability. For example, decision making about long-lived environmental hazards can create challenges because of needs for long-term monitoring (Leach, Sabatier, and Quinn, 2005). Research on common-pool resource management has long emphasized the need for institutional mechanisms involving all resource users (e.g., Ostrom, 1990; National Research Council, 2002a), and there has been increasing interest in public participation in the design of institutional mechanisms for addressing such long-lived problems as ecosystem restoration and management of radioactive waste. For example, a diverse set of civic leaders in Washington State convened a participatory initiative across 14 watersheds following the listing of Puget Sound Chinook salmon as an endangered species. The effort resulted in the adoption of a federal recovery plan and state legislation in 2007 establishing the Puget Sound Partnership, a new state agency with cabinet-level status governed by a seven-member "leadership council."

The literature on public participation has not yet addressed temporal scale explicitly other than to say that representation of the interests of future generations and sustaining collaborative governance mechanisms for problems spanning generations pose obvious difficulties. It is not clear how well public participation processes deal with the difficult trade-offs between short-term and long-term benefits, costs, and risks. Certainly one of the goals of participation processes is to address the beliefs and values of affected publics on just such trade-offs, and, if possible, develop a consensus on such issues that can guide decision making. Many of the tools of conflict resolution, decision science, and economics are intended to aid in such efforts. See Chapter 8 for further discussion of value trade-offs.

Spatial Scale

The scale or scope of the problem plays an important role in defining who the appropriate public is (Markus, Chess, and Shannon, 2005). For some local environmental problems, such as remediating contaminated sites, it can be relatively easy to identify the affected population. Other issues are localized but have aspects that are of broader public concern. For example, a decision to develop mineral resources in a wilderness area may have very localized economic and ecological effects but generate national or global interest, perhaps because of an endangered species. Some environmental issues, such as transport of air pollutants and the management of the Great Lakes, are regional. Still others, such as climate change and ozone depletion, are global in scope. There are some indications that environmental issues with well-defined geographic boundaries can provide a focus that is compelling to participants (Wondolleck and Yaffee, 1997).

The limited available evidence suggests that the success of public participation processes does not depend on whether the issue is local, regional, or national. Processes conducted at large geographic and institutional scales do present potential difficulties for public participation, notably that of ensuring adequate access and representation of the number and range of interested and affected parties. In such situations, practitioners have used a variety of formats, including holding workshops in multiple locations, Internet participation, study circles, deliberative polling, and formal representative processes, such as blue-ribbon commissions or mediated negotiations among organized interests. With sufficient resources, each of these can be designed to achieve representation over large geographic scales.

Achieving inclusive representation when both local and regional or national interests exist may pose the greatest difficulties. When resources permit, multiple processes with different formats can be used so that processes suitable for local participation can be linked to processes that can reach those at a distance who may be concerned about a local action. For example, in its Western Oregon Plan Revision process, the Bureau of Land Management held dozens of local open houses combined with an Internet site for electronic submission of comments, periodic newsletters, and regular meetings with "formal cooperators," including state and federal agencies and representatives of many of the affected western Oregon counties.

As Lubell and Leach (2005) point out, environmental problems that span political boundaries cannot be addressed by individual agencies working in isolation. The more agencies that have jurisdiction over parts of the issue, the more difficult it can be to establish a clear purpose for the process and obtain commitments about how the results will be used. Interagency working groups are sometimes used to address this difficulty, although this approach can present complications of its own.

The overwhelming majority of research on public participation has focused on local and regional issues in which participants live close enough to each other so that face-to-face interaction can be the basis of participation. While there are notable exceptions, most participation practitioners have more experience with local and regional processes than with national or international ones. National policy issues constituted only about 16 percent of the cases of public participation in Beierle and Cayford's (2002) large database. That study compared state and national policy issues with site-specific issues and found a correlation between scale and a composite measure of success of only +0.02. So it would appear that there is little reason to expect differences in success based on geographic scale alone. However, the organizations that convened the large-scale cases included in their database appear to have had adequate resources for handling the access problems raised by scale. Although research is limited on the role that temporal and spatial scales play in the outcomes of public participation processes, scale clearly affects the level of effort required to get adequate participation by the range of relevant parties.

Complexity

Although characteristics of the environmental issue do not strongly and directly affect the likelihood of a successful public participation process, they may affect the motivation of the potential participants, and motivation has a moderate positive correlation with success (Beierle and Cayford, 2002). Lubell and Leach (2005), for example, found that watershed partnerships are most likely to develop in severely degraded watersheds, an association that they attribute to the motivational levels of the participants. However, their overall conclusion is that "features of the watershed itself, such as size and problem severity, influence success much less than do the social and structural characteristics of the process."

Langbein (2005) found that participants' satisfaction with negotiated rule-making procedures was lower when the process was more complex. Complexity of issues may contribute to diversity of perspectives—having many sides with a variety of interests in the outcome rather than only two perspectives: those opposed and those in favor of a particular action. It was this feature of the conflict more than the environmental issue itself that made a difference. Langbein's finding for regulatory negotiations may also apply to many, if not most, other environmental contexts, such as water resource use plans, climate change policy, and transportation projects, in which there may be many sides.

The more sides to an issue, the more likely that there will be multiple views about what the focus of a public participation process should be. This can make clarity of purpose difficult to achieve and requires careful attention to developing shared understandings of how different participants understand the issues. A member of the panel reported a local example. When a mining company with mineral rights near a national wildlife refuge in the Southeast took the initiative to involve interested stakeholders in its permit application process, many sides quickly emerged. Some were interested in the opportunity for new jobs associated with the mine; others brought the knowledge and historic concerns of regional Native American people; still others were concerned about the impact on the wildlife refuge. Some were responsible for implementing state law, others had federal responsibilities, and still others were local elected officials. These sides had differing views about what would constitute a legitimate purpose and scope for the process. Some participants would only participate if the question were whether or not to open the mine, while others would only participate if the question were how to mine in an environmentally appropriate manner. In this case, two processes were conducted in parallel, neither of which required participants to agree in advance that a permit would be approved. In one, the participants discussed the circumstances under which the company would withdraw its proposal. In the other, participants developed the scope and approach to an environmental assessment process to identify potential impacts of the mine, if it were to be permitted. The result of the first process was an agreement not to open a mine.

Qualitative Characteristics of Risks

Considerable research shows that people's judgments and levels of concern about environmental and other risks depend on a number of qualitative characteristics of the hazards in addition to the probability and magnitude of the potential harm (Fischhoff et al., 1978; Slovic, Fischoff, and Lichtenstein, 1979, 1980, 1985, 1986; Slovic, 1987, 2000; McDaniels, Axelrod, and Slovic, 1995, 1996; McDaniels et al., 1997; Lazo et al., 2000; Rosa, Matsuda, and Kleinhesselink, 2000; Morgan et al., 2001; Willis et al., 2004, 2005; Slimak and Dietz, 2006; Willis and DeKay, 2007). For example, things that are perceived as under individual control, such as smoking and driving, tend to be perceived as relatively less risky than things that are perceived as less controllable, such as exposure to toxic substances in the air and flying by commercial jetliner. Hazards that threaten especially catastrophic or dreaded outcomes, such as genetic damage to future generations, are likely to be perceived as especially risky.

Research also indicates that such factors are also associated with people's desires for stricter government regulations or other risk reduction efforts (e.g., Slovic, Fischhoff, and Lichtenstein, 1985; Slovic, 1987; Baron, Hershey, and Kunreuther, 2000; Willis et al., 2005), suggesting that the kinds of risks presented by environmental conditions—and the emotions that those risks evoke—may affect the propensity of otherwise uninvolved individuals to become involved. For example, Fischhoff, Nadai, and Fischhoff (2001) reported that firms that engage in activities involving unknown, dreaded hazards are more likely to be the target of consumer boycotts and more likely to be screened out by socially responsible investment funds (see also Stern, Dietz, and Black, 1986). An intense emotional reaction can act as a signal to take immediate action (e.g., Lazarus, 1991; Cacioppo et al., 1999; Loewenstein et al., 2001), but in some contexts it impairs effective responding (Holloway et al., 1997; DiGiovanni, 1999; Loewenstein et al., 2001). The difference may depend on the nature of the emotion evoked.

Characteristics of the environmental issue that generate involvement can make it easier to secure participation from otherwise silent segments of the public. However, there are cases in which high motivation is accompanied by substantial mistrust of either the agency involved or of other participants, and that mistrust can make effective participation very difficult. The point is that it is not the character of the environmental problem itself that is critical, but, rather, the history of the problem and the psychological and social factors and relationships that emerge from that history (e.g., Peters and Slovic, 1996; Loewenstein et al., 2001).

Collective Action and Common-Pool Resources

Research on collective action and the management of common-pool resources (e.g., Olson, 1965; Ostrom, 1990; National Research Council, 2002a; Dietz, Ostrom, and Stern, 2003) supports the idea that results depend much less on what kind of resource is involved than on the distribution of costs and benefits and other issues related to acquiring information, monitoring the environmental condition and the people involved, and resolving conflict. For example, some issues, such as construction of a new electrical transmission line, subway station, or highway, tend to align many who may benefit or lose a little on one side against a few on the other side who stand to lose or gain a lot. In such cases, effective participation by the many can be difficult to obtain. We return to the issue of involving the parties in Chapter 8.

Motivation to resolve conflict can also be a critical component to a successful participation process. Such motivation may come from the nature and intensity of a person's or group's interests in a resource problem or from a shared recognition of the interdependent nature of participants' interests (Selin and Myers, 1995; Yaffee, Wondolleck, and Lippman, 1997). Motivation may also increase when participants perceive a political stalemate in which there is a lack of viable alternatives to collaboration (U.S.D.A. Forest Service, 2000). These examples suggest some of the ways that the characteristics of an environmental issue or hazard may affect public participation processes by affecting the motivation of important segments of the public to participate.

These findings have two hopeful implications. One is that good public participation practice, in the form of processes designed to meet the potential difficulties that the problem context may create, can improve results for any kind of environmental issue. The other is that what is learned from experiences with public participation in one problem area can be transferred to others: strategies for addressing potential difficulties may be transferable across environmental issues.

THE SCIENCE

Although scientific complexity and uncertainty are often cited as barriers to effective public participation and although scientific knowledge is nearly always limited relative to the question being asked, the available evidence fails to support the contention that incomplete or difficult science precludes effective public participation. How the available knowledge is introduced and used in the process and how new information is generated seem to matter more than the characteristics of the knowledge itself.

In Chapter 6 we review some of the practices that allow for effective integration of science and public participation. These practices are especially important because there has been a tremendous increase over the past decade in the amount of scientific information available to inform environmental policy as well as an improvement in methods for assessing and integrating that information. Despite this growth in the quantity and quality of information brought to environmental decision making, it is rare to face a policy problem for which the information lights the way to a single solution acceptable to all interested and affected parties. In some situations, significant additional information may not further clarify the policy options, and, in most situations, value conflicts are important enough that scientific information alone, no matter what its quality, will not be sufficient to determine a decision. Here we focus on the potential difficulties in public participation that may arise because of the importance of scientific information in environmental assessment and decision making.

As noted in Chapter 6, Bingham (2003) has provided a useful classification of the kinds of problems that can arise in science-intensive disputes. She suggests that there are five "knots" that tie up such disputes: (1) the adequacy of the information for the problem, (2) the clarity of the decision-making process with respect to science, (3) the problems parties have dealing with the data, (4) the problems scientists have among themselves and in communicating with stakeholders, and (5) problems of trust. Each of these is present to some degree in nearly every environmental assessment and decision process. Box 7-1, which draws on this classification, identifies four key sets of questions about the available scientific information that are diagnostic in the sense that the participatory processes can benefit from being designed differently depending on the answers.

Adequacy of Information

The information available to understand an environmental issue can be insufficient in many ways, of which knowledge gaps and scientific uncertainty are only the most obvious ones. Indeed, a very long list of reasons for scientific inadequacy is easy to develop (see Fischhoff, 1989, and Bingham, 2003, for useful compilations). Data may be outdated or from the wrong context. Models may not address key issues of concern, may rely on assumptions that are not accepted by everyone, or may be of limited applicability to the context being considered. Inadequacy of information includes situations in which all participants can readily recognize knowledge gaps as well as situations in which knowledge seems adequate to some participants and inadequate to others because they see the problem differently. Any of

CONTEXT: THE ISSUE

> **BOX 7-1**
> **Key Questions Regarding the Character of the Science Available in Environmental Assessments and Decisions**
>
> 1. Is the information adequate to give a clear understanding of the problem? To what extent do parties define the problem in different ways? Do the various parties agree about the adequacy of the information for the problem(s) defined?
>
> 2. Is the uncertainty associated with the information well characterized, interpretable, and capable of being incorporated into the assessment or decision?
>
> 3. Is the information accessible to and interpretable by interested and affected parties?
>
> 4. Is the information trustworthy?

these contextual factors may make it more difficult to achieve the attributes of good public participation described in Chapters 4-6 unless processes are designed to address them.

The research literature includes numerous examples of problems that can arise from the inadequacy of science in environmental assessment and decision processes. Leach's (2005) review of 25 empirical studies of public participation in the Forest Service from 1960 to the present examines the importance of adequate scientific and technical information. Nine of these studies identified adequate information as critical in facilitating a successful process. Lubell and Leach (2005:23-24) found that adequate scientific expertise, as perceived by participants, was associated with more successful outcomes across the watershed partnerships he examined. He notes that what matters is confidence in expertise rather than mere information and suggests that the key factor may be the embodiment of relevant information in those who can engage in dialogue with participants and perhaps offer judgments in the absence of scientific certainty. We return to the issue of trust below.

If adequacy of information is a matter of participants' judgment, then there may be value in efforts to ensure that participants share in whatever information is available and attempt to reach agreement about what is and is not known. Consistent with this idea, Shindler and Neburka (1997) report more success in processes that selected participants who were already knowledgeable about the issues.[1] Wondolleck and Yaffee (1994) advocate such techniques as "joint fact-finding exercises," in which participants or their designated experts share information and engage in collaborative information generation. Similarly, Daniels and Walker (1997) found evidence that a systems approach to assessing the extent and underlying causes of a problem was useful. Bingham's (2003) recommendations from practitioner experience emphasize processes that are likely to build mutual understanding about the data available, develop new information to fill gaps, and increase confidence that the information is adequate for the decision. And to reiterate a point from Chapters 2 and 6, public participants often bring important context-specific knowledge to the process. Thus, the sharing of relevant information is more than a one-way transmission from scientists to nonscientists.

In some situations, the problem is relatively simple—key information is simply missing. For example, in the case of a power plant in Virginia, all sides agreed that the plant's cooling waters exceeded thermal limits of its discharge permit. What they didn't agree on—and did not have information about—was whether the temperature of the discharge waters had an actual adverse effect on the ecosystem of the lake into which it flowed. Federal and state agencies, the company, and segments of the public were represented by scientists on a steering committee that designed a research project (a joint fact-finding effort) and selected a mutually acceptable research team to answer the question.

Sometimes important information gaps cannot be filled in time for a decision, even if those affected are willing to devote considerable time and resources to joint fact-finding. In one such case reported by a panel member, in which an agency had an unusually short deadline for proposing new regulations for carbon sequestration, the agency organized an iterative process featuring two open workshops, a proposed rule, and plans to use the proposal as a basis for further dialogue. The first workshop was to inform the public about the process, to establish a foundation for ongoing dialogue, and to learn about stakeholders' concerns. In the second workshop, a few months later, the agency informed participants about key questions remaining in the options under consideration and invited thoughts about those questions. Both workshops used a "fishbowl" technique in which experts from the agency and from diverse stakeholder groups discussed the scientific issues in the presence of the other participants, who could use the discussion to improve their understanding rapidly. This combination of practices, along with the opportunity for extensive submission of technical information in response to the proposed rule and for continued iteration of analysis and deliberation through a regulatory mechanism called a notice of data availability, constituted a creative attempt to provide needed information under serious constraints.

Because of the potential for ambiguity, it can be difficult to determine whether or not the information available is adequate for informing an assessment or decision. Most environmental issues can legitimately be addressed from various disciplinary perspectives, in terms of various agencies' responsibilities and in terms of different kinds of public concerns, so there may be competing definitions of the issues.

An example cited a decade ago (National Research Council, 1996) is still telling on this point. The Pennsylvania Environmental Protection Agency, in considering a proposal to site a soil decontamination plant in the low-income, largely African American city of Chester, proposed to make the decision on the basis of an assessment of the incremental health risks to the local population from expected additional hazardous chemical exposures. The Chester city government argued that this assessment would not address the right question. It demanded analyses of the possible synergistic effects of the added exposures and existing exposures from the city's oil refineries, trash incinerator, infectious materials processing center, and Superfund sites; of special risks due to the population's health status and past toxic exposures; and of comparison between the health effects of siting in Chester and in nearby, more affluent communities whose populations were healthier and had lower past exposures.

As this example suggests, legitimate disagreements can arise about how issues for scientific analysis are framed and about which analyses are needed. Scientific information that would be adequate for addressing a policy question when framed in one way would be clearly inadequate under a different framing. Discordant framing or structuring of the issue among the parties can be an important factor underlying conflict about risk and environmental management and can make it more difficult to achieve clarity of purpose for a public participation process (e.g., Miller, 1989; Fisher, 1991; Carnevale and Pruitt, 1992; Fischhoff, 1996a; Kunreuther and Slovic, 1996; Pellow, 1999; Bazerman et al., 2000; Gray, 2004). It can also drive out some affected parties and potentially reduce the inclusiveness of participation and, potentially, the legitimacy of the process, if some parties no longer see the possibility of their concerns being addressed.

When discordant issue framing is a possibility, an explicit and transparent effort to involve participants during the diagnosis and process design in framing or structuring an issue may allow shared frames of reference and definitions of the problem to evolve or, at least, make it possible to come to a shared agreement on the questions or issues to be addressed and a common understanding of the adequacy of information for the decision to be made. Successful negotiations can result in compatible definitions of the problem and ideas about subsequent goals (Kruglanski, Webster, and Klem, 1993). Such successful negotiations were evident in the first U.S. National Assessment of the Potential Consequences of Climate Variability and Change, in which participants in several regions worked together to reframe the focus of the entire assessment from one focused solely on climate change to one focused on climate variability and change (Moser, 2005). Practices commonly used to establish a shared framing or focus for a public participation process include interviews with potential stakeholders to understand their perspectives, drafting a formulation in terms of how to achieve interests at stake rather than whether to accomplish something proposed by one side, publishing and requesting comments on the proposed scope of a process in such publications as the *Federal Register*, holding organizational workshops in the planning phase of a public participation process, and organizing process steering committees.

Disputes about the adequacy of information can also arise because of difficulties over separating the scientific issues involved in understanding and predicting phenomena from the value issues of how to make appropriate trade-offs among goals. At an abstract level, there are serious objections to making a clear-cut division between facts and values (e.g., National Research Council, 1983, 1996; Jasanoff, 1996). More concretely, arguments about facts and arguments about values are often confused with each other in disputes about environmental policy (e.g., Fischhoff, 1989; Dietz, 2003). One problem arises when value assumptions become embedded in analytic methods. For example, in estimating environmental risks, conservative assumptions may be built into the assessment process so that some adverse effects (such as human cancer from pesticide exposure) are much less likely to be underestimated than overestimated (National Research Council, 1983). Generally, the meaning of scientific findings can depend on how they are cognitively framed (Tversky and Kahneman, 1981) and what is regarded as value depends on prior knowledge about the factual implications of different value preferences (Fischhoff, 1975).

The decision relevance of scientific knowledge can also be subject to legitimate dispute. Another decade-old example provides a useful illustration (National Research Council, 1996). A dispute in California in the early 1990s concerned a proposal to spray the insecticide malathion to eradicate populations of Mediterranean fruit flies, believed to have arrived in small isolated groups on infested imported fruit, that threatened the state's $2 billion fruit and vegetable industry. Much attention was given to risk assessments concerning the possible human health effects of malathion exposure. Some critics argued that the fly populations were not isolated and were already established and that therefore targeted malathion spraying would not solve the problem, and the health risk assessment was largely beside the point. They proposed biological pest controls that did not create human health risks, a proposal that required quite a different kind of scientific assessment. In this case there was substantial uncertainty about key facts (whether or not the fly populations were established) that was more important to the controversy than the uncertainty in risks captured in the health risk assessment.

CONTEXT: THE ISSUE

The potential for such fundamental disputes about whether and how available scientific information is decision relevant implies that the adequacy of information cannot be determined without considering participants' perspectives. Issues that are important to some of the parties but are excluded from analysis can create serious conflict, particularly when there is no other forum for addressing these issues (Bradbury, 2005). It is therefore helpful to diagnose whether different understandings of the environmental issue exist among the parties that imply different needs for scientific information. If they do, it is important to look for ways to allow science to address the concerns of all the interested and affected parties and to ensure that scientific information on the full range of these concerns is gathered and presented, so that one issue definition does not dominate because relevant information supporting another definition is missing.

Confidence by participants that they have the best available information, that they have information on issues that matter to them, and that scientists will interpret the available data correctly appears to be important for the integration of incomplete information into a public participation process. Limitations in the available information, including a lack of some information that is seen as desirable by participants, does not by itself preclude effective public participation any more than it precludes effective decision making by government agencies. Rather, what seems crucial is that public participation processes address the inadequacies in the science so as to build mutual trust and understanding between the scientists and the public. Processes for effectively linking scientific analysis and public participation are discussed in Chapter 6.

It is worth noting that disputes about the adequacy or relevance of science can mask other issues. In one example, allocation of Clean Water Act funds was held up by disagreements about whether eutrophication in a bay was being caused by nitrogen or phosphorus. Scientific studies were cited to support both conclusions, leading to the sense that the information was not adequate. In that case, a participatory process was preceded by a workshop of the scientists whose work was being referenced. They issued a report outlining large areas of convergence in the science as well as specific questions that remained unresolved. The scientific uncertainties were relevant, but they were also being used in ways that obscured underlying conflicts of interests. In this case, if phosphorus was the more relevant nutrient, the urbanized counties would get funding for additional sewage treatment capacity; if the culprit was nitrogen, rural counties would get funding for nonpoint source reduction programs. Clarifying what was and was not known, and the decision relevance of what was not known, allowed the disputes to be addressed more directly.

Characterization of Uncertainty

Uncertainty is a form of information inadequacy, but it is so central to environmental assessment and decision making that it deserves separate consideration. We use the term "uncertainty" broadly, to cover various kinds of gaps in knowledge, including those caused by "random" processes for which a statistical distribution is known, those for which the causal factors are known but the probabilities of outcomes are not, and those for which even the causal factors are partly or largely unknown. These latter kinds of unknowns can create more fundamental difficulties than those caused by statistical variation.

The many forms and varieties of scientific uncertainty are themselves a matter of debate (e.g., Funtowicz and Ravetz, 1991, 1993; Wynne, 1992; Rosa, 1998; Yearley, 2000). The issue of uncertainty has not been sufficiently explored in research on public participation. Major international assessments, such as the Intergovernmental Panel on Climate Change and the Millennium Assessment, have offered their authors explicit guidance on how to describe different degrees of scientific uncertainty (Moss and Schneider, 2000). The environmental science community is directing considerable attention to clarifying how scientific uncertainty can be assessed and characterized in support of decision making (e.g., Morgan and Henrion, 1990; Van Asselt, 2000; Kinzig et al., 2003; National Research Council, 2007b).

A growing body of research suggests that the frame used to express uncertainty has a substantial influence on how well members of the public process such information (Gigerenzer and Hoffrage, 1995; Gigerenzer, 1998), a result that suggests that some framings, such as in terms of frequencies, lead to more effective handling of information about uncertainty than others. There have been only a small number of efforts to explore systematically the various formal techniques for characterizing complex and uncertain information about risks in messages for the various participants in environmental deliberative processes (e.g., Johnson and Slovic, 1995; Kuhn, 2000; Florig et al., 2001; Johnson, 2003; Willis et al., 2004; Gregory, Fischhoff, and McDaniels, 2005). Thus, more needs to be learned about how to provide useful information about scientific uncertainty to participants in environmental assessment and decision-making processes, particularly about the potentials of formal techniques and process-based, analytic-deliberative ones.

Ignoring major sources of scientific uncertainty is considered unacceptable practice in most scientific communities because it carries the danger of producing a sense of security and overconfidence that is not justified by the quality or extent of the database (Einhorn and Hogarth, 1978). However, it is often claimed that presenting scientific uncertainty and complexity to nonscientists creates difficulties for participatory processes because the public is not well equipped to understand the science or to deal with uncertainty (see, e.g., Fischhoff, 1995, on the reluctance to acknowledge and communicate uncertainty). Uncertainty about facts and their relationship to values has been highlighted as a possible cause of environmental conflict (e.g., Fischhoff, 1989; Dietz, 2001). In the early 1980s, many risk professionals attributed environmental conflict largely to public ignorance (Dietz, Stern, and Rycroft, 1989), and this view remains widespread (e.g., Frewer et al., 2003; Sweeney, 2004).

Contrary to these views, the evidence from some public participation processes suggests that nonscientists can be quite comfortable with uncertain scientific information. For instance, in the U.S. National Assessment of Climate Change, it was reported that in interactions between scientists and nonscientist stakeholders, the nonscientists were more comfortable in treating uncertainties than the scientists (Moser, 2005). This may be because nonscientists live with uncertainties every day (e.g., deciding what to build, what to buy, what to plant) so that the engaged public does not expect that uncertainties disappear but rather that they be clearly described: for example, what is possible and what is not? What is likely and what is not? What is a 1 in 10 bet versus a 1 in 100 bet (Mahlman, 1998; Frewer et al., 2003)?[2] Indeed, some have suggested that carefully structured public participation processes can help in reducing uncertainty in decision-making processes (Lourenço and Costa, 2007). Much of the analysis we have already reviewed demonstrates that linked analysis and deliberation reveal uncertainties that might otherwise have been missed.

Johnson and Slovic (1995) argued that disclosing uncertainty can be a signal of honesty on the part of the agency presenting the information, but also found that it is a challenge to avoid confusion and outrage (Johnson and Slovic, 1998). It appears that responses to information about uncertainty, like responses to other kinds of risk information, are affected by preexisting views, including views about the credibility of the information source (e.g., Johnson and Slovic, 1995, 1998; Kuhn, 2000; Frewer et al., 2003; Johnson, 2003; Miles and Frewer, 2003). There is some evidence that uncertainty in scientific information increases the salience of concerns with trust and procedural fairness (Van den Bos, 2001; Van den Bos and Lind, 2002), issues we discuss in Chapter 6 and below.

Comfort in assimilating uncertain science may be greater if sources of expertise that stakeholders trust assist in communications between them and representatives of the scientific knowledge base (Association of American Geographers, 2003). In some instances, formal analytic methods to quantify important uncertainties may be helpful (National Research Council, 1983, 1994, 1996). Formal methods for dealing with uncertainty are central to agency traditions in some domains, such as risk analysis of exposures to toxic chemicals (National Research Council, 1983, 1994; Presidential/ Congressional Commission on Risk Assessment and Risk Management, 1997a,b), but not in others, such as natural resource management. Thus, the readiness of scientists to characterize uncertainty for interested and affected parties may be different across environmental issues.

Courts have emphasized process in responding to challenges based on claims of uncertainty. For example, Judge C.L. Dwyer (1994, *Seattle Audubon Society et al. v. James Lyons, Assistant Secretary of Agriculture et al.*), in ruling that the Northwest Forest Plan (NWFP) was a legal exercise of administrative authority, did so based on the clear and transparent discussion of uncertainty both of knowledge and of the likely success of the decision itself. Because the agencies had clearly identified how they would continue to address the uncertainty issue, the court ruled that as long as the agencies met their own process requirements, the plan would continue to be upheld, but if they did not, the essential element of the plan—projecting management options into the future through an adaptive management framework—would be violated. The NWFP envisioned a continuous process of open analytic deliberation among scientists, managers, tribes, governments, other stakeholders, and the public as a mechanism for continuous learning—an expectation only partially met in practice.

Like the difficulties associated with adequacy of information, characterizing uncertainty is a perennial challenge in environmental assessment and decision making. From the research and practitioner experience available, it appears that when uncertainty is at issue, the character of the relationships among the interested and affected parties, the convening agency, and the scientists becomes particularly important and that the way in which information is provided can also have a significant effect on results. As discussed in Chapter 6, there are ways of structuring the participation process that can enhance effective public engagement even in the face of uncertainty.

Accessibility and Comprehensibility of Information

A fundamental requisite of public participants' making effective use of scientific information is that the information is available to them. This requires that participants have access to and are able to critically interpret scientific information. Both of these requirements can be difficult to meet in public participation processes.

Participants may have trouble simply obtaining access to analyses. Scientific research is usually published in the peer-reviewed literature and in the "gray" literature of technical reports. Journals are available online but often only through expensive subscription services. Technical reports are increasingly available online as well and thus potentially accessible to a broad audience. But even if members of the public can view a copy of a scientific analysis, it may be difficult to interpret. Scientific reports are usually written in the dense language of the fields of science that underpin the analysis, so most nonspecialists, and even specialists from different disciplines, find them very time-consuming to interpret or completely opaque.

Some assessment activities have made special effort to make their reports publicly accessible. For example, data collected for the First U.S. National Assessment of the Potential Consequences of Climate Variability and Change were made available on a website. Moser (2005:49) reported that "According to the CCSP staff, the National Assessment constitutes 'the most popular product because it offers information at a scale that people care about.' Between two major access sites . . . it is safe to assume that National Assessment material now sees 400-800 visits every day." In a successor project to the National Assessment, a group of researchers collaborated with private-sector decisions makers (farmers, operators of recreation businesses) to develop a website that provides climate projections (http://www.pileus.msu.edu). The content of the website and the parameters projected were developed through a multiyear collaboration between researchers and the decision makers. Other techniques used in public participation processes to increase accessibility include summarizing technical materials in plain language, providing technical assistance grants to citizen groups, public education workshops or "open houses," and including someone with technical expertise and broad credibility on the staff of the participation process to serve as a translator.

In several advisory committees convened by the U.S. Environmental Protection Agency (EPA) and charged with making recommendations for drinking water regulations, the processes have included a technical work group open to staff or volunteers from all participating organizations to enable broad access to information and broad participation in its analysis. Funding was sometimes made available for groups that did not have their own technical staff. In addition, EPA sometimes hired technical advisors to work for the advisory committee as a whole, playing a leadership role in the technical work group and serving as a translator of the results of the joint technical analyses to the advisory committees.

Other experiments with making scientific information more accessible via the web are under way (e.g., Haklay, 2002, 2003; Harrison and Haklay, 2002; Kellogg and Mathur, 2003). These efforts are likely to yield methods that will greatly increase access to analyses and models. However, making them interpretable by those not trained in environmental science will remain a challenge. When understanding scientific or technical information is important to meaningful participation, it will be important for public participation efforts to invest in meeting the challenge. More generally, research into how to accomplish this will benefit a broad range of environmental public participation efforts.

Trustworthiness of Information

We have alluded to the importance of trust in ensuring effective public participation around issues with a substantial scientific content. Here we elucidate how to anticipate issues of trust that may arise from the application of science in the context of environmental public participation. Chapter 6 discusses processes that have been used to address these issues.

Scientific information can contribute to mistrust in many ways. For example, models and methodologies have often become targets for public wrath when they have been perceived as vehicles for justifying policy decisions with numbers that appear to be scientific but that cannot be verified by critics or that embody assumptions that seem patently incorrect to some of the parties (e.g., Jenni, Merkhofer, and Williams, 1995; Wynne, 1995; Yearley, 2000). They can generate mistrust when it is discovered that an analysis that is presented as comprehensive in fact glosses over important issues by making simplifying assumptions. For example, benefit-cost analyses are often presented without reminding the audience that the estimates are aggregate and do not address who pays the costs and who gets the benefits (e.g., Bentkover, Covello, and Mumpower, 1985; Smith and Desvouges, 1986; Fischhoff, 1989).

On one hand, there are well-recognized difficulties with complex models when applied to public policy (e.g., Hoos, 1973; Van Asselt, 2000; Jaeger et al., 2001). Numerical outputs reflect input data, as well as judgments and assumptions put into the model, all of which can be subject to question and many of which may be hidden from public view. For example, models may focus on what can be quantified easily, which can result in a perhaps unintended but still less than transparent prioritization of certain variables. Possible results include incomplete or inaccurate analyses (and a poorly informed decision), if the variables that are included are not the only factors driving public concern (e.g., Cramer, Dietz, and Johnston, 1980; Dietz, 1987), and the perception of a less open or transparent process and, thus, poor acceptance of a potentially good decision. And the public may want levels of detail and certainty from models that are beyond the scope of the current state of the science. Dietz et al. (2004) found that estuary modelers and local planning officials had incompatible views about what to expect from models intended to guide land use decisions that would affect local estuaries.

On the other hand, there are intriguing experiments with using computer models as tools to aid the deliberative process (e.g., Van den Belt, 2004). In an attempt to resolve water allocation issues on the Truckee and Carson Rivers in western Nevada, experts from the private sector, agricultural interests, environmental groups, and federal, tribal, and state agencies shared water supply forecast models with one another and made multiple

CONTEXT: THE ISSUE   179

runs of these models to enhance transparency and provide their constituencies a shared perspective on similarities and differences in the results produced by the different models. More recently, the Institute for Water Resources at the U.S. Army Corps of Engineers has pioneered a "Shared Vision Planning" process that integrates participation and modeling (see http://www.svp.iwr.usace.army.mil/).

Available evidence indicates that the complexity of scientific issues does not by itself present a significant barrier to effective public participation. Nonscientists can make meaningful use of science when managing complex environmental systems. For example, local communities of fishers can manage the complex ecosystems that produce their fish and can even link effectively to government agencies operating at a larger scale (Berkes, 2002). Complexity can require, however, that special effort be made to organize scientific analyses around questions that are salient to the decision and to participants (e.g., Wilson, 2002) and to ensure that the range of participants understands the science, including its limitations. Although some of the parties may desire a predictive understanding of a complex environmental system, it is important for scientists to be open about the limitations of the available science for producing such an understanding. When the situation is too complex to allow good prediction, assessments and decisions may be better informed by a set of plausible scenarios consistent with scientific knowledge than by poorly grounded predictive models (Brewer, 2007). Complexity can also provide fodder for disputes about what kinds of scientific analyses are needed and about the practical import of available knowledge (see below).

EPA's Total Coliform Rule and Distribution System Advisory Committee, formed in 2007, is a recent example of an effort to make information trustworthy. The advisory committee's task was to make recommendations on revising an existing drinking water regulation to monitor water quality in distribution systems and on data collection and research needs for the future. The advisory committee requested information and analysis from a technical work group at its initial meetings to help define issues, developed options at later meetings and then, prior to evaluating those options, consulted with the technical work group about what information was and was not possible to generate to compare the impacts of those options. This iterative approach included explicit discussions of what information participants said would be relevant to their deliberations and what decision-relevant information would be possible to obtain. These discussions brought disagreements about the practical application of existing information to light and allowed those involved to address them to the extent possible.

It is worth mentioning the possibility of claims of scientific bias or conflict of interest. Such claims can arise when scientists are employed by parties with vested interests or are retained by certain parties to provide them with information but are not trusted by other parties to be fully forthcoming if the information is not advantageous to their clients. A particularly troublesome situation arises when scientific information comes from an agency with a track record of deceit. For example, it was claimed in the 1980s that the Department of Energy's civilian radioactive waste program was still suffering from the reputation of the Atomic Energy Commission, which in the 1950s had attributed radioactive "fallout in St. Louis to Russian sources when it was known to have come from tests in Nevada" (National Research Council, 1989:120).

Difficulties also arise when scientists have personal stakes in the issue, for example, when a scientist's research may be affected by the decision to be made or, when a decision is about a resource or issue that has become the focus of the scientist's professional life. In such cases, a scientist may engage in intentional or unintentional advocacy in the choice of objectives or methods when designing a study. Agency scientists are also sometimes accused of bias toward policies currently in force or that form part of the political agenda of the party in power. Such claims are obvious sources of mistrust in information and are worth looking for. They are best addressed by acknowledging the possibility and opening the scientific discussions to intellectual criticism from any of the parties. If the parties have sufficient resources to participate meaningfully in such discussions, openness is the best way to address bias claims.

Various practices have been used to promote openness. In some situations, agencies have opened the process of nomination to formal peer review panels by consulting interested and affected parties. In other cases, public workshops have included interactive discussions about the science among panels composed of experts nominated by the parties, held in a "fishbowl," a public setting. Technical work groups composed of experts that represent diverse parties can vet analyses prior to their presentation as part of a public participation process, sharing points of agreement and disagreement about methods, analyses, and interpretation along with the analyses. In addition, studies can begin with broad consultations to arrive at a collaboratively designed inquiry or joint fact-finding process.

CONCLUSIONS

Certain issue-related aspects of the context pose difficulties in achieving the goals of effective public participation; others are less consequential. The evidence supports several conclusions.

- The determinants of success are largely the same for participatory environmental assessments and decision-making processes.
- The environmental subject matter has little direct effect on partici-

pation outcomes. What is learned from experiences with public participation in one problem area can generally be transferred to other areas.

- Environmental subject matter can create particular difficulties for participation. For example, long-lived hazards require long-term monitoring and therefore continuing participation over time. Decisions regarding large-scale environmental systems create costs for participation associated with bringing people together over long distances.
- Scientific complexity and uncertainty do not preclude effective public participation. How the available knowledge is introduced and used in the process matters more than the characteristics of the knowledge itself. What matters is how the scientific information is integrated into the process. Concerns about procedural fairness and trust are more salient with scientific uncertainty, and it is therefore important to ensure that public participation processes provide for open and balanced consideration of the scientific issues, including gaps in knowledge, and to provide information in ways that facilitate understanding by nonscientists.
- Some contextual factors can create the potential for serious conflict among the parties. When issues may be framed in competing ways or when there may be credible claims of scientific bias, there is significant potential for conflict over the science. When the parties are polarized at the outset in terms of policy preferences and when some parties expect that other parties, or the responsible agency, may be proceeding in bad faith, there is significant potential for conflict over both science and policy. Such conditions make certain aspects of participatory processes especially important because it is possible to build trust in a process even among parties in fairly strong conflict. Several of the principles of good participation are likely to be especially important when the potential exists for serious conflict. These include transparency of process, inclusiveness, availability of decision-relevant information, explicitness about assumptions and uncertainties, independent review, and iteration.

In sum, although certain characteristics of the issue context can create particular difficulties in public participation and in implementing particular principles of good practice discussed in Chapters 4-6, the difficulties can be addressed and often overcome through the use of various specific practices, tools, and techniques. It may be necessary to collect more information, make special efforts to characterize or discuss areas of incomplete knowledge and scientific uncertainty, or provide some of the parties with resources to allow them to understand the issues and the scientific information well enough to participate meaningfully. The possibility of different framings of the issues and of potentially credible claims of scientific bias call for practices of analytic deliberation that open scientific information to intellectual criticism and thus encourage scientists and others to question their assumptions about what is important to analyze; how physical, biological, and social processes unfold; how to handle uncertainty; and related issues (Renn, 2004, 2008). Opening scientists' judgments to comment by nonspecialists can be unsettling for the scientists. However, deciding which questions are important to ask involves judgments in which scientists and nonscientists alike have a legitimate voice. Transparency in the methods used to gather data and in the assumptions made in the analyses builds trust. Scientists still conduct the science. Involving the public in appropriate ways is consistent with the logic of the scientific process in that all aspects of an analysis are subject to constructive scrutiny and the analysis can be improved as a result.

Table 7-1 provides a diagnostic guide to many of the difficulties associated with aspects of the issue context and to some ways that have been used to try to address them. It identifies particular contextual factors that can make it difficult to implement particular principles of good participation, notes the nature of the likely difficulties, and identifies some practices or techniques that have been used to address the difficulties. It is not meant as an endorsement of any of these practices: evidence is too weak and contexts are too varied for any such endorsement. However, we believe the guide can help agencies, practitioners, and the public anticipate difficulties and begin to think about possible responses. In Chapter 9, we recommend a process for selecting among those responses and addressing context-related difficulties in public participation.

NOTES

[1] It does not follow that it is wise to exclude participants whose scientific knowledge is limited. Limiting the breadth of participation can have serious negative consequences for the overall process, as discussed in Chapter 5. Chapter 6 details practices that can aid in making scientific information useful in a linked process of analysis and deliberation and thus provides guidance on how to make scientific information accessible to those with limited scientific backgrounds.

[2] A difficulty is that while daily life may make people familiar with situations that involve "bets" in which the choice is between odds on the order of 1 in 100 and those on the order of 1 in 10, many environmental health risks require making choices in which the contrasting odds are 1 in 10,000 and 1 in 100,000. The latter numbers are far from the realm of experience of most citizens; however, their comprehension of them may be aided by the use of formal analytical tools.

TABLE 7-1 Diagnostic Guide to Difficulties Related to the Issue in Public Participation

Contextual Factor	Principles That Become More Difficult to Achieve	Difficulties	Illustrative Practices for Addressing Difficulties[a]
Issue Factors			
Long temporal scale	Inclusiveness	Number of interested and affected parties Ability to participate (e.g., future generations)	Create new, longer-lasting institutions
Large spatial scale	Inclusiveness	Number and range of interested and affected parties	Workshops in multiple locations Internet participation Study circles Deliberative polling Blue Ribbon commissions
Large spatial scale	Clarity of purpose Commitment to use the process to inform decisions	Multiple agencies with jurisdiction over parts of the issue	Interagency work groups
Complexity	Clarity of purpose Collaborative problem formulation	Diversity of perspectives can increase with the complexity of the issue	Frame issues in terms of reconciling interests Include multiple issue framings
Collective action and common-pool resources	Inclusiveness	Motivation to participate is lower among those who stand to benefit or lose only a little	
Science Factors			
Inadequate information	Collaborative problem formulation	Missing information may make it difficult either to define the problem clearly or to solve it	Joint fact finding Iterative processes, with steps to solicit information from stakeholders Expert panels or "fishbowl" techniques at workshops

Continued

TABLE 7-1 Continued

Contextual Factor	Principles That Become More Difficult to Achieve	Difficulties	Illustrative Practices for Addressing Difficulties[a]
Inadequate information	Clarity of purpose	Different problem definitions may cause disagreement about what information is needed. Left unresolved, it also can reduce motivation to participate if parties do not see the possibility of their concerns being addressed	Involve participants during diagnosis and design phase in creating an agreement on the scope and objectives Draft a written scope and invite public comment on it Hold workshops during the organizational phase and review the information available and discuss adequacy of the information in terms of the scope Organize a process steering committee to consider problem definition throughout the process
Inadequate information	Availability of decision-relevant information	Disputes about adequacy of science can mask other issues (e.g., when parties select conclusions only from those studies that support an outcome that satisfies their interests)	Joint reports from all scientists about what is known, where areas of disagreement or uncertainty remain, and the decision relevance of what is not known
Scientific uncertainty	Explicitness about analytic assumptions and uncertainties	When uncertainty is poorly characterized, it is hard to estimate the consequences of choices	Formal characterization of uncertainty Adaptive management

Particular difficulty	Desirable attributes	Potential problems	Examples of practices
Accessibility of information	Availability of decision-relevant information	Participants may be unfamiliar with where to find information in peer-reviewed journals or it may be too costly to obtain the information Information may be written in highly technical language and may be difficult for the public or individuals from other disciplines to interpret	Summarize information in plain language Provide technical assistance to participants who need it Organize public education workshops or "open houses" Include a technical expert on the facilitation team to serve as a "translator" Put information on the web
Information may not be trusted	Explicitness about analytic assumptions and uncertainties Good-faith communication	Models and other scientific methods are or are perceived as manipulated to justify a decision made for other reasons Analyses make simplifying assumptions that obscure issues of importance to participants Models focus only on what can be quantified easily, unintentionally prioritizing certain variables, leading to incomplete or inaccurate analyses, or contributing to a perception of bias	Invite scientists to explain limitations of available science Develop scenarios as an alternative to models when predictive models are inadequate Form a technical work group of experts trusted by all sides and develop or vet information and analyses through that group Invite stakeholder nominations for peer review groups Invite stakeholder comments on selection of members of expert panels to ensure confidence that all scientific views are included Engage in joint fact finding

[a] Evidence is inadequate to recommend any of these practices as effective, or as preferable to practices that are not listed. They are listed to suggest some of the practices that might be considered for addressing particular difficulties.

8

Context: The People

This chapter reviews evidence on how the human aspects of the context in which an environmental decision is made—including attributes of the sponsoring agency, as well as its legal and organizational environment; characteristics of the other participants in an assessment or decision; and the dynamics that can occur as people interact—affect the results of public participation. Like Chapter 7, this chapter describes these relationships and provides examples of practices that attempt to overcome difficulties that can emerge from or be exacerbated by the contextual factors. The first section considers the convening and implementing agencies and organizations; the next sections consider the characteristics of the participants and the dynamics of the process.

CONVENING AND IMPLEMENTING AGENCIES

The agencies or other organizations that convene a public participation process and the ones responsible for an environmental assessment or decision are often, but not always, the same.[1] Several diagnostic questions relating to the agency's internal and external context point to challenges for public participation. Participatory processes may need different emphases depending on the answers to these diagnostic questions.

1. Where is the decision-making authority? Who would implement any agreements reached? Are there multiple forums in which the issues are being or could be debated and decided?

Decision-making authority may lie with the agency that convenes the public participation process, in another agency, or be dispersed among several organizations. Generally, the authority to make public decisions lies with governmental agencies, although private entities also convene participatory processes to generate recommendations. Often, given that more than one environmental law or regulation may be applicable to the issue at hand, that stakeholders view other issues as related, or that an issue may be sufficiently controversial that stakeholders raise it in multiple forums, relevant discussions may be taking place in more than one administrative, legislative, or judicial setting. This complexity certainly poses challenges of coordination. Furthermore, because different settings may be advantageous to different parties, it can be difficult to achieve agreement about which forum should be the principal focus of public involvement. The choice of a setting may therefore affect the extent to which certain parties participate or decide instead to be heard in other venues.

Evidence varies about whether public participation is more successfully led by agencies at one level of government or another. A study of health agencies' public participation efforts with contaminated communities found that in some cases, local agencies may provide better, more effective leadership than federal agencies (Henry S. Cole Associates, 1996). Drawing on more cases involving a broader range of issues, Beierle and Cayford (2002) found that outcomes were affected little by whether the convening agency was local, state, or federal but noted that the engagement of multiple agencies does complicate the participation process. Such a complication may be particularly acute when different parties or different parties' incentives to negotiate vary on the basis of the forum in which the dispute is addressed (Bingham, 2003).

Agencies sometimes coordinate their public participation efforts. For example, the Federal Energy Regulatory Commission approved both "integrated" and "alternative" licensing processes for hydroelectric facilities that clarify and coordinate stakeholder involvement with reviews by various agencies with regulatory responsibilities (http://www.ferc.gov/industries/hydropower/gen-info/licensing/licen-pro.asp).[2]

Ashford and Rest (1999) suggest that better interagency coordination not only can save time and money, but also can result in greater agency commitment to public participation. They further suggest that agencies' commitment to increased public involvement is particularly important when interagency coordination presents challenges. Ad hoc efforts to coordinate have not overcome all the difficulties, however, even when different agencies have similar protocols for public participation (Ashford and Rest, 1999).

Coordination sometimes is attempted through formal interagency working groups (e.g., 13 federal agencies collaborate in the Federal Working Group for the Missouri River Recovery Implementation Committee process; http://missouririver.ecr.gov/?link=411). Issues worthy of examination in these efforts include the extent to which resources can be combined, how much of those resources support the public participation process and how much support coordination, and the effectiveness of coordination in terms of the durability of decisions reached. Other practices used to clarify and coordinate stakeholder participation include formal memoranda of understanding between agencies (e.g., to establish cooperating agency status under the National Environmental Policy Act) or written terms of reference (often called protocols) for the public participation process.

2. What are the legal or regulatory mandates or constraints on the convening agency? What laws or policies need to be considered, both in how the process is structured and in defining the scope of the issues that can be addressed?

Applicable laws and regulations or the domain of other agencies affect what can and cannot be done in the participatory process and how agencies with authority to act may use the results of the process. Statutes and regulations shape both the framing of issues and how agencies conduct their work, including the ways they engage in public participation. None of them, however, reduce the complexity that often arises in addressing environmental problems "on the ground."

Open meeting laws, administrative procedure laws, executive directives, judicial rulings, and the procedures and requirements set by senior officials of the agencies are part of the framework for participation. Since the framework varies across agencies, this context must be taken into account, and, in particular, the requirements and limitations under which an agency is operating should be made clear to the participants.

Legislative mandates may either require or constrain public participation. For example, the National Forest Management Act of 1976 requires that the U.S.D.A. Forest Service "hold public meetings or comparable processes . . . that foster public participation" in the "development, review, and revision of forest plans" (Office of Technology Assessment, 1992). The Forest Service may have the most explicit public involvement mandate of all U.S. agencies (Daniels and Walker, 1997).

Other laws help shape public participation practices at the federal level. These include the National Environmental Policy Act (and related guidance from the Council on Environmental Quality concerning involving the public in scoping the issues included in an environmental assessment), the Administrative Procedure Act, the Federal Advisory Committee Act, the Negotiated Rulemaking Act, and the Administrative Dispute Resolution Act. State and local governments often have varying versions of open meeting laws, often referred to as "sunshine" laws, which require announcement of public meetings (see Chapter 2 for a review of the most important statutes).

Even where public participation is encouraged, a mismatch between the interests or concerns of the public and what the convening agency has the authority to do can create misunderstandings and dissatisfaction. Practitioners generally advocate a "situation assessment" prior to convening any significant public participation, to identify whether such differences exist, and explicit discussions with stakeholders about the scope of the process, to establish a clear and agreed-on purpose for the process.

The effect of legislative mandates can depend greatly on how the affected agencies deal with them. For example, the Superfund program of the U.S. Environmental Protection Agency (EPA) has considerable resources devoted to public participation in part because it is required by law. Sometimes, however, significant environmental problems and community concerns may fall outside an agency's legislative mandate, potentially impeding its ability to creatively solve problems and implement solutions. As Ashford and Rest (1999, Part Four, VII-3) conclude, on the basis of seven case studies of hazardous waste sites:

> Agencies may have legal, political, and economic constraints that impede their ability to give the community what it wants—even if the agencies would like to do so. To the extent that the community gets very little of what it wants, it is unlikely to be satisfied with the outcome of a public participation process. This is not to say that governmental agencies should not strive to give the communities what they can. If they have faithfully acted in a trusteeship role for the community, the agencies can feel satisfied—even in the face of articulated dissatisfaction and apparent lack of appreciation—knowing they have done more than resolve a dispute or follow an easy pathway most in line with their narrow mission.

In one case involving environmental justice issues, EPA resisted cleanup of petroleum-contaminated sites because the Superfund legislation did not cover petroleum. This seriously damaged trust with the segments of the community advocating cleanup (Ashford and Rest, 1999). EPA's lack of jurisdiction over certain nuclear issues affected the functioning of boards of the U.S. Department of Defense (DOD) (Branch and Bradbury, 2006).

Although agencies cannot change their legislative mandates in the short term, they can make matters worse with policies that unnecessarily constrain the topics that public participation addresses by treating as a rigid constraint what could be treated as an issue for discussion—how to cope with the limitations of mandates. For example, significant conflict was created between military co-chairs and Restoration Advisory Boards dealing with DOD's nuclear weapons production sites because of the DOD guidance that deliberation be limited to remediation issues funded under the Installation Restoration Program (Bradbury, 2005). In addition, DOD policies did not permit discussion of the reuse of land after the closing of installations. Thus, the participatory process could not address the critical issue of how future use of the lands might affect cleanup decisions (Bradbury, 2005). In some cases, inflexible interpretation made it nearly impossible for community participants to consider the full range of remediation issues (Branch and Bradbury, 2006). In some situations, practitioners try to address such problems by expanding the range of participants to include public or private entities that may have the authority to address issues outside the scope of an agency's authority that are of concern to stakeholders.

One of the complicating factors that comes from having many agencies involved is that there are often substantial differences in the legal mandates and organizational cultures that shape participation practices and in the willingness and ability of agencies to cede influence to public participation. These in turn can have an important influence on the success of the participatory processes. In sum, although legislative mandates may either require or constrain public participation, the effect of these mandates can depend greatly on how the affected agencies deal with them. Agencies should consider them explicitly, communicate them openly to stakeholders, consult with stakeholders about the significance of their constraints, and make efforts to address constraints that could place bounds on public participation that could affect its quality or legitimacy.

3. **What factors in the convening agency influence its willingness or ability to implement principles of public participation?**

Considerations internal to the convening agency can influence its ability to work effectively with its stakeholders. Many of these internal contextual factors relate to the principles of good management generally (see Chapter 4), including clarity of purpose, commitment to use the results of the participation process, and adequate resources. It is important to know whether agency leadership has made, or would be willing to make, specific statements about how the results of the public participation process will be used. Confidence that investing time in the process will have a consequence increases participants' motivation to participate.

Agency leadership commitment is not the only factor relevant to assessing the degree of an agency's commitment. The views of staff about how to use the results of a public participation process also are important. It can be very useful to elicit staff members' views to understand the potential for internal conflicts that may create difficulties in sustaining an agency's commitment to public participation. Limited staff time is a related internal factor, particularly in an era of tight budgets. So it can be important to ask what other responsibilities the staff have and what the implications are for how much time they have to devote to the process. Other internal factors also have an impact on agency commitment, including the level of authority of the individual(s) representing the agency in the process. Another factor is the degree of clarity about how actively engaged the agency will be in the process and whether it will engage directly as a participant in the process, provide technical assistance, or simply receive the results.

Other key questions include: How open are staff and leadership to consulting stakeholders in the design of the process? Is there a clear deadline for a decision and if so, is it functioning as an impetus for action or as a reason to preclude some forms of public participation? What resources does the agency have to invest in the public participation process? Are there personnel available who have training and experience in organizing public participation? The objective in asking these questions is to be realistic, not critical. As stated elsewhere in this report, when circumstances include either internal or external constraints, a more limited process done well may be more effective than trying to do more than can be sustained.

The most critical imperative in meeting the challenges posed by the agency's context is to make clear to participants from the outset what processes and decisions are and are not possible. Yet the extent to which certain factors are within or outside an agency's control can be unclear. Public participation processes can be undermined when an agency uses claims about contextual constraints as cover for internal challenges or resistance to public participation.

WHO PARTICIPATES

Several attributes of participants and potential participants—that is, of the set of interested and affected parties to an assessment or decision—can create challenges for those convening public participation processes. We cannot overstate the importance of finding out from the start who may be affected by an environmental decision, who is interested in the issue, what their positions and interests are, how many perspectives there are, whether the participants are organized, how diverse they are culturally, whether they have worked together successfully or unsuccessfully in the past, the degree of mutual trust, and whether coalitions or oppositional groups have formed, among other factors. As noted in Chapter 3, inadequate representation of interested and affected parties is one of the leading criticisms of public participation processes. Achieving full participation by interested and affected parties can require substantial diligence. Effective communication once participants are engaged also can be affected by characteristics of the participants and their relationships with one another.

Chapter 5 describes two basic approaches to determining who can participate. One is through processes that are bounded, in the sense that representation is based on identified organized parties or on specific stakeholder interests represented by particular individuals. Generally, policy dialogues, advisory committees, and negotiations are bounded processes. These are commonly used in situations in which the type of decision process is formal (e.g., regulatory negotiation) and the outcome is often a joint report or set of recommendations concerning a specific issue or action, such as a set of rules or an adjudication.

Other processes are unbounded, in the sense that they are open to any interested individual and constrained only by who has the interest and the resources to participate. Under certain conditions, such as when an environmental issue has been recently identified and organized groups have not formed or when there may be affected groups that are unorganized, unbounded and open participatory processes are especially appropriate. Unbounded processes are useful for coordinating deliberation to define an issue for assessment or policy, to determine the information needed for action, and to identify the ways in which various parties are affected by or interested in the outcome. Unbounded processes may be formal, as in public hearings, surveys, or public comment processes, or informal, as in study circles, open houses, or other forms of workshops. As participants become self-identified and the needed information and expertise clarified, the process may become more formalized as it coalesces around the need to assess a particular issue or define a policy or program. As this distinction suggests, participatory processes can be tailored to the number of parties, the degree of their organization, the objectives of the process, and time and resource constraints. The injunction to identify and represent "the spectrum of interested and affected parties" (National Research Council, 1996:3) remains a useful guide.

Characteristics of the participants can obviously affect the results of public participation processes. We have identified six diagnostic questions related to the characteristics of the participants, the answers to which should affect the design and conduct of participatory processes; see Box 8-1. They are addressed in turn in the next six sections.

ADEQUACY OF REPRESENTATION

The question of adequate representation, which is often a matter of access, has several dimensions, each of which can affect the likelihood that all parties will be meaningfully represented. The dimensions relate to the scale of the environmental issue; the characteristics of individuals that may reduce their likelihood of participation; difficulties the parties may have in organizing collectively for representation; and disparities among groups in their ability to get to the table.

> **BOX 8-1**
> **Diagnostic Questions Pertaining to Participants**
>
> 1. Are there interested and affected parties who may have difficulty being adequately represented?
>   a. What does the scale of the issue, especially the geographic scale, imply for the range of affected parties?
>   b. Are there disparities in the attributes of individual potential participants that may affect the likelihood of participation?
>   c. Are there diffuse, unorganized, or difficult-to-reach interests?
>   d. Are there disparities across groups of participants in their financial, technical, or other resources that may influence participation?
>
> 2. What are the significant differences in values, interests, cultural views, and perspectives among the parties?
>
> 3. Are the participants polarized on the issue?
>
> 4. Are there substantial disparities across participant groups in their power to influence the process?
>
> 5. To what degree can the individuals at the table act for the parties they are assumed to represent?
>
> 6. To what degree are there problems of trust among the agency, the scientists, and the interested and affected parties? Specifically,
>   a. Are there indications that some participants are likely to proceed insincerely or to breach the rules of the process?
>   b. Are some participants concerned that the convening agency will proceed in bad faith?
>   c. Do some participants view the scientists as partisan advocates and so mistrust them?

Scale of the Issue

As noted in Chapter 7, the scale of an environmental issue may create particular challenges for participation. Many large-scale issues, such as national environmental standards, climate change, regional air quality, water resources, and some transportation issues, make participation difficult for some parties. First, the geographic boundaries of the issue may be unclear, making it difficult to determine who is affected. When an issue extends across political and institutional boundaries, there can be a large number of affected parties. In addition, some local or regional environmental resources (e.g., the Grand Canyon, the Arctic National Wildlife Refuge) have national or global significance because of the value of the resource, its cultural meaning, or the possibility that it will set a precedent.

Many public participation processes rely on repeated face-to-face interaction. For policies with national or global impact, repeated face-to-face interaction is much more expensive, time consuming, and complicated than it is for geographically contained decisions. There are mechanisms to cope with the problem of scale, but their complexities and costs must be taken into account. For example, the cost and time involved in setting up national advisory processes is substantial. Furthermore, federal agencies may invoke the Federal Advisory Committee Act (FACA) in ways that are foreign or unpalatable to participants accustomed to informal local participatory processes. Some agencies treat members of FACA-regulated committees as temporary employees and as a result require that committee members be fingerprinted. Some potential members see this procedure as burdensome and intrusive, and some have refused to participate as a result. In addition, the time and expense of traveling substantial distances for a national decision process favors well-funded organized groups over other parties, so it may be important to hold multiple meetings in diverse locations to allow engagement of those who cannot travel to a national meeting. Such strategies increase costs and the duration of the process. Online participation reduces travel costs, but its effects on who participates and on the quality of deliberation are only beginning to be studied (see Chapter 5).

Determining the relative role of local and national interests can pose significant practical challenges to public involvement regardless of scale. In final decisions, authorities give explicit or implicit weights to national and local interests, but this issue also needs consideration in the design of participatory processes. The logistics problem ("How do we get them to the table?") and the value weighting problem ("How many local versus national interest representatives should we have?") interact. It can be hard for local groups to participate in national processes and for all but the best-funded national groups to participate in local processes far from their offices.

At a relatively local scale, participation can be based on social relationships that extend beyond the responsible organizations and directly involve those affected by a project or policy (Wilbanks, 2003). At larger scales, participation often relies on the involvement of organizations, such as trade or environmental groups, which are presumed to represent interested or affected constituencies. In such circumstances, there is always a concern with the degree to which the representatives share views with their constituencies. For example, there has long been a concern that the major U.S. environmental groups and the foundations that support them do not adequately reflect the concerns of disadvantaged communities (Brulle, 2000; Taylor, 2000; but see Delfin and Tang, 2005, for contrary evidence).

There are other ways to get adequate representation for large-scale assessments or decisions, but data are very limited on their effectiveness compared with the constituency-based approach. As noted in Chapter 7, it is possible to engage representative samples of people in direct deliberation on policy issues. In one example in Texas, such "deliberative polls" led to an increased commitment to renewable energy policy compared with a poll taken without deliberation (Ackerman and Fishkin, 2004). While this approach involves very high costs, other experiments suggest that standard surveys and face-to-face participation can be hybridized effectively (e.g., Pidgeon et al., 2005). Another approach, tried in the U.S. National Assessment of the Potential Consequences of Climate Variability and Change that was completed in 2001, pursues participation at different scales—in this instance, at the national level and in various regions and sectors. Results were summarized both regionally and nationally (Moser, 2005). Such processes are quite expensive and have not been used often. As already noted, there is also the possibility of Internet-based participation (Beierle, 2002), an approach that is only beginning to be explored and studied (see Chapter 7).

Participants' Disparities

Some claim that the potential of public participation to improve decisions is limited because nonspecialists lack the capacity to understand and engage with complex and uncertain scientific information, obscure laws and regulations, and complex value trade-offs (Dietz, Stern, and Rycroft, 1989; Sweeney, 2004). This view attributes failures of public participation to insufficient levels of education, time, or other aspects of "human capital" on the part of the public. Research on public involvement in political decision making across the spectrum of public policy issues presents a more complex picture.

A substantial body of research on processes of deliberation and decision making on public issues of all kinds has shown that individual resources, such as formal education, occupation, social status, and available time and money, condition the likelihood that individuals will participate and participate influentially (e.g., Verba and Nie, 1972; Verba, Schlozman, and Brady, 1994). There are two reasons for this. First, these resources facilitate personal involvement and influence directly. For example, individuals who know more of the arguments about a particular issue tend to be more influential regardless of the quality of their arguments (Kameda, Ohtsubo, and Takezawa, 1997). Those with higher occupational status and educational attainment tend to speak more and are more influential, even if their information is not more accurate than other group members' (Hastie,

Penrod, and Pessington, 1983). Individuals who focus on the merits of an issue tend to have more influence in a group, even though they are also less willing to change their views based on meritorious arguments (Cacioppo et al., 1996). Second, persons who possess these resources are much more likely to be recruited to participate than are their less advantaged peers (Rosenstone and Hansen, 1993; Verba et al., 1993; Verba, Schlozman, and Brady, 1994; Goldstein, 1999; Schier, 2000).

These inequalities contribute to a sense of disconnection and powerlessness among many in the United States, particularly those at the lower end of the socioeconomic continuum. And it is increasingly the case for those in the middle, many of whom have also concluded that government is not much concerned with their interests and aspirations and that the public sphere is open to citizens "by invitation only" (Schier, 2000). In a 1996 national survey, for example, 69 percent of Americans with less than a high school education, 62 percent of high school graduates, and 57 percent of Americans in the bottom two-thirds of household income distribution agreed with the proposition, "People like me don't have a say in what the government does" (Markus, 2002). By way of comparison, only 4 in 10 college graduates or upper-income survey respondents agreed with the statement. Related to, but distinct from, a sense of political inefficacy among many Americans is a judgment by many of them that government is incompetent or untrustworthy (Nye, Zelikow, and King, 1997; Hetherington, 2004). By way of example, a CBS/*New York Times* national survey conducted in July 2007 found that only 24 percent of Americans "trust the government in Washington to do what is right" "just about always" or "most of the time" (http://www.pollingreport.com/institut.htm#Federal). The levels of trust in government among Americans has been low, with only transitory exceptions, for some three decades now. Such chronic levels of disconnection and mistrust may present formidable barriers to participatory processes.

Highly educated, financially comfortable people are much more likely to be active in public affairs than are less educated, lower income people, not because they are more concerned about public matters or more willing to make the effort, but rather because of differences in the control of politically valuable resources (cognitive skills, money, and a sense of political efficacy), embeddedness in social networks that include influential people, and the targeted efforts of political organizations to activate the citizens who control those resources (Rosenstone and Hansen, 1993; Verba et al., 1993; Verba, Schlozman, and Brady, 1994; Putnam, 2000). Data from recent American National Election Studies surveys reveal that college graduates are roughly twice as likely as high school dropouts to be contacted in an election year by party activists urging them to vote. The same odds of being contacted during a campaign distinguish people residing in households in the top one-third of the income distribution from people in the bottom one-third (Markus, 2002). These statistics suggest that efforts to activate political participation in the United States tend to amplify rather than mitigate the effects of resource inequalities among citizens (Powell, 1986; Lijphart, 1994). The situation of socially disadvantaged groups organizing effectively for political change is the exception rather than the rule in American society.

An examination of nearly 1,700 comments filed in the period 1988-1990 as part of EPA's rule-making process regarding 28 "significant" hazardous waste regulations revealed that individual members of the public filed fewer than 6 percent of the comments, whereas corporations and industry groups submitted about 60 percent of the comments, and local, state, and federal government officials submitted approximately 25 percent (Coglianese, 1996). Comment on EPA rule-making is, of course, only one way in which citizens may participate in environmental decision making. The consensus conclusion of research on other common environmental participatory processes (such as public hearings and citizen advisory committees) and on public participation in governance more generally is that the vast majority of the public is uninvolved in, or even unaware of, participatory options that are, in principle at least, available to them (Verba and Nie, 1972; Rosenstone and Hansen, 1993; Verba, Schlozman, and Brady, 1994; Schier, 2000).

These findings imply that unless public participation in environmental assessment and decision making explicitly compensates for these tendencies, the politically disadvantaged and disconnected will be underrepresented among the participants, and the outcomes of such participation are likely to be skewed against their interests (e.g., Bullard, 1990). For example, a review of 30 cases of public participation in the Great Lakes area found that advisory committees were frequently unrepresentative from a socioeconomic perspective (Beierle and Konisky, 1999).

There are notable exceptions to these overall patterns. Numerous cases have been documented in which low-income or minority communities have mobilized very effectively when they see their vital interests as threatened. They have mastered daunting technical analyses, and overcome bureaucratic resistance to have their voices heard in environmental policy processes (Brown and Mikkelsen, 1990; Bullard and Johnson, 2000; Brown et al., 2002, 2003; Pellow, 2002; McCormick, 2006). Numerous studies also show that marginalized people can build on existing local institutions, such as religious congregations, neighborhood associations, schools, and labor unions, to affect issues that concern them, including environmental issues (Piven and Cloward, 1971; Boyte, 1980; Levine, 1982; Evans and Boyte, 1992; Shutkin, 2000; Sirianni and Friedland, 2001; Warren, 2001; Osterman, 2002). But those designing a participation process cannot rely on this to happen spontaneously. Special efforts will usually be required to engage those who are not connected to the policy process and less likely to participate, especially when they do not see their vital interests at stake. As noted above, when a large consequence is shared by large numbers of people, it amounts to a small consequence for each one, reducing their motivation to get involved.

Many practices are used to involve people who are otherwise not likely to participate. They include inviting members of particular groups to participate on a process steering committee; providing resources to existing organizations to send out mailings, organize participation, or host meetings in an affected community; and including members of the community or group on a convening or facilitation team.

Missing Interests

For many environmental issues, well-organized interests, including industries, local political and economic coalitions, and environmental groups, are well prepared to engage in participatory processes. But as already noted, many individuals who may feel substantial effects from a decision may not be organized in a way that facilitates their easy engagement. Those who can expect to receive only modest benefits from a decision may be even less organized. Therefore, if participatory processes are to take public concerns into account equitably, care must be taken to include the voices of those who are not well represented.

In some cases, preliminary fieldwork will reveal that communities presumed to be disorganized are in fact endowed with an array of local organizations and institutions that may be open to collaborating in an effort to foster inclusion of underrepresented interests (Fisher, 1994; Rivera and Erlich, 1998; Boyte, 2004). When such organized groups have few members or are reluctant to participate, some research suggests that the very act of inviting their members into the policy process helps to organize a new constituency, as those groups now have a focus for their organizing efforts (Morone and Kilbreth, 2003). Some research also suggests that certain participatory procedures may be particularly well suited for encouraging involvement of underrepresented or marginalized groups, such as citizen action committees, citizen forums, citizen juries, planning cells, and consensus conferencing (for reviews, see Rowe and Frewer, 2000; Renn, 2004).

Groups' Disparities

As with individuals, groups and organizations differ in the resources that affect their ability to participate meaningfully on behalf of their constituencies. Government agencies and private corporations are generally represented by paid staff or consultants, but citizen groups tend to have fewer financial and technical resources and to rely on representation by unpaid volunteers. Studies focused on public participation in Forest Service processes indicate the importance of community resources in shaping the likelihood of a successful process. External support of the process by the community (Schuett, Selin, and Carr, 2001) and public interest and pressure to move the process along can also increase the likelihood of success (Yaffee, Wondolleck, and Lippman, 1997). A clear assessment of the resources available to all parties can help avoid designing a process in which some do not have an effective voice.

As we have noted, the "general public" is not one amorphous mass but rather many distinct publics, particularly when it comes to environmental matters. The public most concerned about issues related to hazardous waste contamination in urbanized areas may be quite different in many ways from the public most concerned about endangered species and wilderness conservation. However, numerous studies of environmental values and concerns among representative samples have found that concern with the well-being of other humans is strongly positively correlated with concern with other species and the environment itself (Dietz, Fitzgerald, and Shwom, 2005).

The concept of "social capital" is useful for thinking about how to address problems of access to public participation processes. The central idea is that repeated interactions among individuals can give rise to social networks, norms of trust, reciprocity, and empathy, which together increase the possibilities for cooperation (Bourdieu, 1985; Coleman, 1988, 1990; Putnam, 1993; Newton, 1997; Lin, 2001). These linkages are referred to as "social capital." By fostering cooperation, social capital renders possible solutions to classic problems of collective action that arise in the provision and maintenance of public goods, such as clean air and ocean fisheries (Ostrom, 1990; Taylor and Singleton, 1993; Boix and Posner, 1998; National Research Council, 2002a). In the absence of sufficient social capital, powerful incentives exist for individuals to shirk contributing money, time, or other scarce resources, since each individual can benefit from such activities even without contributing (Olson, 1965).

Social capital is relevant to environmental public participation for two reasons. First, participatory processes are themselves a form of collective action: individuals can enjoy whatever environmental benefits accrue from participatory processes without taking part themselves (Lubell, 2002). The extent of social capital among potential participants therefore can be an important influence on the breadth of involvement and the quality of collaboration that a participatory process will have (Warren, 2001; Larsen et al., 2004). Second, public participation changes social capital. Depending on the manner in which it is conducted, public involvement can enhance social capital among participants, thereby increasing the likelihood that future involvement in such processes will be successful—and more generally enriching the group's or community's capacity to cooperate on public matters. Research has documented the ways in which some government programs and policies have nurtured—or in some cases "unraveled" (Skocpol, 1996)—social capital in civil society (see also Berry, Portney, and Thomson, 1993).

Social capital can engender sufficient cooperation to overcome obstacles to collective action (Marwell and Oliver, 1993; Sampson, Raudenbush, and Earls, 1997; Putnam, 2000; Passy, 2003). Indeed, theory and evidence suggest that social capital is implicated in "making democracy work" (Putnam, 1993) by promoting responsive, effective, and efficient government (Boix and Posner, 1998; Putnam, 2000:Chapter 21; Knack, 2002). And research supports the assertion that community-wide social capital, competence, and civic engagement can influence the likelihood of a successful public participation process (Doppelt, Shinn, and John, 2002). Groups that develop social capital can increase their influence relative to less well-organized groups, a result that may be viewed as beneficial or not, depending on one's perspective.

Social capital and the networks that underpin it are usually thought of as having two basic forms: bridging, which involves communications among people from different backgrounds, and bonding, which involves communications among people who share common characteristics, such as social class, nationality, or ethnicity (Putnam, 2000). A potential exists for participatory processes to build "bridging" social capital by facilitating productive relationships among interested and affected parties, including nongovernmental organizations, private business and industry, and governmental entities at the local, state, and national levels. This is an important objective in terms of building capacity for future participatory decision making. It can counter the tendency of policy networks to emphasize connections among those with shared core beliefs and values (Sabatier and Jenkins-Smith, 1993), a network structure that can make compromise and consensus difficult when parties outside the policy network must be involved. Some research indicates that heterogeneous groups tend to be more flexible and innovative. Homogeneous groups often have more positive internal dynamics but poorer performance (Jackson, 1992) and tend to search for information that confirms their beliefs (Schultz-Hardt et al., 2000). Bridging capital can be particularly important for economically and politically marginalized communities because it can help provide the resources they need to achieve effective solutions to the problems they face (Bryant, 1995; Saegert, Thompson, and Warren, 2002).

Research, experience, and common sense suggest that care should be taken to address various other factors that may impede equitable participation, such as time and location of meetings, physical access, availability of public transportation, language diversity, need for child care on site, and participants' varying familiarity with technical information. In some situations, agencies have provided or helped identify resources for community or public interest groups to obtain technical assistance. This approach is obviously subject to resource constraints.

DIFFERING PERSPECTIVES

Policy decisions and the public participation process may be divisive when the effects of an environmental problem and the costs and benefits of potential policy responses affect different groups in different ways. Differences in values, perspectives, and cultural world views can be sources of division among the participants. Such diversity is also a potential source of the strength of participatory processes.

Diversity in experiences, knowledge, values, and perspectives is important for interrogating knowledge claims, assessing the adequacy of problem definitions, and evaluating options for solutions. This process of interrogation is the core characteristic of analytic-deliberative processes (see Chapter 6). The key idea is that the quality and public accountability of a decision are best ensured by engaging a wide variety of participants with diverse perspectives, backgrounds, and experiences to deliberate together to reach a shared understanding of the problem and options for addressing it.

Significant research in communication, problem solving, decision making, and negotiation suggest that good-faith communication, in which parties explain the reasons for their positions in terms of underlying principles or interests, is more likely than other approaches to produce creative solutions. Such research goes back nearly 100 years to the early work of Mary Parker Follett (1918, 1924). It includes, in particular, concepts of "principled" or "integrative" negotiation (Fisher, Ury, and Patton, 1981; Lewicki and Litterer, 1985), understanding of how people engage in escalating and deescalating communication and behavior (Pruitt and Rubin, 1986), and applications of game theory and decision science (Raiffa, 2005, 2007). Many specific concepts and practices (which can be found in the literature) elaborate this basic notion of collaborative problem solving.

Although environmental decision making can benefit substantially from efforts to incorporate and consider a variety of perspectives, it is also the case that differences in values and interests can constitute serious barriers to a productive public participation process. Differences defined by interests are well recognized: there are winners and losers in most policy choices. However, participation in public processes is also connected to people's sense of identity, values, and understandings of norms of appropriate behavior (e.g., Verba, Schlozman, and Brady, 1994; March and Olsen, 1995; Stern et al., 1999; Monroe, 2001; Markus, 2002).

Other important differences are basically cultural and can lead to differences in reasoning and judgments during disputes (Kamenstein, 1996; Triandis, 2000). In many cases involving Native American tribes, difficulties have arisen because of the differences between biocentric values expressed by tribal representatives and the conflicting values of other stakeholders in the process (Lubell and Leach, 2005). In addition, lack of awareness about the importance of sovereignty to tribes or issues associated with a tribe's legal status and lack of understanding of the institutional and cultural norms with which tribes govern themselves may make negotiations between tribal and nonnative stakeholders challenging when nonnative stakeholders do not know the proper tribal etiquette or understand with whom they should negotiate (Jostad, McAvoy, and McDonald, 1996). These cases highlight the ways in which differences in deep core beliefs among participants can shape the likelihood of consensus-based policy agreements (Sabatier and Jenkins-Smith, 1993).

Even within a small geographic area, social classes and racial/ethnic groups may partake of very different cultures that involve different values and assumptions about what are appropriate decisions and what are appropriate processes for reaching decisions. There are also consistent gender differences in risk perceptions and environmental values (Davidson and Freudenburg, 1996; Slovic, 1999; Kalof et al., 2002). In fact, evaluations of the process and even ideas of what success means can vary among such parties as state and federal agency officials, local agencies, resource users, environmentalists, and facilitators (Leach, 2002).

It is important that processes represent and engage the full spectrum of perspectives in planning how a public participation process is conducted so that decision processes are sensitive to them. However, the wider the range of values and norms, the more difficult it may be to come to shared understandings and the more effort may be required to do so. One common strategy is to focus on relationships so that participants get to know one another before considering the issues or even establishing the ground rules for a process. Practices that have been used for this purpose include field trips, social hours at the start of meetings, rotation of meeting locations so that different parties serve as host, story telling, and more formal, shared training on process (Adler and Birkhoff, 2002). The literature on cross-cultural communications also includes concepts and strategies that may be useful.

A large literature and much accumulated experience in using analytical methods from economics and the decision sciences is relevant to addressing issues of value in considering the consequences of possible environmental decisions. Recent reports from the National Research Council (2004, 2005a) and the Millennium Assessment (Reid et al., 2005) provide a useful introduction.

Cost-benefit analysis is one such method. It has been used by many federal agencies to assess and compare different kinds of consequences, such as environmental benefits from ecosystem management and human health benefits from control of environmental toxicants. Its key feature, the comparison of consequences by representing them all in monetary units, is both its strongest advantage and disadvantage. The advantage is that it provides a straightforward method for making difficult comparisons. The disadvantage is it is accomplished by two approaches that are not problem free: (1) transforming all effects into common monetary units (e.g., from lives or species lost to dollars, or from future dollars to present dollars) and (2) making assumptions about social value (e.g., "that social value is nothing more or less than the sum of values individuals express in markets or market-like contexts," National Research Council, 2005a:35), some of which are controversial (Jaeger et al., 2001). These value judgments may be difficult for participants to disentangle from the analytical method. And when they are distentangled, participants may dispute or reject them. These difficulties suggest that any use of these methods should follow the principles outlined in Chapter 6.

Cost-benefit analysis is most likely to be useful for addressing value issues when the consequences to be compared are readily valued in monetary terms (e.g., board feet of timber harvested, cost of emission controls). It is often quite controversial when monetary values are not obvious (e.g., the value of continued viability of an endangered species or of increased visibility with reduced levels of particulate matter in the atmosphere). Such "nonmarket" values are sometimes estimated subjectively (e.g., by eliciting expressions of people's willingness to pay), but that approach is itself a matter of controversy among specialists.

There are other analytical tools for addressing value differences without assuming that they can be measured by a common index. They include multiattribute trade-off analysis and "value-focused thinking" (Keeney, 1992), an analytic hierarchy process (Saaty, 1990), and other methods that focus attention on options and preferred end states without making formal estimates of value or utility (for more detail, see Gregory and McDaniels, 2005). These methods explicitly seek to identify participants' values, goals, or preferred states and, rather than combining and comparing them in a formal analytical framework, structure deliberations to ensure that all such concerns are addressed directly as part of the assessment or decision process.

Value differences among participants sometimes do not affect decisions, either because there are clear legal requirements that specify which values can and cannot be considered (e.g., the Endangered Species Act) or because agreement can be reached on the choice of decision alternatives despite differences among the participants in the values ascribed to environmental consequences. However, disagreements on values are often a sensitive and divisive issue in public participation processes. It is important to diagnose the extent of such disagreement at the outset of the process, to select practices to address them, and to reconsider value issues as the process nears completion. Such practices can reveal paths toward finding common ground among parties with different values as they consider alternatives.

POLARIZATION

The degree of polarization among participants is an important diagnostic factor for determining the need for using specific techniques to help parties deal with different perspectives or conflicting interests as they attempt to achieve the principle of good-faith communication. Some participatory processes begin with participants not being particularly vested in certain desired outcomes, either because positions have not yet formed or because existing positions are relatively flexible and participants acknowledge the need for trade-offs and compromise. However, it is quite common for participation processes to begin with many groups already having strongly held and strongly opposed views. In some cases, participants may be in litigation or engaged in active dispute in other ways. This may not preclude participation, but it does affect it.

Clearly, policy decisions and the public participation process may be divisive when the effects of an environmental problem and the costs and benefits of potential policy responses affect different groups in different ways. A special and critically important challenge arises when some parties believe that they have interests that cannot be met if the interests of another party are served. Sometimes this is the case, but the perception of mutually exclusive interests on the part of some participants may be incorrect. The diagnostic task can be difficult because sometimes the question is initially posed as a choice between mutually exclusive positions. For example, a panel member pointed out that in the first mediated environmental dispute, some participants disagreed in absolute terms about whether to support a specific flood control dam in Washington State. However, the question could have been restated in terms of reconciling opposing groups' interests, that is, as how to reduce flooding while still preserving the whitewater recreational values of that particular reach of river. Thus, initial statements that indicate diametrically opposed interests should not be interpreted as an insurmountable challenge.

Mediators or facilitators often speak to stakeholders in confidence to learn more about the issues of concern and the interests that underlie the positions being articulated, particularly in circumstances in which trust is a barrier but also when parties are not particularly skilled at collaborative problem solving. Other practices for generating solutions when positions appear polarized include brainstorming (also described as "separating inventing from deciding"), tools for applying criteria systematically (including developing models), and adding dimensions such as time or new issues to the scope of the conversation to find solutions that can benefit those involved. Many of these techniques have their roots in game theory or decision science.

Yet, sometimes, interests are truly irreconcilable. For example, in a case in the Rockies, there was no acceptable way to discuss the terms of a permit to mine uranium with parties that were opposed to the use of the uranium for nuclear energy or weapons production. In these cases of irreconcilable interests, particularly when the stakeholders have been consulted and concur in the design of the participatory process, public participation may not reach a consensus but can still be a valuable tool to clarify the roots of disagreement (participants can agree on the points of disagreement), on how to describe the arguments of each side, on how to document the different preferences, or on the forum in which decisions will be made. If the participants believe that the organizers of the process have made all efforts to reach a common understanding of all positions and interests and to document them with openness and transparency, the chances improve for the perceived legitimacy of the process, even if the actions of the authorized decision maker are appealed by those who disagree with the outcome.

Polarization may be seen to present a dilemma for a convening organization: if the participatory process is unlikely to generate consensus, there may be concern that it will make the conflict more intense by giving opponents a platform for debate, increasing the intensity of political opposition and making a decision impossible. As noted in Chapter 3, some observers of public participation (e.g., Sunstein, 2003) argue that it tends to create polarization, although others (e.g., Hamlett and Cobb, 2006) disagree. Increased polarization, some fear, will politicize debate to the detriment of scientific evidence and good judgment in the decision-making process.

These outcomes are possible but, in our judgment, do not provide good justification for curtailing public participation. Agency decisions are inherently exercises of political authority. If political consensus is lacking, the responsible agency should acknowledge this fact, make decisions despite dissent if required, and develop processes whereby public debate and dialogue appropriate to the nature of the situation can take place.

As noted in Chapter 3, there is also the opposite fear, that public participation may be used to co-opt, exhaust, or mislead the public, thereby obstructing the proper role of the public in shaping policy in a democracy and reinforcing existing powerful interests. Although both of these concerns find support in experience, neither outcome is inevitable. The evidence indicates that appropriately structured public participation can serve to reduce both the abuse of science and the dilution of public influence on policy.

The design for the process and the expectations for it should take account of the degree and character of polarization that exists at the outset. In a polarized environment, simply bringing opposing groups together in a forum designed to clarify the points of contention may constitute a constructive first step, one that will be more likely to succeed than something more ambitious yet one that also can provide the basis for additional progress in the future. Such forums can clarify the scientific and political issues and establish useful processes to support good decision making. As noted in Chapter 6, when scientific disagreements are part of what is at issue, an agency can convene a public forum in which scientists with diverse perspectives present their data and conclusions and defend them. In this way, a public agency can create a public forum for decision-focused debate and discussion as well as making the required decisions.

In situations characterized by extreme polarization, which sometimes result from a long history of conflict, extended efforts at trust building also may be necessary to make accommodation possible, even on matters of process. Research and experience in resolution of identity-based conflicts can offer useful insights for such situations (e.g., Saunders, 1999). However, trust-building efforts require time and money, either of which may be in short supply.

POWER DISPARITIES

Disparities that affect influence can play a significant role in who is consulted in the design of a process, who is included as a participant, and, in some cases, in the transparency of the process and the achievement of good-faith communication. In many cases, those with power and influence are also advantaged in terms of other resources related to having an effective voice in a process (time, funding, scientific staff).

As noted above, some parties already participate in environmental decision making very well and effectively. Those parties include large corporations and some professional associations (e.g., Heclo, 1978; Schlozman and Tierney, 1986; Baumgartner and Jones, 1993; Sweeney, 2004). Other parties are much less involved and effective, such as nonunionized workers and their families, the poor, members of racial and ethnic minority groups, and recent immigrants. A small number of prominent national environmental nongovernmental organizations do represent the distinctive interests of noncorporate and nonprofessional constituencies to national agencies, and most of the more populous states have comparable state-level nongovernmental organizations; however, many types of interested and affected parties to environmental decisions are neither involved directly in agency decisions nor represented there by nongovernmental organizations.

A major rationale for public participation is to level the playing field in the sense that everyone should have equal voice in the process, even if outside the process there are vast differentials in resources, power, and influence. This is at the heart of many concepts of legitimacy and capacity building. However, differences in power and influence always exist and those with more power are more likely to have more influence, directly or indirectly, on the choices made about the framing of issues, the nature of participation (e.g., a bounded versus unbounded process), the logistics of meetings, and how the results will be used, mirroring the balance of power in the external playing field. In other words, unless explicitly addressed, collaborative design of a process can be more difficult to achieve the greater the imbalance of power in a situation. This, in turn, can affect the inclusiveness of participation and the transparency of the process, even if all who do participate have the same opportunity to express their views.

Intentional diagnosis and sensitive discussion of the relative power or influence of different groups can promote the principle of collaborative design in a meaningful and realistic way by providing the basis for an informed decision by both organizers and participants as to whether they can convene a process that provides sufficient incentives for inclusive participation. Such understandings are often recorded in ground rules or "terms of reference" for a process. Discussion of relative influence during the diagnostic stage also can enhance a realistic understanding, and sometimes acceptance, of the possibility that some of the parties will seek other forums if they can achieve more of their objectives in that way and, thus promote realistic understandings of the limits of a public participation effort.

Generally, public participation is structured so that a few voices do not dominate the discussion. In some cases, inclusiveness may require subsidies to those with limited resources to compensate for travel costs and time lost from other responsibilities. It may also require providing them with improved access to expertise.

By recognizing existing inequalities and designing and implementing participatory processes so as to minimize their effects, agencies can enhance the quality of input for environmental decision making. The process can be structured to ensure that all stakeholders are motivated to participate and that all parties' voices are given serious consideration in the process. It is also important to be realistic that those involved will be comparing how participation in a process compares with other process alternatives. Those who do not feel they have sufficient influence in the process may seek to increase their power through other strategies, such as community organizing, media outreach, referendums and initiatives, lobbying, and litigation; and those with influence will assess what their influence can accomplish through similar means. Thus, the burden is on the convener to understand the balance of power and influence in a situation and to design a process that motivates participants to work within the process. (Good faith on the part of participants—e.g., in being transparent when they feel they must abandon a process for external politics—is discussed below.)

ROLE OF REPRESENTATIVES

Environmental assessment and decision-making processes typically involve a mix of individuals speaking for themselves and representatives of organizations or groups, among them government agencies, private corporations, trade associations, environmental nongovernmental organizations, and grassroots citizen groups. This means that some participants must get the concurrence of individuals who are not participating directly and therefore who have not experienced the mutual learning that can occur in a good participatory process. The values, internal structures, and dynamics of the organizations that participate in public decision making vary widely and must be recognized in designing a process that successfully accommodates different internal decision-making processes, organizational cultures, conversational styles, potentials for leadership or other organizational change, and the degree to which representatives have access to relevant information, can speak for their organizations or constituencies, make proposals, and support proposed decisions (Pruitt and Carnevale, 1993; O'Conner, 1994). A convening agency's efficacy in creating a good participation process depends on the agency's understanding the parties and the intraorganizational dynamics of their groups and organizations. This is a challenge because of the complexity of environmental issues and the wide variety of ways in which affected parties organize themselves.

An important concern in this regard is whether the individuals at the table are willing and able to make durable agreements. The Forest Service studies point to the importance of the participants' committing to the process, particularly if it is extended in time (Selin and Chavez, 1994; Shindler and Neburka, 1997; Yaffee, Wondolleck, and Lippman, 1997; U.S.D.A. Forest Service, 2000). An agency that engages in a lengthy and formal participatory process may ask participants to agree to continue until the process is completed and to signify by their participation that they accept that the process is fair and that the decisions made as a result of the process will be acceptable to them. However, as experienced negotiators know, participants in a negotiation may choose to cease their participation, or they may assert that a conclusion or choice of decision alternative by the group is unacceptable to them.

Participating organizations may have internal disagreements about which forum should be the principal focus of public involvement. Since participation is voluntary, it is important to consider how the parties' incentives to participate may depend on the forum: What motivates people to give their time and energy to working in a particular forum? These issues also can change over time, as interested and affected parties seek to create new forums in which they can achieve their objectives.

Several practices are commonly used in "bounded" or representative processes to deal with concerns about whether the participants can act for their organizations or constituencies. They include selecting representatives with decision-making authority or strong social networks, being explicit about what authority each representative has and with whom they must consult, planning the time between meetings to allow for consultation, requesting reports about such consultation at meetings, and organizing formal work groups within constituencies. In the Total Coliform Rule advisory committee process, several of the participants established formal working groups of members of the organizations or constituencies they represented.

The amount of time available in any particular context is a related factor in determining what is possible to do. If deadlines are very short, a format that does not depend on representatives consulting with constituencies may be considered. And as noted in Chapter 6, processes based on peer review may help in aligning representatives with their constituencies. The critical element is that onvening agencies need to develop understanding of the parties at the table, in terms of what kinds of commitments they can make on behalf of those they purport to represent. On the basis of such an assessment, they may want to revise the process or their expectations for it.

TRUST

Participants in environmental decisions typically have histories with each other and with the agencies responsible for convening the process and making environmental decisions. These histories form part of the context for decision making and can result in a reservoir of trust or distrust between the agency and the participants, as well as among the participants. For example, research suggests that Native Americans, given their long history of mistreatment at the hands of the U.S. government, may be especially reluctant to participate in watershed partnerships and that partnerships involving Native American tribes may be less successful at achieving policy agreements (Lubell et al., 2002).

It is reasonable to expect that a lack of trust would erode the chances that a participatory process will be successful. In fact, many studies indicate that mistrust among the parties, and between parties and government agencies, has often been a problem for environmental decision making. The strong influence of trust on risk perception is well documented (Siegrist, Earle, and Gutscher, 2007). For example, a major study of risk communication (National Research Council, 1989) identified and presented examples of several sources of mistrust in agencies or their scientists: real or perceived advocacy of unjustified positions; a reputation for deceit, misrepresentation, or coercion; self-serving framing of messages; contradictions between messages from the same source or contradictory messages from other sources; and perceptions of incompetence or impropriety.

Contrary to vivid documented examples, however, Beierle and Cayford (2002) found little correlation between measures of trust and of success in their multicase study (the measures concerned participants' trust in agencies and other participants). They suggest that this is in part because an intense process can overcome initial lack of trust and that such intense processes are more likely to be used when there is a lack of trust. We suggest that what is likely to matter in these more intense processes is how well they implement the principles of good public participation, especially those addressed in Chapters 5 and 6, and how well they address specific difficulties in implementing those principles that arise from issues of trust.

Trust can affect the behavior of organized interests that relate strategically to public participation processes. Depending on their level of trust in the convening agency, they may seek or oppose broad public participation or choose to pursue their ends outside the process, perhaps by litigation, politics, or separate avenues of influence on the responsible agency.

The extent of scientific uncertainty affects the extent to which individuals accept new information or cling to prior beliefs. There is evidence, largely from experimental research, that under conditions of perceived uncertainty, trust and procedural fairness considerations become particularly important to the decision-making process (Van den Bos, 2001; Van den Bos and Lind, 2002) and individuals display a heightened interest in evaluating the credibility of information sources (Halfacre, Matheny, and Rosenbaum, 2000; Brashers, 2001; Van den Bos, 2001). They are more likely to challenge the reliability and adequacy of risk estimates and be less accepting of reassurances (Rich et al., 1995; Couch and Kroll-Smith, 1997). They also tend to become more rigid about beliefs and policy preferences and then to end prematurely the search for facts (e.g., Janis and Mann, 1977; Klein, 1996; Covello et al., 2001). To the extent these findings apply to environmental public participation in real-world settings, they suggest that attention to procedural fairness is especially important for processes that face problems of scientific uncertainty and mistrust (see Chapter 6).

Research on policy networks shows what is termed "biased assimilation"—participants are more likely to accept information that is consistent with prior and deeply held beliefs (Sabatier and Jenkins-Smith, 1993; Sabatier and Weible, 2007). While the degree of uncertainty varies across types of environmental issues, it also varies substantially across specific instances of a single type of environmental problem. For example, hazardous waste sites vary considerably in the degree to which the toxicity of the contaminants is understood and in the dynamics of contaminant movement through air, soil, and water. Problems related to uncertainty are nearly ubiquitous in environmental policy, although the degree of uncertainty varies greatly. This may be one reason that variation in the kind of environmental problem matters relatively little to ultimate outcomes—there is always sufficient uncertainty to entrain the mechanisms described above. As a result, initial levels of trust and the way the participation process deals with trust become critical factors to be considered in process design. Trust or its absence seems likely to be particularly important in cases in which scientific disagreement is an issue or in which adverse effects may be visited on identifiable social groups (Dietz, 2001).

Indications of Lack of Trust

Some parties may use participatory processes to obstruct decisions or may make end runs around the process. Others may simply not be motivated to work toward making the process a success even if they are not actively obstructing it. These challenges may or may not be easy to anticipate in an initial diagnostic assessment, but it is crucial to take a careful and nonjudgmental look at whether the process offers sufficient incentives for good-faith participation.

Beierle and Cayford (2002) found a moderate positive correlation between the motivation of the participants and success of public participation processes. They noted that participant motivation is correlated with several process features and that more intensive processes, which are associated with greater success, require higher levels of motivation. These data do not demonstrate that initial motivation of participants is a causal factor in success. They are also consistent with the proposition that initial success increases participants' motives to stay involved, so that motivation and success reinforce each other.

Motivation and initial success are likely to be related to individuals' levels of resources for engaging effectively in participation. As already noted, those who have financial resources, technical know-how, connections to influential people, and so forth are more likely to be motivated to engage or need less inducement to do so. Public apathy and alienation may simultaneously be a consequence of and a justification for limited meaningful participation (Bowles and Gintis, 1986). Lack of motivation among some parties is a challenge to government agencies that want to draw on the public's experience, insights, and aspirations in crafting and implementing solutions to problems (Schneider and Ingram, 1997).

It is important to note that insincerity is often in the eyes of the beholder. Efforts to achieve objectives through litigation, lobbying, or media attention outside a participatory process that is not meeting a party's needs may be seen from an agency perspective as insincerity. However, those are legitimate political activities and quite different from misrepresenting interests and intentions within a particular process. Excluding participants because of a concern about "insincerity" often backfires because it provides these groups the ammunition for external mobilization. Excluding participants that have repeatedly violated reasonable rules of fairness and joint decision making that have been accepted by all from the beginning of the process seems more likely to be accepted as a legitimate action.

Bad Faith by an Agency

Some writers have claimed that public participation takes place in a climate of greatly diminished public trust in government, particularly in regard to environmental matters (e.g., Nye, Zelikow, and King, 1997; Schier, 2000), and of diminished trust in government among the scientific community (Revkin, 2004). Government agencies with low public credibility and trust are unlikely to enjoy the level of confidence among interested and affected parties that is required to initiate and sustain effective voluntary public participation in agency-sponsored initiatives (Woolcock, 1998).

Beirele and Cayford (2002) consider five indicators of preexisting trust in government agencies: the reputation of the agency with the public, the reputation of the agency with participants, a history of withheld information, a history of unacceptable management, and a history of ignoring management problems. As this list implies, trust in agencies can be built or destroyed. The manner in which participatory processes are conducted can nurture positive relationships among participants (including representatives of government agencies) or erode them. For example, the formal structure of meetings, the forms of discourse in which they are conducted, the timing and location of meetings, and numerous other details of participatory processes can convey messages, intentionally or not, about the relative power and status of participants: whose life circumstances are priorities in setting the agenda and whose facts and knowledge carry the greatest weight in deliberations (Chambers, 1997; Briggs, 1998; Estrella and Gaventa, 1998). If government agencies engage people but that engagement turns out to be ineffectual, it is likely to lead to distrust and cynicism and has the potential to diminish possibilities of future engagement (Halpern, 1995).

Explicit discussions with the convening agency's decision makers during the planning phase of a process can uncover or prevent unanticipated difficulties before expectations are set. In one example in a panel member's experience, plans were being formed to invite the public to discuss alternative ways to expand a city's drinking water supply—a policy choice that had become controversial. When it became clear in individual conversations that the majority of the city council had concluded they had no choice other than to use a new source for drinking water regardless of the opposition, plans for a dialogue about other options were set aside as counterproductive.

Negative Views of Scientists

Some parties may not trust scientists, or agency-sponsored scientists, to produce work that is neutral and nonpartisan. Yet effective public participation processes depend on building a working level of trust in the available issue-relevant information, including an understanding of its limitations. Scientists produce much of this information by applying scientific methods and subjecting their work to scientific peer review. As a result, they may assume that their evidence is neutral or value free, not fully appreciating that world views and assumptions that may be widely shared in their disciplines or fields may be questioned legitimately by outsiders, particularly when analysis is brought into a public policy context in which issues outside their field are central. Thus, we suggest in Chapter 6 that special care must be taken to build trust in science that informs public decisions. As we discuss in that chapter, in most cases the procedures that build trust also improve the science.

In designing environmental public participation processes, it is helpful to make explicit that all scientific analyses have their strengths and limitations and to design deliberative processes to reveal and examine the assumptions that underpin various scientific analyses, clarify where parties differ with regard to assumptions and assessments of facts, highlight why different approaches may lead to different conclusions, and so forth. These efforts become more important the more mistrust or controversy there is about the science (see Chapter 6).

CONCLUSIONS

Certain people-related aspects of the context can pose difficulties in achieving the goals of effective public participation. The evidence supports the following specific conclusions:

- Participatory processes are often constrained by agencies' contexts: their external mandates and internal processes that affect their ability and willingness to use the results of the participatory process. The most critical imperative in meeting such challenges is for conveners to make clear to participants at the outset which outcomes are and are not possible from the process. However, public participation processes can be undermined when an agency uses claims about contextual constraints as cover for resistance to participation.

- Meaningful representation of the public is a major challenge, especially when some parties lack the money, technical expertise, or organization needed for full participation or when there are questions about whether participants can act for the parties they are assumed to represent. Differences among the parties in resources and social influence are not easily addressed in the short run, but special efforts to ensure meaningful access by all the parties are likely to yield benefits in terms of competence and legitimacy.
- Differences among the parties in values and interests, as well as polarization of positions and problems of trust, can pose major challenges in implementing principles of participation in the form of conflicts among the parties or between parties and the responsible agency.
- The above challenges can create significant difficulties for public participation. However, choices can be made in the design of a public participation process to compensate for these difficulties. These choices include the selection of techniques and tools for addressing these difficulties and processes for closing such techniques.

Table 8-1 provides a diagnostic guide to many of the people-related difficulties in public participation and to some ways that have been used to try to address them. Like Table 7-1, it identifies particular contextual factors that can make it difficult to implement particular principles of good participation, describes the difficulties, and identifies practices that have been used to address them. We do not endorse any of these practices; however, we believe the guide can be useful in anticipating difficulties and considering possible responses.

We emphasize that best practice in public participation is a matter of adopting a process for selecting the best techniques and tools for the situation, rather than one of using a preselected set of tools and techniques. There are four main reasons we think it inappropriate to treat certain techniques as "best practices" for overcoming common difficulties in public participation. First, the evidence base is very weak for concluding that any one technique is better than the others, even for a particular context or for addressing a particular difficulty of public participation. Second, the research evidence and practical experience strongly suggest that the best technique is likely to be situation-dependent, so that it is unlikely that any practice will be the best across situations. Third, during the process of an environmental assessment or decision, change often occurs in the state of knowledge, the concerns of participants, or the pressures on the convening agency, such that techniques that had seemed satisfactory at the outset may seem less so later on. And finally, we observe that "best practice" techniques, when adopted in bureaucratic agencies, tend to become standard operating procedures that are implemented formulaically, without monitoring or evaluating their effectiveness, and without providing opportunities to make modifications if they are not working well. We therefore conclude that best practice should not be seen as a matter of adopting particular techniques that have performed well in the past and making them standard operating procedures. Rather, it involves implementing a process that selects techniques for the situation at hand in ways that are informed by evidence, that the participants consider legitimate, and that are open to modification for cause. Selecting "best practice" techniques and implementing them without involving the participants can undermine the legitimacy of public participation processes. In Chapter 9, we recommend such a process.

NOTES

[1] As elsewhere, we use the term *agency* broadly to refer to any entity or group of entities, governmental or not, that convene public participation processes or that may use their results.

[2] An evaluation of the results can be found at http://www.ferc.gov/industries/hydropower/gen-info/licensing/ilp/eff-eva.asp.

TABLE 8-1 Diagnostic Guide to Difficulties Related to the People in Public Participation

Contextual Factor	Principles That Become More Difficult to Achieve	Difficulties	Illustrative Practices for Addressing Difficulties[a]
Agency Factors			
Multiple agencies with decision-making authority	Clarity of purpose Commitment to use the process to inform their actions Appropriate timing in relation to decisions	Different agencies may have different views about the purpose of the process or different degrees of commitment to using results Agencies may have different deadlines for decision making Participants may choose to participate in one forum and not others, making the focus unclear	Interagency work groups for coordinating the decision-making process Memorandums of understanding between agencies Written terms of reference (protocols) for the participation process
Substantive mandates or limits to agency decision-making authority	Inclusiveness of participation Openness of the design	Stakeholders may have concerns that are not within the authority of the convening agency, which may reduce motivation to participate	Situation assessments Openness about external constraints Expand scope of participants to include entities that may be able to implement solutions
Insufficient support or conflict within agency	All management principles		Inform participants of limited possibilities Invite participants to contribute resources

Continued

TABLE 8-1 Continued

Contextual Factor	Principles That Become More Difficult to Achieve	Difficulties	Illustrative Practices for Addressing Difficulties[a]
Participant Factors			
Broad geographic spread	Inclusiveness of participation	Distance can make it difficult for some stakeholders to participate Larger numbers of interested participants can make some formats less inclusive or practical	Deliberative polls Workshops in multiple locations Internet participation Study circles Blue-Ribbon commissions
Differences in formal education, occupation, social status, and available time and money	Inclusiveness of participation	People with fewer resources are less likely to trust that their participation will make a difference and, thus, less likely to participate than those with greater resources	Enlist help of organizations to which these members of the public belong in convening public participation processes Integrate opportunities for participation into events of existing local organizations Provide grants to existing local organizations to send mailings or host meetings
Stakeholders who are diffuse, unorganized, or difficult to reach	Inclusiveness of participation	Interests of those who are less well organized may not be equitably represented	Citizen action committees Citizen forums Citizen juries
Disparities in financial, technical, or other resources?	Inclusiveness of participation	Interests of those who have insufficient time or resources to participate may not be equitably represented	Timing and location of meetings Child care Technical assistance grants

Differences in values, interests, culture and perspectives, and the degree to which parties are polarized	Collaborative problem formulation and process design Good-faith communication	Differences in cultural norms may result in disagreements about what constitutes good-faith communication and/or what constitutes an appropriate process for making a decision Conflicting values can lead to frustration, stalemate, and mistrust of analyses Polarization can make it more difficult to understand the interests that underlie disputes or may have its origins in mutually exclusive interests	Focus on relationships first (e.g., through field trips, social opportunities, story telling) Structured deliberation methods from decision science Confidential conversations with a mediator to identify interests Generating multiple options Systematic application of criteria, including through models Adding issues to the scope of the process to add potential value to the solutions
Disparities of power	Inclusiveness of participation Collaborative design Transparency of the process Good-faith communication	Creating a forum that is attractive both to those with greater and lesser power, compared with pursuing objectives elsewhere	Involve participants during diagnosis and design phase in creating an agreement on the scope and objectives Draft a written scope and invite public comment on it Draft ground rules in which participants agree to inform one another if they intend to pursue their interests in other forums

Continued

TABLE 8-1 Continued

Contextual Factor	Principles That Become More Difficult to Achieve	Difficulties	Illustrative Practices for Addressing Difficulties[a]
Limitations on ability of representatives to act on behalf of their constituency	Good-faith communication	Misperceptions by some participants of other participants' authority to act Internal decision making by some parties may require more time than the process allows	Select participants with authority to represent their constituency Draft ground rules in which participants are specific about their authority and with whom they must consult Allow sufficient time for consultation with decision makers who are not at the table Draft summaries of meetings that are circulated to decision makers and interested others Organize formal working groups within constituencies Arrange briefings for decision makers or broader constituencies being represented

| Significant problems of trust | Good-faith communication | Reduced agency motivation to convene a process
Reduced motivation of the parties to participate | Use situation assessment to identify the nature of the problem
Seek formal agreements
Provide for independent review of scientific analyses
Iterates between analysis and deliberation
Encourage participants to consider possibility of misunderstanding prior to assuming bad faith
Provide incentives for good-faith action and disincentives for acting inconsistently with agreements on either process or substance (e.g., phased processes, phased implementation, contingent agreements) |

[a]Evidence is inadequate to recommend any of these practices as effective, or as preferable to practices that are not listed. They are listed to suggest some of the practices that might be considered for addressing particular difficulties.

9

Overall Conclusions and Recommendations

Engaging stakeholders and the public in environmental assessment and decision making offers clear benefits to all, if done well. The caveat, of course, raises the central questions of this report. What is known about how to engage the public well? What is not known. This chapter presents the panel's overall conclusions, based on the specific conclusions established in Chapters 3-8, and offers recommendations for good public participation practice and for research.

Our recommendations for practice are organized around three sets of principles—for good management, for organizing the process, and for integrating the science—as described in Chapters 4, 5, and 6. They also draw on the diagnostic questions developed in Chapters 7 and 8, which help identify characteristics of the context that tend to create particular challenges or difficulties with respect to achieving successful results (see Box 9-1). The principles are consistent with those from past studies at the National Research Council (1996, 1999a, 2007a) and those recently offered by the Office of Management and Budget and the President's Council on Environmental Quality (2005; also see Table 1-2).

We reiterate such previously stated principles for two reasons. First, it is important to recognize principles that are supported by a convergence of evidence: from practitioners' experience, careful case-study research, case-comparison studies, and basic social science knowledge. Second, the principles bear repeating because they are so often violated in practice. Because public participation in environmental assessments and decision making is a new area for systematic research and much is yet to be learned, we also offer suggestions for advancing knowledge in the field.

BOX 9-1
Diagnostic Questions to Assess the Challenges to Public Participation in a Particular Context

Questions About Scientific Context

1. What information is currently available on the issues? How adequate is available information for giving a clear understanding of the problem? Do the various parties agree about the adequacy of the information?

2. Is the uncertainty associated with the information well characterized, interpretable, and capable of being incorporated into the assessment or decision?

3. Is the information accessible to and interpretable by interested and affected parties?

4. Is the information trustworthy?

Questions About Convening and Implementing Agencies

1. Where is the decision-making authority? Who would implement any agreements reached? Are there multiple forums in which the issues are being or could be debated and decided?

2. Are there legal or regulatory mandates or constraints on the convening agency? What laws or policies need to be considered?

Questions About the Abilities of and Constraints on the Participants

1. Are there interested and affected parties who may have difficulty being adequately represented?

    a. What does the scale of the problem, especially its geographic scale, imply for the range of affected parties?

    b. Are there disparities in the attributes of individual potential participants that may affect the likelihood of participation?

    c. Are there interests that are diffused, unorganized, or difficult to reach?

    d. Are there disparities across groups of participants in terms of their financial, technical, or other resources that may influence participation?

2. What are the differences in values, interests, cultural views, and perspectives among the parties? Are the participants polarized on the issue?

3. Are there substantial disparities across participant groups in their power to influence the process?

4. To what degree can the individuals at the table act for the parties they are assumed to represent?

5. Are there significant problems of trust among the agency, the scientists, and the interested and affected parties?

    a. Are there indications that some participants are likely to proceed insincerely or to breach the rules of the process?

    b. Are some participants concerned that the convening agency will proceed in bad faith?

    c. Do some participants view the scientists as partisan advocates and so mistrust them?

We also offer advice for implementing the principles. For reasons discussed in Chapter 8, we do not consider it advisable to recommend specific techniques as "best practices" for general use. Instead, we recommend a best process for selecting such techniques and for monitoring their effectiveness and adjusting them to achieve the desired purposes.

We remind the reader that there are many goals for public participation processes and thus many criteria for what constitutes a "good" or "effective" outcome and a "good" or "effective" process. Goals include both those focused on the quality of environmental assessments and decisions and those focused on the relationships among the participants. Participation ideally should improve the *quality* of assessments and decisions and their *legitimacy* among those involved and potentially affected. It should lead to increased understanding and decision-making *capacity* among agency officials, scientists, and the interested and affected parties involved and the interests they represent.[1] And it should enhance the ability to implement decisions once they are made both by producing better decisions and by producing legitimate, credible, and well-understood decisions.

The evidence suggests that in most cases, these three kinds of desired results are complementary rather than contradictory: achieving one goal of participation usually accompanies success in reaching other goals. A substantial portion of this chapter presents our conclusions and recommendations regarding how best to proceed. Thus, we present a series of principles for public participation that, if implemented in a way that is sensitive to context, can aid in achieving desirable outcomes, and we recommend a process for finding ways to implement those principles in the context at hand.

Our conclusions and recommendations are based on the convergence of multiple lines of evidence, including studies with one or a few cases; statistical analyses of many cases; systematic case comparisons, including several conducted for this study; our review of basic social science research relevant to public participation; analysis of the legal framework for participation; an assessment of practitioner experience embedded in handbooks and agency guidance; and the expertise of the panel members. Systematic research on public participation is still relatively new, and although the literature is growing rapidly, we had to use our judgment to evaluate different forms of evidence. We think our conclusions and recommendations are reasonably robust given the state of knowledge and practice but, as with any statements based on an emerging field of research, we will not be surprised if further work suggests modifications to and elaborations of them.

THE VALUE OF PUBLIC PARTICIPATION

CONCLUSION 1: When done well, public participation improves the quality and legitimacy of a decision and builds the capacity of all involved to engage in the policy process. It can lead to better results in terms of environmental quality and other social objectives. It also can enhance trust and understanding among parties. Achieving these results depends on using practices that address difficulties that specific aspects of the context can present.

RECOMMENDATION 1: Public participation should be fully incorporated into environmental assessment and decision-making processes, and it should be recognized by government agencies and other organizers of the processes as a requisite of effective action, not merely a formal procedural requirement.

Substantial evidence shows that good public participation not only helps fulfill norms of popular sovereignty in democratic societies, but also improves the substantive quality, legitimacy, and accountability of environmental assessments and decisions. In other words, the reason to engage the public is not simply because laws, regulations, and habit require it, nor is it only because public participation makes decisions more legitimate in the eyes of the public. Rather, substantial evidence shows that effective public participation can help agencies do a better job in achieving public purposes for the environment by ensuring better decisions and increasing the likelihood that they will be implemented effectively. Good public participation also helps build capacity in agencies and among participants and the scientific community for future environmental decision making.

As Chapter 3 shows, innumerable studies of one or a few cases demonstrate the positive results that often come with public participation. Systematic comparisons of larger numbers of cases show that the results of public participation have been positive far more often than they have been negative. The same conclusions can be drawn from examination of practitioner experience and from focused case-comparison studies. These results apply across a wide range of well-documented public participation processes for making many kinds of assessments and decisions and across the United States.[2] Thus, multiple sources of evidence give strong confidence that public participation, done well, can be effective in achieving multiple desired benefits in a wide variety of settings and that it can be effective even within the resource limitations that commonly exist in federal, state, and local governments.

It is also true that public participation, if not done well, may not provide any of these benefits—in some circumstances, participation has done more harm than good. A poorly designed process that lacks adequate support and engagement by the agency or that fails to meet major challenges posed by the specific context can decrease, rather than increase, the quality and legitimacy of an assessment or decision and damage capacity for future processes. Conclusions and Recommendations 2 through 5 present principles that can be used to shape successful participation practice and a process for designing participation processes to implement those principles.

MANAGEMENT

CONCLUSION 2: Basic principles of program management apply to environmental public participation. When government agencies engage in public participation processes without careful prior planning, adequate resources, and organizational commitment, the results may fall short of the potential of public participation.

RECOMMENDATION 2: When government agencies engage in public participation, they should do so with

1. clarity of purpose,
2. a commitment to use the process to inform their actions,
3. adequate funding and staff,
4. appropriate timing in relation to decisions,
5. a focus on implementation, and
6. a commitment to self-assessment and learning from experience.

We caution that although public participation often provides multiple benefits, the available evidence also shows that it is possible to conduct public participation processes that are counterproductive and that may be worse than not including the public at all. Participatory processes convened as a superficial formality or without adequate support by decision makers increase the public's distrust of government when, almost inevitably, the results have little impact. Some participatory processes have functioned as a political tactic to divert the energy of the public away from engaging in dissent on important differences and into activities that are considered safer by an agency, such as projects based on shared goals that ignore important conflicts. This use of public participation is counterproductive in the long run. Choices about whom to involve can also be problematic, as when an agency involves parties that share the agency's basic premises about a decision to be made, while excluding those whose views differ more fundamentally. The power to define the questions to be addressed and to shape the public participation approach—how it is used and by whom—is critical. We return to this point below.

The above recommendation embodies six basic principles of good management that offer practical guidance to agencies for achieving the goals of environmental public participation. The principles describe actions that agencies can take to enhance the quality and effectiveness of the public participation processes in both the short and the long run. It is vital for managers at all levels of government and in the private sector to learn how to involve the public well, if the benefits are to be realized.

Clarity of Purpose The process should be designed with a clear purpose in mind and be organized to meet the objectives. When the responsible agency develops a clear set of objectives, integrated with a plan for how the outcomes of the participatory process will be used and serious efforts to share that understanding with the participants, it increases the likelihood of acceptance of agency decisions and of public willingness to engage in future participation efforts. From the outset, the convening organization and the particpants should develop a clear agreement about the objectives of the process, taking account of the objectives of all parties involved, the scope of legally possible actions, and the constraints on the process.

A Commitment to Use the Process to Inform Actions Public participation processes are more likely to be successful when the agency responsible for the relevant environmental decisions is committed to supporting the process and taking seriously the results. This is in part because the more committed a decision-making agency is to act on the results of a public participation process, the more likely the parties are to engage seriously. As the objectives of the process are codetermined by the agency and the participants, the support of agency leadership and staff at all levels for the objectives of the process should be confirmed. At the beginning of the process, it is essential to clarify how and by whom the outputs of the participatory process will be used and that the responsible organization is committed to open-minded consideration of those outputs. These commitments should be updated periodically, as both the participation process and the context evolve.

Adequate Funding and Staff Public participation processes are more likely to be successful when agencies have adequate capacity and resources including skilled staff and deploy them appropriately to the scale, complexity, and difficulty of the issues involved. If resources are too constrained to support a desired public participatory process, the diagnosis needed to plan an effective process also should be used to target public participation strategically. It is better to do only what can be adequately supported than to provide inadequate support for a more ambitious process.

It is important to match the objectives and scope of the participatory process to the resources available. Diagnosis of the situation can help scope the complexity and difficulty of the tasks required and provide a sense of what sort of investments are required to achieve various goals. If resource limitations make it unlikely that all goals can be achieved, it is critical to invest in meeting the most important challenges or obstacles that have been identified. And it is equally critical to understand what can and cannot be accomplished within the resource constraints that exist. It is often useful to be creative in looking for additional resources, including from participants and the public.

Appropriate Timing in Relation to Decisions Public participation processes are more likely to have good results when planned so that they can be informed by emerging analysis and so that their outputs are timely with regard to the decision process. In designing the participatory process, it is critical to conform to agency decision-making timetables so that closure is achievable and outcomes are available to decision makers in a timely manner. This often requires adjusting the intensiveness of participation and the scope of issues to be covered so that the time is realistic—time is a resource constraint. It is also important not to rush the process unneccessarily, as changes in context or in available analysis might obviate the value of the outcomes of the process if these changes occur after the process is completed. Sometimes, time constraints can be addressed by using a participatory process as part of an adaptive management strategy. The process can be convened to inform a provisional decision with the understanding that it will be reconstituted at a later time to revisit that decision for the purpose of revising it as appropriate.

A Focus on Implementation Participation processes tend to be more successful when designed so as to relate in clear ways to policy decision making and implementation. Increasingly, public participation is viewed as an element of adaptive governance rather than as a one-time, one-way flow of information. So the design of the process should consider implementation: how the process can inform both initial assessment and decision making and ongoing analysis and action. It is useful to identify roles and responsibilities following the public involvement process and to involve those who are needed for implementation. By anticipating difficulties in implementation and discussing contingencies, the public participation process will be better informed and produce more useful results. In many cases, the process can support successful implementation by considering partnerships, monitoring and oversight mechanisms, and incentives and disincentives to implementation.

A Commitment to Self-Assessment and Learning from Experience Public participation processes, as well as the larger assessment and decision processes in which they are embedded, benefit from engaging in self-assessment and design correction as they proceed. The design of a participatory process should create opportunities for participants and sponsors to assess the process both as it is under way and at the end. The design must be flexible enough to allow for mid-course adjustments and to generate lessons learned that can be incorporated into future public participation efforts.

The self-assessment process should include evaluation by an external reviewer or review body whenever possible, as well as by the participants and the sponsors. Careful evaluation research often reveals knowledge that does not emerge from intuitive judgments of what works and what does not. While some systematic studies of public participation now exist, the state of knowledge would be much advanced if organizers of participation supported evaluation studies. Even when resources are limited, expenditures on systematic evaluation deserve high priority, as this is the only valid means to ensure institutional learning and constant improvement.

ORGANIZING THE PROCESS

CONCLUSION 3: The outcomes of a public participation process depend strongly on the way the process is organized and carried out. Although contextual factors can create difficulties for achieving principles of good practice, choices about key aspects of effective participatory process can do much to overcome these difficulties.

RECOMMENDATION 3: Agencies undertaking a public participation process should, considering the purposes of the process, design it to address the challenges that arise from particular contexts. Process design should be guided by four principles:

1. inclusiveness of participation,
2. collaborative problem formulation and process design,
3. transparency of the process, and
4. good-faith communication.

These elements of design are appropriate to all participatory processes, although the way they are implemented will vary across contexts. There is no single best format or set of procedures for achieving good outcomes in all situations.

Inclusiveness of Participation The process should include credible representatives of the full spectrum of parties who are interested in or will be affected by a decison. It should be structured to encourage their voluntary commitment to it. At the outset, care should be taken to identify and engage all such parties. As the process proceeds and more information becomes available, the scope of participation may have to be expanded to reflect enhanced understanding of who may be affected or concerned. The process should be designed to give all involved a fair voice, so as to benefit from differences in perspectives, approaches, backgrounds, and culture.[3] When in doubt, it is preferable to err on the side of too much inclusiveness than too little, although there are often practical constraints on how large a process can be that may require special care in ensuring inclusiveness in a group of restricted size. Care should be given to understanding what would motivate members of the public, whether organized into interest groups or not, to engage seriously in a participatory process. The design of the process should maximize the incentives to participate and minimize disincentives and obstacles.

Collaborative Problem Formulation and Process Design Public participation processes should, to the greatest extent feasible, be designed collaboratively by those convening them and those participating in them. Of particular importance in achieving quality and legitimacy is engaging the spectrum of interested and affected parties in *formulating the problem* for assessment or decision to the extent the agency's context allows.

This principle requires developing the process by collaboration among all who will be engaged in it, with a particular emphasis on engaging members of the public in problem formulation, including defining the scope of the assessment process or policy and diagnosing the obstacles to effective participation. To effectively engage the capabilities and needs of all participants, it is essential that participants co-invent and govern the process. This means that the potential participants should be identified and brought into the planning process as early as possible. To the extent possible, they should participate in defining the issues to be examined (problem formulation), as a mismatch between the scope of the problem as defined by the agency and the scope as defined by participants can be a source of serious problems. The participants should also consider the barriers to achieving effective participation by various groups and other obstacles to an effective process (see Chapters 7 and 8 for diagnostic questions). Participants should co-design the formats and decision rules (process design) to ensure that the process is effective and trusted. Participation specialists can make recommendations and advise all parties on what is likely to work best in the given context, but the final decision should be made in a collaborative effort with the main parties involved.

Transparency of the Process The process should be clear to those involved in it and to those observing it.

All participants and the public should be informed of the purpose and objectives of the process and of agency authorities, requirements, and constraints. Mechanisms should be built into the design for ongoing communication about the process and for public access to information about the process and information being used in it. The Internet provides powerful tools for this purpose, but not everyone has access, and maintaining and using websites requires dedicated resources. We are still in the early stages of understanding the dynamics of Internet-based communication.

Good-Faith Communication All parties must commit to act in good faith and to maintain communications with those they represent. The process should be structured to encourage this.

From the start, it is important to have in place mechanisms for communication to and from decision makers or other constituencies in organizations involved in the process, including agency sponsors and interest groups as well as the public. These groups should be kept informed of progress and encouraged to report back actions taken or changes that may affect the process and the reasons why such changes occurrred. It is advisable to negotiate a consensus on rules for communication, deliberation, and decision making before the substantive issues are discussed. If all parties agree to a common set of procedural rules, it is easier for the moderator to enforce these rules and to ensure fair play among all participants.

It is worth noting two considerations that we do not include among the principles of good participation: the format of the process and its intensity. With regard to format, the public participation literature is replete with proposed formats for conducting the overall participatory process and techniques within the process, including methods for conducting meetings and other interactions among participants. This literature provides a rich toolkit for those designing participatory processes (including members of the public involved in codesign of the process). However, it does not demonstrate that any of these methods is universally superior to the others.

Various public participation formats have been successful in achieving the goals of high quality and widely acceptable assessments and decisions, and each format has also failed at times in achieving these goals. There is no single best format or set of procedures for achieving good outcomes in all situations. However, whatever format is used, the best practices for public participation we recommend should be followed.

As with much else in the practice of public participation, the most successful practice will involve considering the goals for participation and the context in which the participation will take place and designing the process so as to best achieve those goals in that context. The use of the Internet for public participation deserves special note. A research literature on Internet-based public participation is just beginning to emerge, and the technology available for such interaction is evolving and presents new formats and techniques for participation not yet carefully studied (e.g., on-line video conferencing, websites, e-mail). It is inevitable that agencies will experiment with such technologies, and we consider it imperative that such experiments be accompanied by careful research to build a knowledge base to guide future efforts. We suggest that the biggest advantages of electronic participation at present may be found when an environmental issue has broad geographic impact, and the biggest disadvantage may be the difficulty of engaging groups who do not regularly use the Internet.

With regard to the intensity of the public participation process, the evidence suggests that the proper level of intensity is context dependent. The most effective participatory processes are those whose intensity is dictated by responding to context-specific challenges with appropriate participation strategies. For example, contexts that involve serious potential for conflict can benefit more from high-intensity processes than contexts that do not present such challenges. However, when the context calls for intense interactions, results will be highly dependent on how those interactions are organized.

By intensity of deliberation, we mean both the amount of time during which participants are engaged in focused discussion and other activities and the structure of the interactions. If well designed (i.e., following established principles), appropriately intense deliberation often can overcome barriers of trust and can help develop common understandings of scientific and other important information. Relatively intense deliberation may be essential when the issue is especially contentious or complicated or when the goal of the process is ongoing comanagement or adaptive management rather than advice on a specific assessment or policy.

More intense deliberation is more costly for all involved, and that cost may reduce the breadth of participation. More intensity is not always better; rather, the intensity of a participatory process should be designed so that it is appropriate to the context and takes account of the costs and benefits of intensity. An appropriately intense, well-organized deliberative process will influence agencies that have committed to taking the results seriously and followed the other principles of practice. It should also influence the thinking and positions of the other participants. There are two key elements in making a process influential, aside from agency commitment: transparency and good-faith communication.

INTEGRATING SCIENCE

CONCLUSION 4: Processes that iterate between analysis and broadly based deliberation, as recommended in *Understanding Risk* (National Research Council, 1996) and subsequent National Research Council reports (National Research Council, 1999a, 2005, 2007a), have the greatest chance of being effective in linking participation and scientific analysis. In contrast, processes that treat analysis and deliberation in isolation from each other impede both analysis and deliberation.

RECOMMENDATION 4: Environmental assessments and decisions with substantial scientific content should be supported with collaborative, broadly based, integrated, and iterative analytic-deliberative processes, such as those described in *Understanding Risk* and subsequent National Research Council reports. In designing such processes, the responsible agencies can benefit from following five key principles for effectively melding scientific analysis and public participation:

    1. ensuring transparency of decision-relevant information and analysis
    2. paying explicit attention to both facts and values
    3. promoting explicitness about assumptions and uncertainties
    4. including independent review of official analyses and/or engage in a process of collaborative inquiry with interested and affected parties
    5. allowing for iteration to reconsider past conclusions on the basis of new information

As noted in *Understanding Risk* (National Research Council, 1996:3):

> [S]uccess depends critically on systematic analysis that is appropriate to the problem, responds to the needs of the interested and affected parties, and treats uncertainties of importance to the decision problem in a comprehensible way. Success also depends on deliberations that formulate the decision problem, guide analysis to improve decision participants' understanding, seek the meaning of analytic findings and uncertainties, and improve the ability of interested and affected parties to participate effectively in the risk decision process. The process must have an appropriately diverse representation of the spectrum of interested and affected parties, and of specialists in risk analysis, at each step.

This formulation from *Understanding Risk* applies to environmental assessments and decisions more generally.

Special care is needed to integrate science into public participation processes because of three kinds of potential obstacles to effective use of science in assessment and decision-making processes that involve interested and affected parties. First, the science required is inherently complex and uncertain, and the data available are nearly always less than ideal. Consequently, scientists must be explicit about the extent and limits of knowledge, develop understanding of which knowledge participants consider most decision relevant, and possibly reconsider standard approaches to handling uncertainty.

Second, many participants in environmental assessment and decision processes lack sufficient scientific and technical background to easily interpret complex scientific information. Moreover, in the absence of structured decision processes, people tend to consider less than the full range of relevant information in making decisions. And there is not just one view among participants. Rather, there are diverse values, interests, and concerns. Third, there are substantial challenges in communication between scientists and the public. Scientific models are difficult to translate into forms that are transparent to scientists across fields and even more difficult to translate for the public. In addition, debates about scientific uncertainty can be hard for nonspecialists to follow, and the rules for validating facts may be different for scientists than for many segments of the public and may even differ in significant ways across scientific disciplines. All of this can make the public skeptical of the neutrality of scientific analyses and the scientists skeptical of local experience-based knowledge of the public.

Formidable as these challenges may be, there are effective tools available for meeting them. Research in the decision sciences, research on environmental assessment and decision-making processes, and insights garnered from the practice of science and from analyses of public policy processes all converge on five key points of guidance about how to integrate science and public participation into analytic-deliberative processes.

Ensuring Transparency of Information and Analysis We have already identified transparency of the overall participation process as an important principle of process design. Making scientific analyses transparent is especially important. As noted, members of the public will generally not be aware of the assumptions that are embedded in an analysis, especially if the analysis uses complex models. Nor is a lack of transparency a problem only for the public—many scientists are not aware of what are assumptions in specialties outside their own. Processes to ensure that decision-relevant information is accessible and interpretible to all participants and that analyses are available in open sources and presented in enough detail to allow for independent review not only build public trust in the science, they also ensure the open discussion of assumptions and uncertainty that is central to improving scientific analyses.

Paying Explicit Attention to Both Facts and Values An effective analytic-deliberative process must deal with both facts and values and in particular with how anticipated changes in the world will affect the things people value. However, facts will always be uncertain, and some facts may be sharply contested. Values also may be uncertain, in the sense that members of the public may not see how anticipated changes will affect the things they care about. In addition, there is usually substantial diversity in values among the interested and affected public, and different ways of formulating the problem to be analyzed may embody different values or concerns. A variety of tools and processes can help characterize uncertainty about facts, examine the implications of analyses and their uncertainties for decision making, elicit the diversity of public values, and guide individuals and groups through the examination of value trade-offs. Experimentation with using these tools to enhance participation processes is warranted.

Promoting Explicitness About Assumptions and Uncertainties Uncertainties about facts and values will always be present, and all analyses must rely on assumptions. Trust, understanding, and constructive criticism can emerge only when there is awareness of uncertainty and assumptions. Careful analysis linked to effective deliberation can identify assumptions and uncertainties, examine how much they matter, and thus tighten the focus of further analysis and allow honest discussion about what underpins conclusions and decisions.

Including Independent Review of Official Analyses or Collaborative Inquiry Since all analyses are fallible, independent peer review has become the gold standard for judging scientific analysis and is also enshrined in the concept of adversarial argument in the context of trial by a jury of peers. While independent review itself is never perfect, the progress of science demonstrates the power of the method to improve the quality of analysis. Effective independent review of scientific analyses will enhance the ability of a process to adhere to the other guidelines for integrating science. Because the interested and affected parties are rarely able to conduct an independent review themselves, it is important that the analysts who conduct reviews are credible to the parties. Collaborative inquiry that involves the range of interested and affected parties can achieve some of the benefits of independent review.

Allowing for Iteration Between Analysis and Deliberation Iteration is essential to allow for reconsideration of past conclusions on the basis of new information. Practical constraints may limit the amount of iteration that is possible. But some iteration should always be built into the process to refine both the questions being asked and the answers being offered.

IMPLEMENTATION

CONCLUSION 5: Contextual factors—attributes of the environmental issue, the state of knowledge, the agency and its environment, and the participants—can present a variety of difficulties in implementing the principles of good public participation. However, choices made in the design of a public participation process can compensate for the difficulties that specific attributes of the context may pose. The best choices are likely to be situation dependent. It is counterproductive to define "best practice" in terms of any specific techniques to be routinely used.

RECOMMENDATION 5: Public participation practitioners, working with the responsible agency and the participants, should adopt a best-process regime consisting of four elements:

1. diagnosis of the context to identify likely difficulties;
2. collaborative choice of techniques to address those difficulties;
3. monitoring of the process to see how well it is working; and
4. iteration, including changes in tools and techniques if needed, to overcome difficulties.

As discussed in Chapter 8, it would be a mistake to name certain techniques as "best practices" for several reasons: the evidence is very weak for such a recommendation; the research evidence and practical experience strongly suggest that the best technique is likely to be situation dependent; practices need to be sensitive to changes that occur during the process; and recommended "best practices" too easily turn into standard operating procedures that are implemented formulaically, without sensitivity to their effectiveness, which may be less than hoped and may vary over time. Selecting "best practice" techniques and implementing them without involving the participants can also undermine the legitimacy of public participation processes.

Given these considerations, we recommend a best process regime for selecting and adjusting tools and techniques to meet the challenges of particular public participation settings as they evolve. Best process for public participation has four elements: diagnosis, collaborative choice, monitoring, and iteration.

1. Diagnosis of the Context Practitioners, the agency, and participants should identify potentially significant difficulties or challenges in the situation at hand with respect to implementing the principles of good public participation. Chapters 7 and 8 provide a set of diagnostic questions that can be useful for this purpose (see Box 9-1).

2. Collaborative Choice of Techniques Practitioners, working with the agency and the participants, should collaboratively design the process, selecting tools and techniques from among those available for addressing the anticipated difficulties or challenges.

3. Monitoring of the Process Practitioners, participants, and the convening agency should agree to monitor the process to see whether it is in fact meeting anticipated and emerging challenges. The monitoring procedure may be informal or may involve formal evaluation, integrated by agreement of the participants into the public participation process design.

4. Iteration Practitioners, participants, and the agency should establish procedures that allow for adaptation and change in the public participation process when needed.

The four elements of best process and their relationships to each other and to the principles of public participation practice and to contextual factors are presented schematically in Figure 9-1. In this process, those involved begin with diagnosis to identify the important contextual factors and the difficulties they are likely to create. They then collaboratively select specific techniques and tools to use to address the difficulties and agree on how to monitor the process. The results of monitoring can lead to a decision to adopt new or different techniques for continuing the participatory process.

Diagnosis, collaborative choice, monitoring, and iteration are all critical to finding effective ways to implement public participation. Diagnosis highlights the key issues the process must address. Collaborative choice is important for legitimacy, but cannot alone address the possibility that important questions, perspectives, or participants might be inadvertently left out or that some of the participants may be duplicitous. Monitoring helps create accountability to address these problems. Iteration allows for their correction.

Accountability through monitoring, evaluation, and iteration can be very costly when organized in the context of bureaucratic standard procedure or adversarial legal interaction. Getting broad initial acceptance of a process that has these features can greatly reduce these potential costs. Thus, systems that provide for iteration can be part of a collaboratively chosen process and also provide a check on it. They allow any of the parties to raise questions during the process about whether the practices in use are actually solving the problems and implementing principles of good participation, and they provide a legitimate place for making and addressing claims of failure to solve problems or failure to implement principles of good participation.

NEEDED RESEARCH

CONCLUSION 6: Research on the public participation process has lagged far behind the need. However, both a community of researchers and a set of appropriate research methods are available for developing a scientifically grounded understanding of public participation.

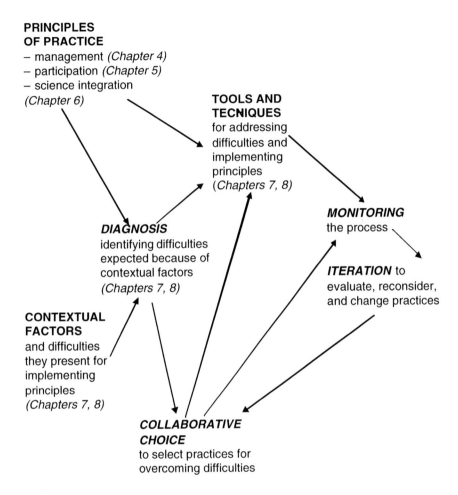

FIGURE 9-1 Elements of best process for public participation in relation to the principles of good public participation and variations in context.
NOTE: The four elements of best process are indicated in italics. Arrows indicate lines of influence: principles and contextual factors contribute to diagnosis; principles, diagnosis, and collaborative choice influence the selection of tools and techniques; the tools and collaborative choice determine what is monitored and how; monitoring leads to iteration; and iteration, via collaborative choice, feeds back to the selection of tools and techniques.

RECOMMENDATION 6: Agencies that involve interested and affected parties in environmental assessments and decision making should invest in social science research to inform their practice and build broader knowledge about public participation. Routine, well-designed evaluation of agency public participation efforts is one of the most important contributions they can make. Because public participation makes a useful test bed for examining basic social science theory and methods, the National Science Foundation should partner with mission agencies in funding such research, following the model of the successful Partnership for Environmental Research of the National Science Foundation and the Environmental Protection Agency.

Our conclusions and recommendations for the practice of environmental public participation flow from the available empirical evidence and are consistent with the judgments of experienced practitioners and basic social science knowledge. However, further and more rigorous research will be necessary to test and build on these judgments. For example, further research may determine that particular tools or techniques are efficacious with regard to meeting specific challenges that arise with public participation in certain contexts. The inability to provide solidly supported conclusions of this type at present should not be surprising. Empirical research on environmental public participation is recent and is still dominated by studies that examine only one or a few cases. True experimental research on the model of case-control clinical trials is almost nonexistent. To arrive at robust general conclusions will require, at a minimum, evidence from detailed comparisons across many public participation cases that differ in terms of the challenges they present and that are observed over time. Knowledge in other similarly complex areas of environmental practice, such as the management of common-pool resources, took decades of work using multiple research methods to get to the point of providing broad and robust guidance to practitioners (National Research Council, 1986, 2002a; Dietz et al., 2003). Research to date has yielded useful findings, as noted throughout this report and as reflected in our conclusions and recommendations. But much more remains to be done.

The National Research Council report, *Decision Making for the Environment*, identified the creation of effective analytic-deliberative processes as one of the important priority research areas and identified several important research questions in this area. We agree with that panel's assessment. The following key questions should guide future research on public participation (National Research Council, 2005a:38-39):

- What are good indicators for key attributes of success for analytic-

deliberative processes, such as decision quality, legitimacy, and improved decision capacity?
  • How are these outcomes affected by the ways in which the processes are organized, the range and diversity of people involved, the rules used for deliberating and reaching conclusions, the ways technical information is organized and made available, and the environmental, social, organizational, and legal contexts of the decision at hand?
  • What are effective ways to make technical analyses transparent to a wide range of decision participants, some of whom lack technical training?
  • How can decision-analytic techniques for preference elicitation, characterizing uncertainty, and aggregating preferences be used to best advantage in broadly based analytic-deliberative processes?
  • How can decision processes be organized to ensure that all sources of relevant information, including the local knowledge of nonscientists, are gathered and appropriately considered?
  • How can analytic-deliberative decision processes be organized to reach closure effectively and with broad acceptance, especially when the processes involve a diversity of perspectives and interests? What tests could be applied to decisions and decision processes to support claims that they are ready for closure?

The analyses in this volume suggest the need for research on some further refinements of these questions, such as on ways in which the effects of certain aspects of practice may depend on contextual variables or on the phase of the decision-making process, as well as on ways to overcome common imperfections of small-group decision making that have been observed in experimental research and on ways to combine analytic and deliberative methods for addressing value trade-offs.

Such research questions can be answered only by an interacting community of scholars and by an increased level of rigor in research design. It is clear that a community of researchers interested in theoretical and empirical examination of what happens in public participation is emerging. If this community grows and matures, it will provide the evidence needed for more effective and efficient public participation practice and, in the end, better environmental assessment and decision making.

But as is often the case in the emergence of a new field, there are important obstacles to developing a strong and coherent body of knowledge. The network of researchers is not well connected across disciplines and across classes of environmental decisions, so that the literature is not as interconnected, self-critical, and cumulative as it might be. In addition, much of the research is conducted with little or no funding, so that the overwhelming majority of studies are based on one or a few cases. While the case-study approach allows for a detailed understanding of the situations examined, it does not lend itself to disentangling the myriad and interacting effects of context and process. In preparing this report we have particularly valued the few studies that have been able to deploy larger sample sizes. The continued use of case studies will help advance the field, but it must be complemented with more multicase comparative and longitudinal studies that allow a stronger assessment of generality and causality. To this end, agencies should be open to having their participation procedures evaluated prospectively by researchers and to ensuring cooperation and funding for such evaluation studies.

We note that two important social science methodologies, formal experiments and modeling, are notable by their near absence from the public participation literature. It will be useful to explore the utility of formal controlled experiments to contribute to understanding of public participation processes. Their ability to produce strong evidence about causal effects can make them very useful, even though concerns regarding the external validity of results generated in artificial settings must be taken into account. Field research using experimental and quasi-experimental designs hold great promise but are absent from the available knowledge base. Since there have been virtually no attempts to model public participation processes using analytic or simulation methods (we know of only one, by Howarth and Wilson, 2006), it is less clear what such approaches might contribute, but they surely deserve exploration.

In addition to individual research projects, it is important that agencies that use public participation routinely invest in building the community of public participation researchers so that, over time, agency efforts will benefit from an improved base of systematic knowledge. This emerging community is highly interdisciplinary and, to realize its full potential, needs to be more strongly interconnected and better linked to public participation practitioners. The community of researchers and practitioners focused on management of common-pool resources is exemplary in this regard and might serve as a useful model for building public participation research.

The scientific study of public participation in environmental assessment and decision making is still very new and the methods employed in available studies are usually less than ideal. Therefore, an investment in more and stronger research will almost certainly yield further insights that in turn will have substantial payoff in the form of improved environmental assessments and decisions and enhanced capacity for sound environmental policy.

NOTES

[1] We use the term *agency* generically throughout our conclusions and recommendations. An environmental public participation process may be convened by a federal, state, or local government agency, by a group of agencies, by a business or nonprofit nongovernmental organization, or even by a previously unorganized group of affected individuals. We sometimes use the term *agency* to refer broadly to any entity or group of entities that may convene a public participation process, provide the resources for it to proceed, or take action based on its results.

[2] Our reading of the literature indicates that this conclusion also applies outside the United States. However, to keep our task manageable, we have focused our analysis on the literature grounded in U.S. experience to ensure that our results are relevant to the agencies that sponsored the study.

[3] We use the term *culture* broadly to refer to characteristics of belief, thought, or practice that are shared within a social group, whether that group is defined by ethnicity, race, gender, religion, occupation, scientific discipline, or some other characteristic. The cultural differences that are important to public participation are those that affect the ways people understand the policy issues or the information offered to aid in assessment or decision making. The most important cultural factors are likely to vary with the issue at hand.

References

Abel, R.L.
   1974   A comparative theory of dispute institutions in society. *Law and Society Review, 8,* 217-347.
   1982   *The Politics of Informal Justice: Volume 1.* New York: Academic Press.

Abels, G.
   2007   Citizen involvement in public policy-making: Does it improve democratic legitimacy and accountability? The Case of pTA. *Interdisciplinary Information Sciences, 13*(1), 103-116.

Abelson, J., Forest, P.G., Eyles, J., Casebeer, A., Martin, E., and Mackean, G.
   2007   Examining the role of context in the implementation of a deliberative public participation experiment: Results from a Canadian comparative study. *Social Science & Medicine, 64*(10), 2115-2128.

Abelson, J., Forest, P.-G., Eyles, J., Smith, P., Martin, E., and Gauvin, F.-P.
   2003   Deliberations about deliberative methods: Issues in the design and evaluation of public participation processes. *Social Science & Medicine, 57,* 239-251.

Acheson, D.
   1941   *Final Report of the Attorney General's Committee on Administrative Procedure.* Washington, DC: U.S. Government Printing Office.

Ackerman, B., and Fishkin, J.S.
   2004   *Deliberation Day.* New Haven, CT: Yale University Press.

Adler, P.S., and Birkhoff, J.
   2002   *Building Trust: When Knowledge from Here Meets Knowledge from Away.* Portland, OR: National Policy Consensus Center. Available: http://www.keystone.org/spp/documents/Building_trust.pdf [accessed June 2008].

Advisory Commission on Intergovernmental Relations
   1979   *Citizen Participation in the American Federal System.* Washington, DC: Author.

Agrawal, A.
  2002 Common resources and institutional sustainability. In National Research Council, *The Drama of the Commons* (pp. 41-85). Committee on the Human Dimensions of Global Change. E. Ostrom, T. Dietz., N. Dolsak, P.C. Stern, S. Stonich, and E.U. Weber (Eds.). Division of Behavioral and Social Sciences and Education. Washington, DC: National Academy Press.

Andresen, S., Skodvin, T., Underdal, A., and Wettestad, J. (Eds.)
  2000 *Science and Politics in International Environmental Regimes: Between Integrity and Involvement.* New York: Manchester University Press.

Applegate, J.S., and Dycus. S.
  1998 Institutional controls or emperor's clothes? Long-term stewardship on the nuclear weapons complex. *The Environmental Law Reporter News & Analysis, 28*(11), 10631-10652.

Arctic Climate Impact Assessment
  2004 *Impacts of a Warming Arctic: Arctic Climate Impact Assessment.* Cambridge, England: Cambridge University Press.

Argryris, C.
  1982 The executive mind and double-loop learning. *Organizational Dynamics*, Autumn, 5-22.

Arkes, H.R.
  1991 Costs and benefits of judgment errors: Implications for debiasing. *Psychological Bulletin, 110,* 486-498.
  2003 The nonuse of psychological research at two federal agencies. *Psychological Science, 14,* 1-6.

Arnstein, S.
  1969 A ladder of participation. *Journal of the American Institute of Planners, 5,* 216-224.

Aron, J.B.
  1979 Citizen participation at government expense. *Public Administration Review, 39,* 477-485.

Aronoff, M., and Gunter, V.
  1992 Defining disaster: Local constructions for recovery in the aftermath of chemical contamination. *Social Problems, 39,* 345-365.
  1994 A pound of cure: Facilitating participatory processes in technological hazard disputes. *Society and Natural Resources, 7,* 235-252.

Arrow, K.J.
  1951 *Social Choice and Individual Values.* New Haven, CT: Yale University Press.

Arvai, J.L.
  2003 Using risk communication to disclose the outcome of a participatory decision-making process: Effects on the perceived acceptability of risk-policy decisions. *Risk Analysis, 23,* 281-289.

Ashford, N.A., and Rest, K.M.
  1999 *Public Participation in Contaminated Communities.* Cambridge: Massachusetts Institute of Technology Center for Technology, Policy, and Industrial Development. Available: http://web.mit.edu/ctpid/www/tl/TL-pub-PPCC.html [accessed June 2008].

Association of American Geographers
  2003 *Global Change and Local Places: Estimating, Understanding, and Reducing Greenhouse Gases.* New York: Cambridge University Press.

REFERENCES

Attorney General's Commission on Administrative Procedure
    1941   *Final Report of the Attorney General's Commission on Administrative Procedure.* Washington, DC: Department of Justice.

Aven, T.
    2003   *Foundations of Risk Analysis: A Knowledge and Decision-Oriented Perspective.* Chichester, England: Wiley.

Avery, M., Streibel, B.J., and Auvine, B.
    1981   *Building United Judgment: A Handbook for Consensus Decision Making.* Madison, WI: Center for Conflict Resolution.

Bacow, L.S., and Wheeler, M.
    1984   *Environmental Dispute Resolution.* New York: Plenum Press.

Baland, J., and Platteau, J.
    1996   *Halting Degradation of Natural Resources: Is There a Role for Rural Communities?* Oxford, England: Clarendon Press.

Barber, B.R.
    1984   *Strong Democracy: Participatory Politics for a New Age.* Berkeley: University of California Press.

Barke, R.P., and Jenkins-Smith, H.C.
    1993   Politics and scientific expertise: Scientists, risk perception, and nuclear waste policy. *Risk Analysis, 13,* 425-439.

Baron, J., and Hershey, J.C.
    1988   Outcome bias in decision evaluation. *Journal of Personality and Social Psychology, 54,* 569-579.

Baron, J., and Spranca, M.
    1997   Protected values. *Organizational Behavior and Human Decision Processes, 70,* 1-16.

Baron, J., Hershey, J.C., and Kunreuther, H.
    2000   Determinants of priority for risk reductions: The role of worry. *Risk Analysis, 20,* 413-427.

Bartos, O.J.
    1974   *Process and Outcome of Negotiations.* New York: Columbia University Press.

Baughman, M.
    1995   Mediation. In O. Renn, T. Webler, and P. Wiedemann (Eds.), *Fairness and Competence in Citizen Participation: Evaluating New Models for Environmental Discourse* (pp. 253-266). Boston, MA: Kluwer Academic.

Baumgartner, F., and Jones, B.D.
    1993   *Agendas and Instability in American Politics.* Chicago: University of Chicago Press.

Baumol, W.J.
    1986   *Superfairness: Applications and Theory.* Cambridge, MA: MIT Press.

Bayley, C., and French, S.
    2008   Designing a participatory process for stakeholder involvement in a societal decision. *Group Decision and Negotiation, 17,* 195-210.

Bazerman, M.H., Curhan, J.R., Moore, D.A., and Valley, K.L.
    2000   Negotiation. *Annual Review of Psychology, 51,* 279-314.

Bechtel, R.B., Verdugo, V.C., and Pinheiro, J.D.Q.
    1999   Environmental belief systems: United States, Brazil, and Mexico. *Journal of Cross-Cultural Psychology, 30,* 122-128.

Behn, R.D., and Vaupel, J.V.
    1982   *Quick Analysis for Busy Decision Makers.* New York: Basic Books.

Beierle, T.C.
2000 *The Quality of Stakeholder-Based Decisions: Lessons from Case Study Record.* (Discussion paper #00-56.) Washington, DC: Resources for the Future.
2002 *Democracy On-Line: An Evaluation of the National Public Dialogue on Public Involvement in EPA Decisions.* Washington, DC: Resources for the Future.

Beierle, T.C., and Cayford, J.
2002 *Democracy in Practice: Public Participation in Environmental Decisions.* Washington, DC: Resources for the Future.

Beierle, T.C., and Konisky, D.M.
1999 *Public Participation in Environmental Planning in the Great Lakes Region.* (Discussion paper #99-50.) Washington, DC: Resources for the Future.
2000 Values, conflict, and trust in participatory environmental planning. *Journal of Policy Analysis and Management, 19,* 587-602.

Benhabib, S.
1992 Autonomy, modernity, and community: Communitarianism and critical theory in dialogue. In A. Honneth, T. McCarthy, C. Offe, and A. Wellmer (Eds.), *Cultural-Political Interventions in the Unfinished Project of Enlightenment* (pp. 39-61). Cambridge, MA: MIT Press.

Bentkover, J.D., Covello, V.T., and Mumpower, J. (Eds.)
1985 *Benefits Assessment: The State of the Art.* Dordrecht, the Netherlands: D. Reidel.

Bercovitch, J.
1984 *Social Conflict and Third Parties: Strategies of Conflict Resolution.* Boulder, CO: Westview Press.

Berkes, F.
2002 Cross-scale institutional linkages: Perspectives from the bottom up. In National Research Council, *The Drama of the Commons* (pp. 292-321). Committee on the Human Dimensions of Global Change. E. Ostrom, T. Dietz, N. Dolsak, P.C. Stern, S. Stonich, and E.U. Weber (Eds.). Division of Behavioral and Social Sciences and Education. Washington, DC: National Academy Press.

Berkes, F., Mahon, R., McConney, P., Pollnac, R.C., and Pomeroy, R.S.
2001 *Managing Small-Scale Fisheries: Alternative Directions and Methods.* Ottawa: International Development Research Centre.

Berry, J.B., Portney, K.E., and Thomson, K.
1993 *The Rebirth of Urban Democracy.* Washington, DC: Brookings Institution Press.

Besley, J.C., and McComas, K.A.
2005 Framing justice: Using the concept of procedural justice to advance political communication research. *Communication Theory, 15,* 414-436.

Bingham, G.
1986 *Resolving Environmental Disputes: A Decade of Experience.* Washington, DC: The Conservation Foundation.
2003 *When the Sparks Fly: Building Consensus When the Science Is Contested.* Washington, DC: Resolve.

Birkhoff, J., and Bingham, G.
2004 *Conflict Management and Dispute Resolution.* Paper prepared for the National Research Council Panel on Public Participation in Environmental Assessment and Decision Making, April, Washington, DC.

Black, D.T.
1948 On the rationale of group decision making. *Journal of Political Economy, 56,* 23-34.

Blackstock, K.L., Kelly, G.J., and Horsey, B.L.
  2007   Developing and applying a framework to evaluate participatory research for sustainability. *Ecological Economics*, 60, 726-742.
Blader, S.L., and Tyler, T.R.
  2003   A four-component model of procedural justice: Defining the meaning of a "fair" process. *Personality & Social Psychology Bulletin*, 29, 747-758.
Bleiker, A., and Bleiker, H.
  2000   *Citizen Participation Handbook for Public Officials and Other Professionals Serving the Public* (13th ed.). Monterey, CA: Institute for Participatory Management and Planning.
Blundel, R.
  2004   *Effective Organizational Communication*. London, England: Pearson.
Bobbio, N.
  1987   *The Future of Democracy*. Cambridge, England: Polity Press.
Bohm, G., and Pfister, H.R.
  2000   Action tendencies and characteristics of environmental risks. *Acta Psychologica*, 104, 317-337.
Bohman, J.
  1997   Deliberative democracy and effective social freedom: Capabilities, resources, and opportunities. In J. Bohman and W. Rehg (Eds.), *Deliberative Democracy: Essays on Reason and Politics* (pp. 321-348). Cambridge, MA: MIT Press.
Boholm, A.
  1998   Comparative studies of risk perception: A review of twenty years of research. *Journal of Risk Research*, 1(2), 135-163.
Boix, C., and Posner, D.N.
  1998   Social capital: Explaining its origins and effects on government performance. *British Journal of Political Science*, 28(4), 686-693.
Bora, A., and Hausendorf, H.
  2006   Participatory science governance revisited: Normative expectations versus empirical evidence. *Science and Public Policy*, 33, 478-488.
Borcherding, K., Rohrmann, B., and Eppel, T.
  1986   A psychological study on the cognitive structure of risk evaluations. In B. Brehmer, H. Jungermann, P. Lourens, and G. Sevon (Eds.), *New Directions in Research on Decision Making* (pp. 245-262). Amsterdam, the Netherlands: North-Holland.
Bord, R.J., and O'Connor, R.E.
  1997   The gender gap in environmental attitudes: The case of perceived vulnerability to risk. *Social Science Quarterly*, 78, 830-840.
Bornstein, G.
  1992   The free-rider problem in intergroup conflicts over step-level and continuous public goods. *Journal of Personality and Social Psychology*, 62, 597-606.
Bourdieu, P.
  1985   The forms of capital. In J.G. Richardson (Ed.), *Handbook of Theory and Research for the Sociology of Education* (pp. 241-258). New York: Greenwood.
Bowles, S., and Gintis, H.
  1986   *Democracy and Capitalism: Property, Community, and the Contradictions of Modern Social Thought*. New York: Basic Books.
Boyer, B., and Meidinger, E.
  1985   Privatizing regulatory enforcement: A preliminary assessment of citizen suits under federal environmental laws. *Buffalo Law Review*, 35, 834-965.

Boyte, H.C.
- 1980 *The Backyard Revolution: Understanding the New Citizen Movement.* Philadelphia: Temple University Press.
- 2004 *Everyday Politics: Reconnecting Citizens and Public Life.* Philadelphia: University of Pennsylvania Press.

Bradbury, J.
- 1989 The policy implications of differing concepts of risk. *Science, Technology, and Human Values,* 14(4), 380-399.
- 2005 *Evaluating Public Participation in Environmental Decisions.* Paper prepared for the National Research Council Panel on Public Participation in Environmental Assessment and Decision Making, Feb. 3-5, National Academy of Sciences, Washington, DC. Available: http://www7.nationalacademies.org/hdgc/Bradbury%20Text.pdf [accessed June 2008].

Bradbury, J., and Branch, K.M.
- 1999 *An Evaluation of the Effectiveness of Local Site-Specific Advisory Boards for the U.S. Department of Energy Environmental Restoration Programs.* (PNNL-12139.) Richland, WA: Pacific Northwest National Laboratory. Available: http://www.osti.gov/bridge/servlets/purl/4269-c1tCdi/webviewable/4269.PDF [accessed June 2008].

Bradbury, J., Branch, K., and Malone, E.
- 2003 *An Evaluation of DOE-EM Public Participation Programs.* (Report prepared for the DOE Environmental Management Program, PNNL-14200.) Richland, WA: Pacific Northwest National Laboratory.

Bradbury, J., Branch, K., Santos, S., and Chess, C.
- 2005 *An Evaluation and Recommendations for Enhancing U.S. Army Restoration Advisory Boards.* (Revised final report.) Fort Belvoir, VA: Department of Defense, Defense Technical Information Center.

Bramwell, R., West, H., and Salmon, P.
- 2006 Health professionals' and service users' interpretation of screening test results: Experimental study. *British Medical Journal,* 333(7562), 284-286.

Branch, K., and Bradbury, J.
- 2006 Comparison of DOE and Army advisory boards: Application of a conceptual framework for evaluating public participation in environmental risk decision making. *Policy Studies Journal,* 34(4), 723-754.

Brashers, D.E.
- 2001 Communication and uncertainty management. *Journal of Communication,* 51, 477-497.

Brewer, G.D
- 2007 Inventing the future: Scenarios, imagination, mastery and control. *Sustainability Science,* 2, 159-177.

Breyer, S.
- 1993 *Breaking the Vicious Circle: Toward Effective Risk Regulation.* Cambridge, MA: Harvard University Press.

Briggs, X. de Sousa
- 1998 Doing democracy up-close: Culture, power, and communication in community building. *Journal of Planning Education and Research,* 18(1), 1-13.

Brouwer, R., Powe, N., Turner, R.K., Bateman, I.J., and Langford, I.H.
- 1999 Public attitudes to contingent valuation and public consultation. *Environmental Values,* 8(3), 325-347.

Brown, P.
- 1992 Popular epidemiology and toxic waste contamination: Lay and professional ways of knowing. *Journal of Health and Social Behavior,* 33, 267-281.

Brown, P., and Mikkelsen, E.J.
  1990   *No Safe Place: Toxic Waste, Leukemia, and Community Action.* Berkeley: University of California Press.
Brown, P., Mayer, B., Zavetoski, S., Luebke, T., Mandlebaum, J., and McCormick, S.
  2003   The politics of asthma suffering: Environmental justice and the social movement transformation of illness experience. *Social Science and Medicine,* 57(3), 453-464.
Brown, P., McCormick, S., Mayer, B., Zavetoski, S., Morello-Frosch, R., Altman, R.G., and Senier, L.
  2006   "A lab of our own": Environmental causation of breast cancer and challenges to the dominant epidemiological paradigm. *Science, Technology, and Human Values,* 31(5), 499-536.
Brown, P., Zavetoski, S., Mayer, B., McCormick, S., and Webster, P.
  2002   Policy issues in environmental health disputes. *Annals of the American Academy of Political and Social Science,* 584, 175-202.
Brown, P., Zavetoski, S., McCormick, S., Mayer, B., Morello-Frosch, R., and Gasior, R.
  2004   Embodied health movements: New approaches to social movements in health. *Sociology of Health and Illness,* 26(1), 50-80.
Brulle, R.J.
  1992   Jürgen Habermas: An exegesis for human ecologists. *Human Ecology Bulletin,* 8(Spring/Summer), 29-40.
  1994   Power, discourse, and social problems: Social problems from a rhetorical perspective. *Current Perspectives in Social Problems,* 5, 95-121.
  2000   *Agency, Democracy, and Nature: The U.S. Environmental Movement from a Critical Theory Perspective.* Cambridge, MA: MIT Press.
Brun, F., and Buttoud, G.
  2003   *The Formulation of Integrated Management Plans (IMPs) for Mountains Forests.* Quaderni of the Department of Economics and Agricultural Engineering, Forestry and Environmental University of Turin.
Bryan, F.M.
  2004   *Real Democracy: The New England Town Meeting and How It Works.* Chicago: University of Chicago Press.
Bryant, B. (Ed.)
  1995   *Environmental Justice: Issues, Policies, and Solutions.* Washington, DC: Island Press.
Bullard, R.D.
  1990   *Dumping in Dixie: Race, Class and Environmental Quality.* Boulder, CO: Westview Press.
Bullard, R.D., and Johnson, G.S.
  2000   Environmentalism and public policy environmental justice: Grassroots activism and its impact on public policy decision making. *Journal of Social Issues,* 56(3), 555-578.
Burgess, J., and Clark, J.
  2006   Evaluating public and stakeholder engagement strategies in environmental governance. In A.G. Peirez, S.G. Vaz, and S. Tognetti (Eds.), *Interfaces Between Science and Society* (Chapter 13). Sheffield, England: Greenleaf.
Burgess, J., Stirling, A., Clark, J., Davies, G., Eames, M., Staley, C., and Williamson, S.
  2007   Deliberative mapping: A novel analytic-deliberative methodology to support contested science-policy decisions. *Public Understanding of Science,* 16, 299-322.
Burns, T.R., and Überhorst, R.
  1988   *Creative Democracy: Systematic Conflict Resolution and Policymaking in a World of High Science and Technology.* New York: Praeger.

Cacioppo, J.T., Petty, R.E., Feinstein, J., and Jarvis, W.B.G.
  1996 Dispositional differences in cognitive motivation: The life and times of individuals varying in need for cognition. *Psychological Bulletin, 119*, 197-253.
Caldwell, L.K.
  1998 *The National Environmental Policy Act: An Agenda for the Future*. Bloomington: Indiana University Press.
Campbell, D.T.
  1969 Reforms as experiments. *American Psychologist, 24*, 409-429.
Campbell, M.C.
  2003 Intractability in environmental disputes: Exploring a complex construct. *Journal of Planning Literature, 17*, 360-371.
Campbell, S., and Currie, G.
  2006 Against Beck: In defense of risk analysis. *Philosophy of the Social Sciences, 36*(2), 149-172.
Canadian Round Tables
  1993 *Building Consensus for a Sustainable Future: Guiding Principles* (1st ed.). Ottawa: National Round Table on the Environment and Economy.
Capella, J., Price, V., and Nir, L.
  2002 Argument repertoire as a reliable and valid measure of opinion quality: Electronic dialogue in campaign 2000. *Political Communication, 19*, 73-93.
Carnes, S.A., Schweitzer, M., Peelle, E.B., Wolfe, A.K., and Munro, J.F.
  1998 Measuring the success of public participation on environmental restoration and waste management activities in the U.S. Department of Energy. *Technology in Society, 20*(4), 385-406.
Carnevale, P.J.D., and Pruitt, D.G.
  1992 Negotiation and mediation. *Annual Review of Psychology, 43*, 511-582.
Carpenter, S., and Kennedy, W.J.D.
  1988 *Managing Public Disputes*. San Francisco, CA: Jossey-Bass.
Carr, D.S., Selin, S.W., and Schuett, M.A.
  1998 Managing public forests: Understanding the role of collaborative planning. *Environmental Management, 22*(5), 767-776.
Cestero, B.
  1999 *Beyond the Hundredth Meeting: A Field Guide to Collaborative Conservation on the West's Public Lands*. Tucson, AZ: The Sonoran Institute.
Chakraborty, J., and Bosman, M.M.
  2005 Measuring the digital divide in the United States: Race, income, and personal computer ownership. *The Professional Geographer, 57*(3), 395-410.
Chambers, R.
  1997 *Whose Reality Counts?: Putting the First Last*. London, England: Intermediate Technology Publications.
  2005 *Ideas for Development*. London, England: Earthscan.
Chambers, S.
  1996 *Reasonable Democracy*. Ithaca, NY: Cornell University Press.
  2003 Deliberative democratic theory. *Annual Review of Political Science, 6*, 307-326.
Charnley, G.
  2003 How the Risk Commission evolved from the Red Book. *Human and Ecological Risk Assessment, 9*, 1213-1218.
Chess, C.
  2001 Organizational theory and the stages of risk communication. *Risk Analysis, 21*(1), 179-188.

Chess, C., and Johnson, B.B.
   2006   Organizational learning about public participation: "Tiggers" and "Eeyores." *Human Ecology Review, 13*(2), 182-192.
Chess, C., Dietz, T., and Shannon, M.
   1998   Who should deliberate when? *Human Ecology Review, 5*, 45-48.
Chess, C., Tamuz, M., and Greenberg, M.
   1995   Organizational learning about environmental risk communication: The case of Rohm and Haas' Bristol plant. *Society and Natural Resources, 8*, 57-66.
Chilvers, J.
   2005   Towards analytic-deliberative forms of risk governance in the UK? Reflections on learning in radioactive waste. *Journal of Risk Research, 10*(2), 197-222.
   2008   Deliberating competence. Theoretical and practitioner perspectives on effective participatory appraisal practice. *Science, Technology, and Human Values, 33*(2), 155-185.
Chrislip, D.
   1994   *Collaborative Leadership: How Citizens and Civic Leaders Can Make a Difference.* San Francisco, CA: Jossey-Bass.
Clark, J., Burgess, J., and Harrison, C.M.
   2000   "I struggled with this money business": Respondents' perspectives on contingent valuation. *Ecological Economics, 33*(1), 45-62.
Clark, W.C., Mitchell, R.B., and Cash, D.W.
   2006   Evaluating the influence of global environmental assessments. In R.B. Mitchell, W.C. Clark, D.W. Cash, and N.M. Dickson (Eds.), *Global Environmental Assessments: Information and Influence* (pp. 1-28). Cambridge, MA: MIT Press.
Clarke, J.N., and McCool, D.
   1985   *Staking out the Terrain: Power Differentials Among Natural Resource Management Agencies.* Albany: State University of New York.
Clayton, S., and Opotow, S.
   2003   Justice and identity: Perspectives on what is fair. *Personality and Social Psychology Review, 7*, 298-310.
Clemen, R.
   1996   *Making Hard Decisions: An Introduction to Decision Analysis.* Belmont, CA: Duxbury Press.
Cobb, S., and Rifkin, J.
   1991   Practice and paradox: Deconstructing neutrality in mediation. *Law and Social Inquiry, 16*, 35-62.
Coburn, J.
   2005   *Street Science: Community Knowledge and Environmental Health Justice.* Cambridge, MA: MIT Press.
Coglianese, C.
   1996   Litigating within relationships: Disputes and disturbance in the regulatory process. *Law and Society Review, 30*(4), 735-766.
   1997   Assessing consensus: The promise and performance of negotiated rulemaking. *Duke Law Journal, 46*(6), 1255-1349.
   1999   Limits of consensus. *Environment, 41*, 28-33.
   2000   *Is Consensus an Appropriate Basis for Regulatory Policy?* (KSG Faculty Research Working Paper Series RWP01-012.) Harvard University, Kennedy School of Government, Boston, MA. Available: http://ksgnotes1.harvard.edu/Research/wpaper.nsf/rwp/RWP01-012/$File/rwp01_012_coglianese.pdf [accessed June 2008].

2003a Does consensus work? A pragmatic approach to public participation in the regulatory process. In A. Morales (Ed.), *Renascent Pragmatism: Studies in Law and Social Science* (pp. 180-195). Burlington, VT: Ashgate Press.
2003b Is satisfaction success? Evaluating public participation in regulatory policymaking. In R. O'Leary and L. Bingham (Eds.), *The Promise and Performance of Environmental Conflict Resolution* (pp. 69-89). Washington, DC: Resources for the Future.
2003c *The Internet and Public Participation in Rulemaking.* Cambridge, MA: Harvard University, Center for Business and Government.

Coleman, J.S.
1957 *Community Conflict.* New York: Free Press.
1988 Social capital in the creation of human capital. *American Journal of Sociology,* 94, 95-120.
1990 *Foundations of Social Theory.* Cambridge, MA: Harvard University Press.

Collins, H.M., and Evans, R.
2002 The third wave of science studies: Studies of expertise and experience. *Social Studies of Science,* 32, 235-296.

Condorcet, M.J.A.N. de C., Marquis de
1785 Essai sur l'application de l'analyse à la probabilité des decisions rendues à la pluralité des voix. Paris: l'Imprimerie Royale. [facsimile ed., New York: Chelsea House, 1972.]

Consensus Building Institute
1999 *Using Assisted Negotiation to Settle Land Use Disputes: A Guidebook for Public Officials.* Washington, DC: Author.

Coppock, R.
1985 Interactions between scientists and public officials: A comparison of the use of science in regulatory programs in the United States and West Germany. *Policy Sciences,* 18, 371-390.

Cormick, G.W.
1980 The "theory" and practice of environmental mediation. *The Environmental Professional,* 2(1), 24-33.

Corrigan, P., and Joyce, P.
1997 Five arguments for deliberative democracy. *Political Studies,* 48, 947-969.

Cortner, H.J.
1996 Public Involvement and Interaction. In A.W. Ewert (Ed.), *Natural Resource Management: The Human Dimension* (pp. 167-180). Boulder, CO: Westview Press.

Cortner, H.J., and Shannon, M.A.
1993 Embedding public participation in its political context. *Journal of Forestry,* 91(7), 14-16.

Coser, L.A.
1956 *The Function of Social Conflict.* New York: Free Press.

Cosmides, L., and Tooby, J.
1996 Are humans good intuitive statisticians after all? Rethinking some conclusions from the literature on judgment under uncertainty. *Cognition,* 58, 1-73.

Couch, S.R., and Kroll-Smith, S.
1994 Environmental controversies, interactional resources, and rural communities: Siting versus exposure disputes. *Rural Sociology,* 59, 25-44.
1997 Environmental movement and expert knowledge: Evidence of a new populism. *International Journal of Contemporary Sociology,* 34(2), 185-211.

Covello, V.T.
1983 The perception of technological risks: A literature review. *Technological Forecasting and Social Change,* 23, 285-297.

Covello, V.T., Peters, R.G., Wojtecki, J.G., and Hyde, R.C.
  2001   Risk communication, the West Nile virus epidemic, and bioterrorism: Responding to the communication challenges posed by the intentional or unintentional release of a pathogen in an urban setting. *Journal of Urban Health: Bulletin of the New York Academy of Medicine*, 78, 382-391. Available: http://www.centerforriskcommunication.com/publications.htm [accessed June 2008].

Cox, R.T.
  1961   *The Algebra of Probable Inference*. Baltimore, MD: Johns Hopkins University Press.

Cramer, J.C., Dietz, T., and Johnston, R.
  1980   Social impact assessment of regional plans: A review of methods and a recommended process. *Policy Sciences*, 12, 61-82.

Cramton, R.C.
  1971   The why, where, and how of broadened participation in the administrative process. *Georgetown Law Journal*, 60, 525-550.

Creighton, J.L.
  1992   *Involving Citizens in Community Decision Making: A Guidebook*. Washington, DC: Program for Community Problem Solving, National League of Cities.
  1999   *How to Design a Public Participation Program*. Washington, DC: U.S. Department of Energy, Office of Intergovernmental and Public Accountability.
  2005   *The Public Participation Handbook: Making Better Decisions Through Citizen Involvement*. San Francisco: Jossey-Bass.

Cross, F.B.
  1992   The risk of reliance on perceived risk. *Risk Issues in Health and Safety*, 3, 59-70.
  1998   Facts and values in risk assessment. *Reliability Engineering and Systems Safety*, 59, 27-45.

Crowfoot, J., and Wondolleck, J.
  1990   *Environmental Disputes: Community Involvement in Conflict Resolution*. Washington, DC: Island Press.

Cupps, D.S.
  1977   Emerging problems of citizen participation. *Public Administration Review*, 37, 478-487.

Dahl, R.A.
  1989   *Democracy and Its Critics*. New Haven, CT: Yale University Press.
  1998   *On Democracy*. New Haven, CT: Yale University Press.

Dana, D.A.
  1994   The promise of a bureaucratic solution: Breaking the vicious circle toward effective risk regulation. *Boston University Law Review*, 74, 365-372.

Daneke, G.A.
  1983   Public involvement: Why, what, how? In G.A. Daneke, M.W. Garcia, and J. Delli Priscoli (Eds.), *Public Involvement and Social Impact Assessment* (pp. 11-33). Boulder, CO: Westview Press.

Daniels, S.E., and Walker, G.B.
  1997   *Rethinking Public Participation in Natural Resource Management: Concepts from Pluralism and Five Emerging Approaches*. Paper prepared for the Food and Agriculture Organization Working Group on Pluralism and Sustainable Forestry and Rural Development, December 9-12, Rome, Italy. Available: http://www.mtnforum.org/oldocs/260.doc [accessed April 2008].
  2001   *Working Through Environmental Conflict: The Collaborative Learning Approach*. Westport, CT: Praeger.

Davidson, D.J., and Freudenburg, W.R.
    1996   Gender and environmental risk concerns: A review and analysis of available research. *Environment and Behavior, 28*, 302-339.

Dawes, R.M., van de Kragt, A.J.C., and Orbell, J.M.
    1990   Cooperation for the benefit of us: Not me, or my conscience. In J. Mansbridge (Ed.), *Beyond Self-Interest* (pp. 97-110). Chicago: University of Chicago Press.

DeKay, M., and Vaughan, E.
    2005   *Individual Judgment and Decision-Making Processes: Influences on Public Participation in Environmental Assessment and Decision Making.* Paper prepared for the National Research Council Panel on Public Participation in Environmental Assessment and Decision Making, Feb. 3-5, National Academy of Sciences, Washington, DC.

DeKay, M.L., Small, M.J., Fischbeck, P.S., Farrow, R.S., Cullen, A., Kadane, J.B., Lave, L.B., Morgan, M.G., and Takemura, K.
    2002   Risk-based decision analysis in support of precautionary policies. *Journal of Risk Research, 5*, 391-417.

Delfin, F.G., and Tang, S.-Y.
    2005   Elitism, pluralism, or resource dependency: Patterns of environmental philanthropy among private foundations in California. *Environment and Planning A, 39*(9), 2167-2186.

Delli Carpini, M.X., Cook, F.L., and Jacobs, L.R.
    2004   Public deliberation, discursive participation, and citizen engagement: A review of the empirical literature. *Annual Review of Political Science, 7*, 315-344.

Delli Priscoli, J.
    1983   The citizen advisory group as an integrative tool in regional water resources planning. In G. Daneke, M. Garcia, and J. Delli Priscoli (Eds.), *Public Involvement and Social Impact Assessment* (pp. 79-87). Boulder, CO: Westview Press.

Deutsch, M.
    1973   *The Resolution of Conflict: Constructive and Destructive Processes.* New Haven, CT: Yale University Press.

Dewey, J.
    1923   *The Public and Its Problems.* New York: Henry Holt.

Diani, M., and McAdam, D. (Eds.)
    2003   *Social Movements and Networks: Relational Approaches to Collective Action.* New York: Oxford University Press.

Dienel, P.C., and Renn, O.
    1995   Planning cells: A gate to "fractal mediation." In O. Renn, T. Webler, and P. Wiedemann (Eds.), *Fairness and Competence in Citizen Participation: Evaluating New Models for Environmental Discourse* (pp. 117-140). Boston, MA: Kluwer Academic.

Dietz, T.
    1987   Theory and method in social impact assessment. *Sociological Inquiry, 57*, 54-69.
    1988   Social impact assessment as applied human ecology: Integrating theory and method. In R. Borden, J. Jacobs, and G.R. Young (Eds.), *Human Ecology: Research and Applications* (pp. 220-227). College Park, MD: Society for Human Ecology.
    1994   What should we do? Human ecology and collective decision making. *Human Ecology Review, 1*, 301-309.
    2001   Thinking about environmental conflict. In L. Kadous (Ed.), *Celebrating Scholarship* (pp. 31-54). Fairfax, VA: George Mason University.
    2003   What is a good decision? Criteria for environmental decision making. *Human Ecology Review, 10*(1), 60-67.

Dietz, T., and Pfund, A.
   1988   An impact identification method for development program evaluation. *Policy Studies Review, 8*, 137-145.

Dietz, T., and Stern, P.C.
   1995   Toward realistic models of individual choice. *Journal of Socio-Economics, 24*, 261-279.
   1998   Science, values, and biodiversity. *BioScience, 48*, 441-444.
   2005   *Further Analysis of the Democracy in Practice Database.* Paper prepared for the National Research Council Panel on Public Participation in Environmental Assessment and Decision Making, Feb. 3-5, National Academy of Sciences, Washington, DC.

Dietz, T., Fitzgerald, A., and Shwom, R.
   2005   Environmental values. *Annual Review of Environment and Resources, 30*, 335-372.

Dietz, T., Ostrom, E., and Stern, P.C.
   2003   The struggle to govern the commons. *Science, 301*, 1907-1912.

Dietz, T., Stern, P.C., and Dan, A.
   In press   How deliberation effects stated willingness to pay for mitigation of carbon dioxide emissions: An experiment. Submitted to *Land Economics*.

Dietz, T., Stern, P.C., and Rycroft, R.W.
   1989   Definitions of conflict and the legitimization of resources: The case of environmental risk. *Sociological Forum, 4*, 47-70.

Dietz, T., Tanguay, J., Tuler, S., and Webler, T.
   2004   Making computer models useful: An exploration of expectations by experts and local officials. *Coastal Management, 32*(3), 307-318.

DiGiovanni, C.
   1999   Domestic terrorism with chemical or biological agents: Psychiatric Aspects. *American Journal of Psychiatry, 156*, 1500-1505.

Dingwall, R., and Greatbatch, D.
   1993   Who is in charge? Rhetoric and reality in the study of mediation. *Journal of Social Welfare and Family Law*, 367-385.

Donohue, W.
   1991   *Communication, Marital Dispute and Divorce Mediation.* Hillsdale, NJ: Lawrence Erlbaum Associates.

Doppelt, B., Shinn, C., and John, D.
   2002   *Review of USDA Forest Service Community-Based Watershed Restoration Partnerships.* Portland, OR: Portland State University, Center for Watershed and Community Health, Mark O. Hatfield School of Government. Available: http://www.fs.fed.us/largewatershedprojects/DoppeltReport/index.html [accessed June 2008].

Downs, A.
   1957   *An Economic Theory of Democracy.* London, England: HarperCollins.

Doyle, M., and Straus, D.
   1993   *How to Make Meetings Work: The New Interaction Method.* Berkeley, CA: Berkeley Publishing Group.

Dryzek, J.S.
   1990   *Discursive Democracy: Politics, Policy, and Political Science.* Cambridge, England: Cambridge University Press.
   1994a   *Discursive Democracy. Politics, Policy, and Political Science* (2nd ed.). Cambridge, England: Cambridge University Press.
   1994b   Ecology and discursive democracy: Beyond liberal capitalism and the administrative state. In M. O'Connor (Ed.), *Is Capitalism Sustainable? Political Economy and the Politics of Ecology* (pp. 176-197). New York: Guilford Press.

Dryzek, J., and List, C.
2003 Social choice theory and deliberative democracy: A reconciliation. *British Journal of Political Science, 33*(1), 1-28.

Dukes, E.F.
1996 *Resolving Public Conflict: Transforming Community and Governance.* New York: Manchester University Press.
2004 What we know about environmental conflict resolution: An analysis based on research. *Conflict Resolution Quarterly, 22*(1-2), 191-220.

Dukes, F., and Firehock, K.
2001 *Collaboration: A Guide for Environmental Advocates.* (A publication of the University of Virginia's Institute for Environmental Negotiation, the Wilderness Society, and the National Audubon Society.) Charlottesville: University of Virginia Press.

Duram, L.A., and Brown, K.G.
1998 Assessing public participation in U.S. watershed planning initiatives. *Society and Natural Resources, 12*(5), 455-467.

Durant, R.F., Fiorino, D.J., and O'Leary, R.
2004 *Environmental Governance Reconsidered: Challenges, Choices, and Opportunities.* Cambridge, MA: MIT Press.

Durodie, B.
2003 The true cost of precautionary chemicals regulation. *Risk Analysis, 23*, 389-398.

Dwyer, W.L.
1994 *Seattle Audubon Society et al. v. James Lyons, Assistant Secretary of Agriculture et al.* Order on Motions for Summary Judgment. RE 1994 Forest Plan, U.S. District Court, Western District of Washington at Seattle.

Edwards, H.T.
1986 Alternative dispute resolution: Panacea or anathema? *Harvard Law Review, 99*, 668-681.

Edwards, P.N., and Schneider, S.H.
2001 Self-governance and peer review in science-for-policy: The case of the IPCC Second Assessment Report. In C. Miller and P.N. Edwards (Eds.). *Changing the Atmosphere: Expert Knowledge and Environmental Governance.* Cambridge, MA: MIT Press. Available: http://stephenschneider.stanford.edu/Mediarology/Mediarology.html [accessed September 2008].

Einhorn, H.J., and Hogarth, R.M.
1978 Confidence in judgment: Persistence of the illusion of validity. *Psychological Review, 85*, 395-416.

English, M., Gibson, A., Felman, D., and Tonn, B.
1993 *Stakeholder Involvement: Open Processes for Reaching Decisions About the Future Use of Contaminated Sites.* Final Report to the D.O.E. Knoxville, TN, Waste Management Research and Education Institute, University of Tennessee.

Epstein, S.
2000 The rationality debate from the perspective of the cognitive-experiential self-theory. *Behavioral & Brain Sciences, 23*, 671-672.

Estrella, M., and Gaventa, J.
1998 *Who Counts Reality? Participatory Monitoring and Evaluation: A Literature Review.* (IDS working paper #70.) Brighton, England: University of Sussex, Institute of Development Studies.

Ethridge, M.E.
1987 Procedures for citizen involvement in environmental policy: An assessment of policy effects. In J. DeSario and S. Langton (Eds.), *Citizen Participation in Public Decision Making* (pp. 115-132). Westport, CT: Greenwood Press.

REFERENCES

Evans, S.M., and Boyte, H.C.
- 1992 *Free Spaces: The Sources of Democratic Change in America.* Chicago: University of Chicago Press.

Farrar, C., Fishkin, J.S., Green, D.P., List, C., Luskin, R.C., and Paluck, E.L.
- 2003 *Experimenting with Deliberative Democracy: Effects on Policy Preferences and Social Choice.* Paper presented at the European Consortium for Political Research General Conference, Sept., Marburg, Germany. Available: http://cdd.stanford.edu/research/papers/2003/experimenting.pdf [accessed June 2008].
- 2006 *Disaggregating Deliberation's Effects: An Experiment Within a Deliberative Poll.* Center for Deliberative Democracy, Stanford University. Available: http://cdd.stanford.edu/research/papers/2006/nh-disaggregating.pdf [accessed June 2008].

Farrell, A.E., and Jäger, J. (Eds.)
- 2006 *Assessments of Regional and Global Environmental Risks: Designing Processes for the Effective Use of Science in Decisionmaking.* Washington, DC: Resources for the Future.

Feldman, M.S., and Khademian, A.M.
- 2002 To manage is to govern. *Public Administration Review, 62*(5), 541-554.

Feldman, M.S., Khademian, A.M., Ingram, H., and Schneider, A.S.
- 2006 Ways of knowing and inclusive management practices. *Public Administration Review, 66*(Supp. 1), 89-99.

Felstiner, W., Abel, R., and Sarat, A.
- 1980-1981 The emergence and transformation of disputes: Naming, blaming and claiming. *Law and Society Review, 15*(3-4), 631-654.

Ferreira, M.B., Garcia-Marques, L., Sherman, S.J., and Sherman, J.W.
- 2006 Automatic and controlled components of judgment and decision making. *Journal of Personality & Social Psychology, 91,* 797-813.

Festinger, L.B.
- 1957 *A Theory of Cognitive Dissonance.* Palo Alto, CA: Stanford University Press.

Fetterman, D.M.
- 1994 Steps of empowerment evaluation: From California to Capetown. *Evaluation and Program Planning, 17,* 305-313.
- 1996 Empowerment evaluation: An introduction to theory and practice. In D.M. Fetterman, S.J. Kafterian, and A. Wandersman (Eds.), *Empowerment Evaluation: Knowledge and Tools for Self-Assessment and Responsibility* (pp. 3-45). Thousand Oaks, CA: Sage.

Finer, H.
- 1941 Administrative responsibility in democratic government. *Public Administration Review, 1*(4), 335-350.

Fiorino, D.J.
- 1989 Environmental risk and democratic process: A critical review. *Columbia Journal of Environmental Law, 14*(2), 501-547.
- 1990 Citizen participation and environmental risk: A survey of institutional mechanisms. *Science, Technology, and Human Values, 15*(2), 226-243.

Fischer, F.
- 2000 *Citizens, Experts, and the Environment: The Politics of Local Knowledge.* Durham, NC: Duke University Press.
- 2005 Participative governance as deliberative empowerment. The cultural politics of discursive space. *The American Review of Public Administration, 36,* 19-40.

Fischer, F., and Forester, J. (Eds.)
- 1993 *The Argumentative Turn in Policy Analysis and Planning.* Durham, NC: Duke University Press.

Fischhoff, B.
- 1975 Hindsight ≠ foresight: The effect of outcome knowledge on judgment under uncertainty. *Journal of Experimental Psychology: Human Perception and Performance, 1*, 288-299.
- 1982 Debiasing. In D. Kahneman, P. Slovic, and A. Tversky (Eds.), *Judgment under Uncertainty: Heuristics and Biases* (pp. 422-444). New York: Cambridge University Press.
- 1985 Managing risk perceptions. *Issues in Science and Technology, 2*(1), 83-96.
- 1989 Risk: A guide to controversy. In National Research Council, *Improving Risk Communication* (pp. 211-319). Committee on Risk Perception and Communication. Commission on Behavioral and Social Sciences and Education and Commission on Physical Sciences, Mathematics, and Resources. Washington, DC: National Academy Press.
- 1991 Value elicitation: Is there anything in there? *American Psychologist, 46*, 835-847.
- 1995 Risk perception and communication unplugged: Twenty years of process. *Risk Analysis, 15*(2), 137-145.
- 1996a Public values in risk research. *Annals of the American Academy of Political and Social Science, 545*, 75-84.
- 1996b The real world: What good is it? *Organizational Behavior and Human Decision Processes, 65*, 232-248.

Fischhoff, B., and Furby, L.
- 1988 Measuring values: A conceptual framework for interpreting transactions with special reference to contingent valuation of visibility. *Journal of Risk and Uncertainty, 1*, 147-184.

Fischhoff, B., Lichtenstein, S., Slovic, P., Derby, S.L., and Keeney, R.L.
- 1981 *Acceptable Risk*. New York: Cambridge University Press.

Fischhoff, B., Nadai, A., and Fischhoff, I.
- 2001 Investing in Frankenfirms: Predicting socially unacceptable risks. *The Journal of Psychology and Financial Markets, 2*, 100-111.

Fischhoff, B., Slovic, P., Lichtenstein, S., Read, S., and Combs, B.
- 1978 How safe is safe enough? A psychometric study of attitudes towards technological risks and benefits. *Policy Sciences, 8*, 127-152.

Fishburn, P.
- 1981 Subjective expected utility: A review of normative theories. *Theory and Decision, 13*, 139-199.

Fisher, A.
- 1991 Risk communication challenges. *Risk Analysis, 11*, 173-179.

Fisher, R.
- 1994 *Let the People Decide*. New York: Twayne.

Fisher, R., and Ury, W.
- 1981 *Getting to Yes: Negotiating Agreement Without Giving In*. New York: Penguin Books.

Fisher, R., Ury, W., and Patton, B.
- 1991 *Getting to Yes: Negotiating Agreement Without Giving In* (2nd ed.). New York: Penguin Books.

Fishkin, J.S.
- 1991 *Democracy and Deliberation: New Directions for Democratic Reform*. New Haven, CT: Yale University Press.
- 1997 *The Voice of the People: Public Opinion and Democracy*. New Haven, CT: Yale University Press.
- 2006 Strategies of public consultation. *The Integrated Assessment Journal, 6*(2), 57-72.

REFERENCES

Fishkin, J.S., and Luskin, R.C.
2005 Experimenting with a democratic ideal: Deliberative polling and public opinion. *Acta Politica, 40*, 284-298.
Fleck, L.M.
2007 Can we trust "democratic deliberation?" *Hastings Center Report, 37*(4), 22-25.
Florig, H.K., Morgan, M.G., Jenni, K.E., Fischoff, B., Fischbeck, P.S., and DeKay, M.L.
2001 A deliberative method for ranking risks(I): Overview and test-bed development. *Risk Analysis, 21*, 913-921.
Floyd, D.W., Germain, R.H., and ter Horst, K.
1996 A model for assessing negotiations and mediation in forest resource conflicts. *Journal of Forestry, 94*(5), 29-33.
Flynn, J., Slovic, P., and Mertz, C.K.
1994 Gender, race, and perception of environmental health risks. *Risk Analysis, 14*, 1101-1108.
Follett, M.P.
1918 *The New State: Group Organization, the Solution of Popular Government.* New York: Longmans, Green.
1924 *Creative Experience.* New York: Longmans, Green.
Forester, J.
1989 *Planning in the Face of Power.* Berkeley: University of California Press.
Forester, J., and Stitzel, D.
1989 Beyond neutrality: The possibilities of activist mediation in public sector conflicts. *Negotiation Journal, 5*(July), 251-264.
Franklin, M.N.
2004 *Voter Turnout and the Dynamics of Electoral Competition in Established Democracies Since 1945.* Cambridge, England: Cambridge University Press.
Franklin, M.N., Lyons, P., and Marsh, M.
2004 Generational basis of turnout decline in established democracies. *Acta Politica, 39*(2), 115-151.
Frentz, I.C., Voth, D.E., Burns, S., and Sperry, C.W.
2000 Forest Service community relationship building: Recommendations. *Society and Natural Resources, 13*, 549-566.
Frewer, L., Hunt, S., Brennan, M., Kuznesof, S., Ness, M., and Ristons, C.
2003 The views of scientific experts on how the public conceptualize uncertainty. *Journal of Risk Research, 6*(1), 75-85.
Frisch, D., and Clemen, R.T.
1994 Beyond expected utility theory: Rethinking behavioral decision research. *Psychological Bulletin, 116*, 46-54.
Fung, A.
2005 *Empowered Participation: Reinventing Urban Democracy.* Princeton, NJ: Princeton University Press.
2006 Varieties of participation in complex governance. *Public Administration Review, 66*(Supp. 1), 66-75.
Fung, A., and Wright, E.O.
2001 Deepening democracy: Innovations in empowered local governance. *Politics and Society, 29*, 5-41.
Funtowicz, S.O., and Ravetz, J.R.
1991 A new scientific methodology for global environmental issues. In R. Costanza (Ed.), *Ecological Economics* (pp. 137-152). New York: Columbia University Press.
1993 Science for the post-normal age. *Futures, 7*, 739-755.

Furger, F., and Fukuyama, F.
    2007   A proposal for modernizing the regulation of human biotechnologies. *Hastings Center Report, 37*(4), 16-20.

Futrell, R.
    2003   Technical adversarialism and participatory collaboration in the U.S. chemical weapons program. *Science, Technology, and Human Values, 28,* 451-482.

Gaertner, S.L., Dovidio, J.F., Rust, M.C., Nier, J.A., Banker, B.S., Ward, C.M., Mottola, G.R., and Houlette, M.
    1999   Reducing intergroup bias: Elements of intergroup cooperation. *Journal of Personality and Social Psychology, 76*(3), 388-402.

Gamson, W.A.
    1990   *The Strategy of Social Protest.* Belmont, CA: Wadsworth.
    1992   The social psychology of collective action. In A.D. Morris and C.M. Mueller (Eds.), *Frontiers in Social Movement Theory* (pp. 53-76). New Haven, CT: Yale University Press.

Gastil, J.
    1993   *Democracy in Small Groups.* Gabriola Island, BC: New Society.
    2008   *Political Communication and Deliberation.* Thousand Oaks, CA: Sage.

Gastil, J., and Levine, P. (Eds.)
    2005   *The Deliberative Democracy Handbook: Strategies for Effective Citizen Engagement in the 21st Century.* San Francisco, CA: Jossey-Bass.

Gehrlein, W.V.
    2004   *Probabilities of Election Outcomes with Two Parameters: The Relative Impact of Unifying and Polarizing Candidates.* Working paper, Department of Business Administration, University of Delaware.

George, A.L., and Bennett, A.
    2005   *Case Studies and Theory Development in the Social Sciences.* Cambridge, MA: MIT Press.

Gericke, K.L., and Sullivan, J.
    1994   Public participation and appeals of Forest Service plans: An empirical examination. *Society and Natural Resources, 7,* 125-135.

Germain, R.H., Floyd, D.W., and Stehman, S.V.
    2001   Public perceptions of the USDA Forest Service public participation process. *Forest Policy and Economics, 3,* 113-124.

Geys, B.
    2006   Explaining voter turnout: A review of aggregate level research. *Electoral Studies, 25*(4), 637-663.

Gigerenzer, G.
    1998   Ecological rationality: An adaptation for frequencies. In D.D. Cummins and C. Allen (Eds.), *The Evolution of Mind* (pp. 9-29). New York: Oxford University Press.

Gigerenzer, G., and Hoffrage, U.
    1995   How to improve Bayesian reasoning without instruction: Frequency formats. *Psychological Review, 102,* 684-704.

Gigone, D., and R. Hasti
    1993   The common knowledge effect: Information sharing and group judgment. *Journal of Personality and Social Psychology, 65,* 959-974.

Goldberg, S.B., Green, E.D., and Sander, F.E.A.
    1985   *Dispute Resolution.* Boston, MA: Little Brown.

Goldschmidt, R., and Renn, O.
  2006 *Meeting of Minds—European Citizens' Deliberation on Brain Sciences*. (Final report of the external evaluation, vol. 5.) Stuttgart, Germany: University of Stuttgart, Social Science Department.
Goldstein, K.
  1999 *Interest Groups, Lobbying, and Participation in America*. New York: Cambridge University Press.
Graham, J.D., and Wiener, J.B.
  1995 *Risk vs. Risk: Tradeoffs in Protecting Health and the Environment*. Cambridge, MA: Harvard University Press.
Graham, S.D.N.
  1996 Flight to the cyber suburbs. *The Guardian*, April 18, p. 2-3.
Gray, B.
  2004 Strong opposition: Frame-based resistance to collaboration. *Journal of Community and Applied Psychology*, 14(3), 166-176.
Greenstone, D., and Peterson, P.E.
  1973 *Race and Authority in Urban Politics: Community Participation and the War on Poverty*. New York: Russell Sage Foundation.
Gregory, R., and McDaniels, T.
  1987 Valuing environmental losses: What promise does the right measure hold? *Policy Sciences*, 20, 11-26.
  2005 Improving environmental decision processes. In National Research Council, *Decision Making for the Environment: Social and Behavioral Science Research Priorities* (pp. 175-199). G.D. Brewer and P.C. Stern (Eds.). Panel on Social and Behavioral Science Research Priorities for Environmental Decision Making. Committee on the Human Dimensions of Global Change. Washington, DC: The National Academies Press.
Gregory, R., Fischhoff, B., and McDaniels, T.R.
  2005 Acceptable input: Using decision analysis to guide public policy deliberations. *Decision Analysis*, 2(1), 4.
Gregory, R., Lichtenstein, S., and Slovic, P.
  1993 Valuing environmental resources: A constructive approach. *Journal of Risk and Uncertainty*, 7, 177-197.
Grunig, J.E., and Grunig, L.A.
  1992 Models of public relations and communication. In J.E. Grunig, D.M. Dozier, W.P. Ehling, L.A. Grunig, F.C. Repper, and J. White (Eds.), *Excellence in Public Relations and Communication Management* (pp. 285-326). Hillsdale, NJ: Lawrence Erlbaum Associates.
Guba, E., and Lincoln, Y.
  1989 *Fourth Generation Evaluation*. Newbury Park, CA: Sage.
Gulliver, P.H.
  1979 *Disputes and Negotiations: A Cross-Cultural Perspective*. New York: Academic Press.
Gutmann, A., and Thompson, D.F.
  1996 *Democracy and Disagreement*. Cambridge, MA: Harvard University Press.
Habermas, J.
  1970 *Towards a Rational Society*. Boston, MA: Beacon Press.
  1975 *Legitimation Crisis*. (T. McCarthy, Trans.) Boston, MA: Beacon Press.
  1984 *The Theory of Communicative Action. Volume One, Reason and the Rationalization of Society*. Boston, MA: Beacon Press.

1987 *The Theory of Communicative Action. Volume Two, Lifeworld and System: A Critique of Functionalist Reason.* Boston, MA: Beacon Press.
1989 The public sphere: An encyclopedia article. (S. Lennox and F. Lennox, Trans.) In S.E. Bronner and D. MacKay Kellner (Eds.), *Critical Theory and Society: A Reader* (pp. 136-144). London, England: Routledge.
1991 *Moral Consciousness and Communicative Action.* Boston, MA: Beacon Press.
1996 *Between Facts and Norms: Contributions to a Discourse Theory of Law and Democracy.* (W. Rehg, Trans.) Cambridge, MA: MIT Press.

Hagendijk, R., and Irwin, A.
2006 Public deliberation and governance: Engaging with science and technology in contemporary Europe. *Minerva, 44,* 167-184.

Hajer, M.
1997 *Politics of Environmental Discourse: Ecological Modernization and the Policy Process.* Oxford, England: Oxford University Press.

Hajer, M., and Wagenaar, H.
2003 *Deliberative Policy Analysis: Understanding Governance in the Network Society.* Boston, MA: Cambridge University Press.

Haklay, M.E.
2002 Public environmental information: Understanding requirements and patterns of likely public use. *Area, 34*(1), 17-28.
2003 Public access to environmental information: Past, present and future. *Computers, Environment, and Urban Systems, 27*(2), 163-180.

Halfacre, A.C., Matheny, A.R., and Rosenbaum, W.A.
2000 Regulating contested local hazards: Is constructive dialogue possible among participants in community risk management? *Policy Studies Journal, 28,* 648-667.

Halpern, R.
1995 *Rebuilding the Inner City.* New York: Columbia University Press.

Halvorsen, K.E.
2003 Assessing the effects of public participation. *Public Administration Review, 63,* 535-543.

Hamlett, P.W., and Cobb, M.D.
2006 Potential solutions to public deliberation problems: Structured deliberations and polarization cascades. *Policy Studies Journal, 34*(4), 629-648.

Hammond, J., Keeney, R., and Raiffa, H.
1999 *Smart Choices: A Practical Guide to Making Better Decisions.* Cambridge, MA: Harvard Business School Press.

Hammond, K.R., Harvey, L.O., Jr., and Hastie, R.
1992 Making better use of scientific knowledge: Separating truth from justice. *Psychological Science, 33*(2), 80-87.

Harrison, C., and Haklay, M.
2002 The potential of public participation geographic information systems in U.K. environmental planning: Appraisals by active publics. *Journal of Environmental Planning and Management, 45*(6), 841-863.

Harter, P.J.
2000 Assessing the assessors: The actual performance of negotiated rulemaking. *New York University Environmental Law Journal, 9,* 32-59. Available: http://www.law.nyu.edu/journals/envtllaw/issues/vol9/1/v9n1a2.pdf [accessed June 2008].

Hastie, R., Penrod, S.D., and Pennington, N.
1983 *Inside the Jury.* Cambridge, MA: Harvard University Press.

Healey, P.
 2006 *Collaborative Planning: Shaping Places in Fragmented Societies* (2nd ed.). New York: Palgrave Macmillian.
Heclo, H.
 1978 Issue networks and the executive establishment. In A. King (Ed.), *The New American Political System*. Washington, DC: American Enterprise Institute.
Held, D.
 1987 *Models of Democracy*. Stanford, CA: Stanford University Press.
 1996 *Models of Democracy* (2nd ed.). Stanford, CA: Stanford University Press.
Henry S. Cole Associates
 1996 *Learning from Success: Health Agency Efforts to Improve Community Involvement in Communities Affected by Hazardous Waste Sites*. Atlanta, GA: Agency for Toxic Substances and Disease Registry.
Hershey, J.C., and Baron, J.
 1992 Judgment by outcomes: When is it justified? *Organizational Behavior and Human Decision Processes, 53*, 89-93.
Hetherington, M.J.
 2004 *Why Trust Matters: Declining Political Trust and the Demise of American Liberalism*. Princeton, NJ: Princeton University Press.
Hibbing, J.R., and Theiss-Moore, E.
 2001 Process preferences and American politics: What the people want government to be. *American Political Science Review, 95*, 145-153.
Holloway, H., Norwood, A., Fullerton, C., Engel, C., and Ursano, R.
 1997 The threat of biological weapons: Prophylaxis and mitigation of psychological and social consequences. *Journal of the American Medical Association, 278*, 425-427.
Hoos, I.
 1973 Systems techniques for managing society: A critique. *Public Administration Review, 33*(2), 157-164.
Horlick-Jones, T., Rowe, G., and Walls, J.
 2007 Citizen engagement processes as information systems: The role of knowledge and the concept of translation quality. *Public Understanding of Science, 16*, 259-278.
Howard, R.A.
 1966 Decision analysis: Applied decision theory. In D.B. Hertz and J. Melese (Eds.), *Proceedings of the Fourth International Conference on Operational Research*. New York: Wiley-Interscience. (Reprinted in 1989.)
 1968 The foundations of decision analysis. *IEEE Transactions on Systems Science and Cybernetics, SSC-4*(3), 211-219.
Howard, R.A., Matheson, J.E., and North, D.W.
 1972 The decision to seed hurricanes. *Science, 176*, 1191-1202. Available: http://www.northworks.net/hurricanes.pdf [accessed June 2008].
Howarth, R.B., and Wilson, M.A.
 2006 A theoretical approach to deliberative valuation: Aggregation by mutual consent. *Land Economics, 82*(1), 00781-007816.
Imperial, M.T.
 1998 Analyzing institutional arrangements for ecosystem-based management: Lessons from the Rhode Island Salt Ponds SAM Plan. *Coastal Management, 27*(1), 31-56.
 2005 Using collaboration as a governance strategy: Lessons from six watershed management programs. *Administration & Society, 37*(3), 281-320.
Ingram, H., and Smith, S.R. (Eds.)
 1993 *Public Policy for Democracy*. Washington, DC: Brookings Institution Press.

Insko, C.A., Schopler, J., Drigotas, S.M., Graetz, K.A., Kennedy, J., Cox, C., and Bornstein, G.
    1993   The role of communication in interindividual-intergroup discontinuity. *Journal of Conflict Resolution,* 37(1), 108-138.

Institute for Environmental Negotiation
    2001   *Collaboration: A Guide for Environmental Advocates.* (A publication of the University of Virginia's Institute for Environmental Negotiation, the Wilderness Society, and the National Audubon Society, June.)

Intergovernmental Panel on Climate Change
    2001   *IPCC Third Assessment Report.* Available: http://www.grida.no/climate/ipcc_tar/ [accessed June 2008].

International Association for Public Participation (IAP2)
    2006   *Public Participation Toolbox.* Available: http://www.iap2.org/associations/4748/files/06Dec_Toolbox.pdf [accessed June 2008].

International Finance Corporation
    2006   *Stakeholder Engagement: A Good Practice Handbook for Companies Doing Business in Emerging Markets.* Washington, DC: Author. Available: http://www.ifc.org/ifcext/enviro.nsf/AttachmentsByTitle/p_Stakeholder Engagement_Full/$FILE/IFC_StakeholderEngagement.pdf [accessed June 2008].

International Risk Governance Council
    2005   *Risk Governance: Towards an Integrative Approach.* (White Paper No. 1.) Geneva, Switzerland: Author.

Irwin, A.
    1995   *Citizen Science: A Study of People, Expertise, and Sustainable Development.* London, England: Routledge.

Iyengar, S., Luskin, R.C., and Fishkin, J.S.
    2003   Facilitating informed public opinion: Evidence from face-to-face and online deliberative polls. In *Annual Meeting of the American Political Science Association,* Philadelphia, PA. Available: http://pcl.stanford.edu/common/docs/research/iyengar/2003/facilitating.pdf [accessed June 2008].

Jackson, S.E.
    1992   Team composition in organizational settings: Issues in managing an increasingly diverse work force. In S. Worchel, W. Wood, and J.A. Simpson (Eds.), *Group Process and Productivity* (pp. 138-173). Newbury Park, CA: Sage.

Jaeger, C.C., Renn, O., Rosa, E.A., and Webler, T.
    2001   *Risk, Uncertainty and Rational Action.* London, England: Earthscan.

Janis, I.L.
    1972   *Victims of Groupthink: A Psychological Study of Foreign Policy Decisions and Fiascos.* Boston, MA: Houghton-Mifflin.

Janis, I.L., and Mann, L.
    1977   *Decision Making: A Psychological Analysis of Conflict, Choice, and Commitment.* New York: Free Press.

Jasanoff, S.
    1996   The dilemma of environmental democracy. *Issues in Science & Technology, 13,* 63-70.
    2005   *Designs on Nature: Science and Democracy in Europe and the United States.* Princeton, NJ: Princeton University Press.

Jasanoff, S., and Martello, M.L. (Eds.)
    2004   *Earthly Politics: Local and Global in Environmental Governance.* Cambridge, MA: MIT Press.

Jaynes, E.T.
    2003   *Theory of Probability: The Logic of Science.* New York: Cambridge University Press.

Jeffreys, H.
  1961 *Theory of Probability* (3rd ed.). Oxford, England: Clarendon Press.
Jenni, K.E., Merkhofer, M.W., and Williams, C.
  1995 The rise and fall of a risk-based priority system: Lessons from DOE's Environmental Restoration Priority System. *Risk Analysis, 15*, 397-410.
Johnson, B.B.
  2003 Further notes on public response to uncertainty in risks and science. *Risk Analysis, 23*, 781-789 [also see erratum, *Risk Analysis* (2004), 24, 781].
Johnson, B.B., and Slovic, P.
  1995 Presenting uncertainty in health risk assessment: Initial studies of its effects on risk perception and trust. *Risk Analysis, 15*, 485-494.
  1998 Lay views on uncertainty in environmental health risk assessment. *Journal of Risk Research, 1*, 261-279.
Johnson, K.N., Swanson, F., Herring, M., and Greene, S.
  1999 *Bioregional Assessments: Science at the Crossroads of Management and Policy*. Washington, DC: Island Press.
Joss, S.
  2005 Lost in translation? Challenges for participatory governance of science and technology. In A. Bogner and H. Torgersen (Eds.), *Wozu Experten? Ambivalenzen der Beziehung von Wissenschaft und Politik* (pp. 197-219). Wiesbaden, Germany: VS Verlag fur Sozialwissenschaften.
Jostad, P.M., McAvoy, L.H., and McDonald, D.
  1996 Native American land ethics: Implications for natural resource management. *Society and Natural Resources, 9*, 564-581.
Kahneman, D.
  2003 A perspective on judgment and choice: Mapping bounded rationality. *American Psychologist, 58*, 697-720.
Kahneman, D., Slovic, P., and Tversky, A. (Eds.)
  1982 *Judgment Under Uncertainty: Heuristics and Biases*. New York: Cambridge University Press.
Kalof, L., Dietz, T., Guagnano, G.A., and Stern, P.C.
  2002 Race, gender, and environmentalism: The atypical values and beliefs of white men. *Race, Gender & Class, 9*(2), 1-19.
Kameda, T., Ohtsubo, Y., and Takezawa, M.
  1997 Centrality in sociocognitive networks and social influence: An illustration in a group decision-making context. *Journal of Personality and Social Psychology, 73*, 296-309.
Kamenstein, D.S.
  1996 Persuasion in a toxic community: Rhetorical aspects of public meetings. *Human Organization, 55*(4), 458-464.
Kaner, S.
  2007 *Facilitator's Guide to Participatory Decision Making*. San Francisco: Jossey-Bass.
Kasemir, B., Jager, J., Jaeger, C., and Gardner, M.T. (Eds.)
  2003 *Public Participation in Sustainability Science*. Cambridge, England: Cambridge University Press.
Kaufman, A.S.
  1960 Human nature and participatory democracy. In C.J. Friedrich (Ed.), *Responsibility: NOMOS III* (pp. 266-289). New York: Liberal Arts Press. (Reprinted in W.E. Connolly (Ed.), 1969, *The Bias of Pluralism*. New York: Atherton.)
Keeney, R.L.
  1980 Equity and public risk. *Operations Research, 28*, 527-534.

1992 *Value-Focused Thinking: A Path to Creative Decision Making.* Cambridge, MA: Harvard University Press.
1996 The role of values in risk management. In H. Kunreuther and P. Slovic (Eds.), *Annals of the American Academy of Political and Social Science, Special Issue: Challenges in Risk Assessment and Risk Management* (pp. 126-134). Thousand Oaks, CA: Sage.

Keller, R.L., and Sarin, R.K.
1988 Equity in social risk: Some empirical observations. *Risk Analysis, 8,* 135-146.

Kellogg, W.A., and Mathur, A.
2003 Environmental justice and information technologies: Overcoming the information-access paradox in urban communities. *Public Administration Review, 63*(5), 573-585.

Kelly, J.R., and Karau, S.J.
1999 Group decision making: The effects of initial preferences and time pressure. *Personality and Social Psychology Bulletin, 25,* 1342-1354.

Kemmis, D.
2002 Science's role in natural resource decisions. *Issues in Science and Technology 18*(Summer), 31-34.

Kerwin, C., and Langbein, L.
1995 *An Evaluation of Negotiated Rulemaking at the Environmental Protection Agency, Phase I.* Washington, DC: Administrative Conference of the United States.

King, C.S., Feltey, K.M., and O'Neil Susel, B.
1998 The question of participation: Toward authentic public participation in public administration. *Public Administration Review, 58*(4), 317-326.

Kinney, P.L, Northridge, M.E., Chew, G.L., Groming, E., Joseph, E., Correa, J.C., Prakash, S., Goldstein, I., and The Reducing Indoor Allergens Study Team
2002 On the front lines: An environmental asthma intervention in New York City. *American Journal of Public Health, 92,* 24-26.

Kinzig, A., Starrett, D., Arrow, K., Aniyar, S., Bolin, B., Dasgupta, P., Ehrlich, P., Folke, C., Hanneman, M., Heal, G., Hoel, M., Jansson, B.-O., Jansson, A., Kautsky, N., Levin, S., Lubchencko, J., May, R.M., Mäler, K., Pacala, S.W., Schneider, S.H., Siniscalco, D., and Walker, B.
2003 Coping with uncertainty: A call for a new science-policy forum. *Ambio, 32*(5), 330-335.

Klein, G.
1996 The effect of acute stressors on decision-making. In J.E. Driskall and E. Salas (Eds.), *Stress and Human Performance* (pp. 49-88). Mahwah, NJ: Lawrence Erlbaum Associates.

Kleinhesselink, R.R., and Rosa, E.A.
1991 Cognitive representation of risk perceptions: A comparison of Japan and the United States. *Journal of Cross-Cultural Psychology, 22*(1), 11-28.

Knack, S.
2002 Social capital and the quality of government: Evidence from the states. *American Journal of Political Science, 46*(4), 772-785.

Knight, J., and Johnson, J.
1994 Aggregation and deliberation: On the possibility of democratic legitimacy. *Political Theory, 22*(2, May), 277-296.

Knopman, D.S., Susman, M.M., and Landy, M.K.
1999 Civic environmentalism: Tackling tough land-use problems with innovative governance. *Environment, 41*(10), 24-32.

Kolb, D.M.
1994 *When Talk Works: Profiles of Mediators.* San Francisco: Jossey-Bass.

Koontz, T.M., and Thomas, C.W.
  2006   What do we know and need to know about the environmental outcomes of collaborative management? *Public Administration Review*, 66(6), 109-119.
Koopmans, R.
  1996   New social movements and changes in political participation in Western Europe. *European Politics*, 19(1), 28-50.
Kraft, M.E
  2000   Environmental policy in Congress: From consensus to gridlock. In N.J. Vig and M.E. Kraft (Eds.), *Environmental Policy* (4th ed.).Washington, DC: Congressional Quarterly Press.
Kramer, R.
  1969   *Participation of the Poor: Comparative Community Case Studies in the War on Poverty*. Englewood Cliffs, NJ: Prentice-Hall.
Kraus, N., Malmforms, T., and Slovic, P.
  1992   Intuitive toxicology: Expert and lay judgments of chemical risks. *Risk Analysis*, 12, 215-232.
Kriesberg, L.
  1973   *The Sociology of Social Conflicts*. Englewood Cliffs, NJ: Prentice-Hall.
  1982   *Social Conflicts*. Engelwood Cliffs, NJ: Prentice-Hall.
Kroll-Smith, J.S., and Couch, S.R.
  1991   As if exposure to toxins were not enough: The social and cultural system as a secondary stressor. *Environmental Health Perspectives*, 95, 61-66.
  1993   Technological hazards: Social responses as traumatic stressors. In J.P. Wilson and B. Raphael (Eds.), *International Handbook of Traumatic Stress Syndromes* (pp. 79-91). New York: Plenum Press.
Kruger, L.E., and Shannon, M.A.
  2000   Getting to know ourselves and our places through participation in civic social assessment. *Society and Natural Resources*, 13, 461-478.
Kruglanski, A.W., and Webster, D.M.
  1996   Motivated closing the mind: "Seizing" and "freezing." *Psychological Review*, 103(2), 263-283.
Kruglanski, A.W., Bar-Tal, D., and Klar, Y.
  1993   A social-cognitive theory of conflict. In K.S. Larsen (Ed.), *Conflict and Social Psychology* (pp. 45-57). London, England: Sage.
Kruglanski, A.W., Webster, D.M., and Klem, A.
  1993   Motivated resistance and openness to persuasion in the presence or absence of prior information. *Journal of Personality and Social Psychology*, 65, 861-876.
Krutilla, J., and Haigh, J.
  1978   An integrated approach to national forest management. *Environmental Law*, 8, 373-415.
Kuhn, K.M.
  2000   Message format and audience values: Interactive effects of uncertainty information and environmental attitudes on perceived risk. *Journal of Environmental Psychology*, 20, 41-51.
Kunreuther, H., and Slovic, P.
  1996   Science, values and risk. *Annals of the American Academy of Political and Social Science*, 545(1), 116-125.
Laird, F.N.
  1993   Participatory analysis, democracy, and technological decision making. *Science, Technology, and Human Values*, 18, 341-361.

Lampe, D., and Kaplan, M.
   1999   *Resolving Land-Use Conflicts Through Mediation: Challenges and Opportunities.* Cambridge, MA: Lincoln Institute of Land Policy.
Landy, M.
   1993   Public policy and citizenship. In H. Ingram and S. Rathgeb Smith (Eds.), *Public Policy for Democracy* (pp. 19-44). Washington, DC: Brookings Institution Press.
Langbein, L.I.
   2005   *Negotiated and Conventional Rulemaking at EPA: A Comparative Case Analysis.* Paper prepared for the National Research Council Panel on Public Participation in Environmental Assessment and Decision Making, January, National Academy of Sciences, Washington, DC.
Langford, I.H., Georgiou, S., Bateman, I.J., Day, R.J., and Turner, R.K.
   2000   Public perceptions of health risks from polluted coastal bathing waters: A mixed methodological analysis using cultural theory. *Risk Analysis, 20*, 691-704.
Langton, S.
   1978   Citizen participation in America: Current reflections on state of the art. In S. Langton (Ed.), *Citizen Participation in America* (pp. 1-12). Lanham, MD: Lexington Books.
Larsen, L., Harlan, S.L., Bolin, B., Hackett, E.J., Hope, D., Kirby, A., Nelson, A., Rex, T.R., and Wolf, S.
   2004   Bonding and bridging: Understanding the relationship between social capital and civic action. *Journal of Planning Education and Research, 24*(1), 64-77.
Lauber, B.T., and Knuth, B.A.
   1999   Measuring fairness in citizen participation: A case study of moose management. *Society and Natural Resources, 11*, 19-37.
Lawrence, R.L., and Deagen, D.A.
   2001   Choosing public participation methods for natural resources. A context specific guide. *Society & Natural Resources, 14*, 859-874. Available: http://remotesensing.montana.edu/pdfs/lawrenceDeagen2001.pdf [accessed August 2008].
Lax, D.A., and Sebenius, J.K.
   1986   *The Manager as Negotiator: Bargaining for Cooperation and Competitive Gain.* New York: Free Press.
Lazarus, R.
   1991   *Emotion and Adaptation.* New York: Oxford University Press.
Lazo, J.K., Kinnell, J.C., and Fisher, A.
   2000   Expert and layperson perceptions of ecosystem risk. *Risk Analysis, 20*, 179-193.
Leach, W.D.
   2002   Surveying diverse stakeholder groups. *Society and Natural Resource, 15*(7), 641-649.
   2005   *Public Involvement in the USDA Forest Service Policy Making: A Literature Review.* Paper prepared for the National Research Council Panel on Public Participation in Environmental Assessment and Decision Making, January, National Academy of Sciences, Washington, DC. Available: http://www7.nationalacademies.org/hdgc/Tab%20_10%20Public%20Involvement.pdf [accessed August 2008].
   2006   Public involvement in USDA Forest Service policymaking: A literature review. *Journal of Forestry, 104*(1), 43-49.
Leach, W.D., and Pelkey, N.W.
   2001   Making watershed partnerships work: A review of the empirical literature. *Journal of Water Resources Management & Planning, 127*, 378-385.

Leach, W.D., and Reza, K.S.
  2004 *The Collaborative Capacity of American Tribes: Sovereignty, History, and Culture in Watershed Management.* Draft manuscript, Center for Collaborative Policy California State University, Sacramento.
Leach, W.D., Pelkey, N.W., and Sabatier, P.
  2002 Stakeholder partnerships as an emergent form of collaborative policymaking: Evaluation criteria applied to watershed management in California and Washington. *Journal of Policy Analysis and Management, 21*(4), 645-670.
Leach, W.D., Sabatier, P.A., and Quinn, J.F.
  2005 *Watershed Partnerships' Pursuit and Neglect of Scientific Monitoring.* Draft manuscript, Center for Collaborative Policy California State University, Sacramento.
Leighninger, M.
  2006 *The Next Form of Democracy: How Expert Rule Is Giving Way to Shared Governance . . . and Why Politics Will Never Be the Same.* Nashville, TN: Vanderbilt University Press.
Lemos, M.C., and Morehouse, B.J.
  2005 The co-production of science and policy in integrated climate assessments. *Global Environmental Change-Human and Policy Dimensions, 15,* 57-68.
Lerner, J.S., and Tetlock, P.E.
  1999 Accounting for the effects of accountability. *Psychological Bulletin, 125,* 255-275.
  2003 Bridging individual, interpersonal, and institutional approaches to judgment and decision making: The impact of accountability on cognitive bias. In S. Schneider and J. Shanteau (Eds.), *Emerging Perspectives on Judgment and Decision Research* (pp. 431-457). New York: Cambridge University Press.
Leventhal, G.S.
  1980 What should be done with equity theory? New approaches to the study of fairness in social relationships. In K. Gergen, M. Greenberg, and R. Willis (Eds.), *Social Exchange: Advances in Theory and Research* (pp. 27-55). New York: Plenum Press.
Levine, A.
  1982 *Love Canal: Science, Politics, and People.* Lexington, MA: Lexington Books.
Levine, J.M., and Moreland, R.L.
  1998 Small groups. In D.T. Gilbert, S.T. Fiske, and G. Lindzey (Eds.), *The Handbook of Social Psychology* (vol. 2, 4th ed., pp. 415-469). New York: McGraw-Hill.
Levins, R.
  1966 The strategy of model building in population biology. *American Scientist, 54,* 421-431.
Levitt, B., and March, J.G.
  1988 Organizational Learning. *Annual Review of Sociology, 14,* 319-340.
Lewicki, R.J., and Litterer, J.A.
  1985 *Negotiation.* Homewood, IL: R.D. Irwin.
Lewicki, R.J., Gray, B., and Elliot, M.
  2003 *Making Sense of Intractable Environmental Conflicts: Concepts and Cases.* Washington, DC: Island Press.
Liberatore, A., and Funtowicz, S.
  2003 Democratising expertise, expertising democracy: What does this mean, and why bother? *Science and Public Policy, 30*(3), 146-150.
Lijphart, A.
  1994 *Electoral Systems and Party Systems: A Study of Twenty-Seven Democracies, 1945-1990.* New York: Oxford University Press.

1997   Unequal participation: Democracy's unresolved dilemma. *American Political Science Review, 91,* 1-14.
Lin, N.
2001   *Social Capital: A Theory of Social Structure and Action.* Cambridge, England: Cambridge University Press.
Lind, E.A., and Tyler, T.R.
1988   *The Social Psychology of Procedural Justice.* New York: Plenum Press.
Linnerooth-Bayer, J., and Fitzgerald, K.B.
1996   Conflicting views on fair siting processes: Evidence from Austria and the U.S. *Risk: Health, Safety and Environment, 7,* 119-134.
List, C.
2001   *Mission Impossible? The Problem of Democratic Aggregation in the Face of Arrow's Theorem.* Doctorate of Philosophy thesis, University of Oxford.
2006   The discursive dilemma and public reason. *Ethics, 116,* 362-402.
List, C., Luskin, R.C., Fishkin, J.S., and McLean, I.
2006   Deliberation, Single-Peakedness, and the Possibility of Meaningful Democracy: Evidence from Deliberative Polls. London School of Economics, England. Available: http://personal.lse.ac.uk/list/PDF-files/DeliberationPaper.pdf [accessed August 2008].
Liu, J., Dietz, T., Carpenter, S.R., Alberti, M., Folke, C., Moran, E., Pell, A.N., Deadman, P., Kratz, T., Lubchencko, J., Ostrom, E., Ouyang, Z., Provencher, W., Redman, C.L., Schneider, S.H., and Taylor, W.W.
2007a   Complexity of coupled human and natural systems. *Science, 317,* 1513-1516.
Liu, J., Dietz, T., Carpenter, S.R., Folke, C., Alberti, M., Redman, C.L., Schneider, S.H., Ostrom, E., Pell, A.N., Lubchencko, J., Taylor, W.W., Ouyang, Z., Deadman, P., Kratz, T., and Provencher, W.
2007b   Coupled human and natural systems. *Ambio, 36,* 639-649.
Loewenstein, G.F., Weber, E.U., Hsee, C.K., and Welch, E.S.
2001   Risk as feelings. *Psychological Bulletin, 127,* 267-286.
Löfstedt, R.
1999   The role of trust in the North Blackforest: An evaluation of a citizen panel project. *Risk: Health, Safety and Environment, 10,* 7-30.
Lopes, L.L.
1983   Some thoughts on the psychological concept of risk. *Journal of Experimental Psychology: Human Perception and Performance, 9,* 137-144.
Lourenço, R.P., and Costa, J.P.
2007   Incorporating citizens' views in local policy decision making processes. *Decision Support Systems, 43,* 1499-1511.
Lubell, M.
2000   Cognitive conflict and consensus building in the National Estuary Program. *American Behavioral Scientist, 44,* 629-648.
2002   Environmental activism as collective action. *Environment and Behavior, 34*(4), 431-454.
2004a   Collaborative watershed management: A view from the grassroots. *Policy Studies Journal, 32,* 341-361.
2004b   Resolving conflict and building cooperation in the National Estuary Program. *Environmental Management, 33,* 677-691.
Lubell, M., and Leach, W.D.
2005   *Watershed Partnerships: Evaluating a Collaborative Form of Public Participation.* Paper prepared for the National Research Council Panel on Public Participation in Environmental Assessment and Decision Making, Feb. 3-4, National Academy of Sciences, Washington, DC.

Lubell, M., Schneider, M., Scholz, J.T., and Mete, M.
   2002   Watershed partnerships and the emergence of collective action institutions. *American Journal of Political Science*, 46(1), 148-163.
Lynn, F.M.
   1990   Public participation in risk management decisions: The right to define, the right to know, and the right to act. *Risk Issues in Health and Safety*, 1, 95-101.
Lynn, F.M., and Kartez, J.D.
   1995   The redemption of citizen advisory committees: A perspective from critical theory. In O. Renn, T. Webler, and P. Wiedemann (Eds.), *Fairness and Competence in Citizen Participation: Evaluating New Models for Environmental Discourse* (pp. 87-102). Boston, MA: Kluwer Academic.
   1998   Science and nonscience concerning human-caused climate warming. *Annual Review of Energy and the Environment*, 23, 83-105.
Macedo, S.
   1999   *Deliberative Politics: Essays on Democracy and Disagreement*. New York: Oxford University Press.
Mahlman, J.D.
   1998   Science and nonscience concerning human-caused climate warming. *Annual Review of Energy and the Environment*, 23, 83-105.
Mansbridge, J.
   1983   *Beyond Adversary Democracy*. Chicago: University of Chicago Press.
March, J.G., and Olsen, J.P.
   1995   *Democratic Governance*. New York: Free Press.
Markus, G.B.
   2002   *Institutional and Individual Origins of Civic Engagement*. Paper presented at the Annual Meeting of the American Political Science Association, August, Boston, MA.
Markus, G.B., Chess, C., and Shannon, M.A.
   2005   *Political Perspectives on Public Participation in Environmental Assessment and Decision Making*. Paper prepared for the National Research Council Panel on Public Participation in Environmental Assessment and Decision Making, Feb. 3-5, National Academy of Sciences, Washington, DC.
Marshall, B.K., and Picou, J.S.
   2008   Post-normal science, the precautionary principle, and worst cases: Managing scientific uncertainty in the 21st Century. Submitted to *Sociological Inquiry*.
Martin, S.P., and Robinson, J.P.
   2007   The income digital divide: Trends and predictions for levels of Internet use. *Social Problems*, 54(1), 1-22.
Marwell, G., and Oliver, P.E.
   1993   *The Critical Mass in Collective Action: A Micro-Social Theory*. New York: Cambridge University Press.
Mather, L., and Yngvesson, B.
   1980-
   1981   Language, audience, and the transformation of disputes. *Law and Society Review*, 15(3-4).
McCloskey, M.
   1996   The skeptic: Collaboration has its limits. *High Country News*, May 13.
McComas, K.A., Trumbo, C.W., and Besley, J.C.
   2007   Public meetings about suspected cancer clusters: The impact of voice, interactional justice, and risk perception on attendees' attitudes in six communities. *Journal of Health Communication*, 12, 527-549.

McCormick, S.
2006 The Brazilian anti-dam movement: Knowledge contestation as communicative action. *Organization and Environment, 19*(30), 321-346.
2007a Democratizing science movements: A new framework for contestation. *Social Studies of Science, 37*, 1-15.
2007b Governing hydroelectric dams in Brazil. *Journal of Latin American Studies, 39*(2), 227-261.

McCormick, S., Brody, J., Brown, P., and Polk, R.
2004 Public involvement in breast cancer research: An analysis and model for future research. *International Journal of Health Services, 34*(4), 625-646.

McCormick, S., Brown, P., and Zavestoski, S.
2004 The personal is scientific, the scientific is political: The public paradigm of the environmental breast cancer movement. *Sociological Forum, 18*(4), 545-576.

McDaniels, T.
1996 The structured value referendum: Eliciting preferences for environmental policy alternatives. *Journal of Policy Analysis and Management, 15*, 227-251.
1998 Ten propositions for untangling descriptive and prescriptive lessons in risk perception findings. *Reliability Engineering and System Safety, 59*, 129-134.

McDaniels, T.L., Axelrod, L.J., and Slovic, P.
1995 Characterizing perception of ecological risk. *Risk Analysis, 15*(5), 575-588.
1996 Perceived ecological risks of global change: A psychometric comparison of causes and consequences. *Global Environmental Change, 6*(2), 159-171.

McDaniels, T.L., Axelrod, L.J., Cavanagh, N.S., and Slovic, P.
1997 Perception of ecological risk to water environments. *Risk Analysis, 17*(3), 341-352.

McKeown, R., Hopkins, C.A., and Chrystalbridge, M.
2002 *Education for Sustainable Development Toolkit* (version 2). Available: http://www.esdtoolkit.org/about.htm [accessed June 2008].

McKeown, T.
2004 Case studies and the limits of the quantitative worldview. In H.E. Brady and D. Collier (Eds.), *Rethinking Social Inquiry: Diverse Tools, Shared Standards* (pp. 139-167). Lanham, MD: Rowman and Littlefield.

Mendelberg, T.
2002 The deliberative citizen: Theory and evidence. In M.X. Delli Carpini, L. Huddy, and R. Shapiro (Eds.), *Research in Micropolitics: Political Decision Making, Deliberation, and Participation* (vol. 6, pp. 159-193). Greenwich, CT: JAI Press.

Mesquita, B., and Frijda, N.H.
1992 Cultural variations in emotions: A review. *Psychological Bulletin, 112*, 179-204.

Metzger, E.S., and Lendvay, J.M.
2006 Seeking environmental justice through public participation: A community-based water quality assessment in Bayview Hunters Point. *Environmental Practice, 8*, 104-114.

Mikula, G., Scherer, K.R., and Athenstaedt, U.
1998 The role of injustice in the elicitation of differential emotional reactions. *Personality & Social Psychology Bulletin, 24*, 769-783.

Miles, S., and Frewer, L.
2003 Public perception of scientific uncertainty in relation to food hazards. *Journal of Risk Research, 6*(3), 267-283.

Miller, C.E.
1989 The social psychological effects of group decision rules. In P. Paulus (Ed.), *Psychology of Group Influence* (pp. 327-355). Hillsdale, NJ: Lawrence Erlbaum Associates.

REFERENCES

Miller, D.
 1992 Deliberative democracy and social choice. *Political Studies*, 40(Special Issue), 54-67.
Miller, J.G.
 1999 Cultural psychology: Implications for basic psychological theory. *Psychological Science, 10*, 85-91.
Mitchell, R.B., Clark, W.C., and Cash, D.W.
 2006 Information and influence. In R.B. Mitchell, W.C. Clark, D.W. Cash, and N.M. Dickson (Eds.), *Global Environmental Assessments: Information and Influence* (Chapter 11, pp. 307-338). Cambridge, MA: MIT Press.
Mitchell, R.B., Clark, W.C., Cash, D.W., and Dickson, N.M. (Eds.)
 2006 *Global Environmental Assessments: Information and Influence*. Cambridge, MA: MIT Press.
Modavi, N.
 1996 Mediation of environmental conflicts in Hawaii: Win-win or co-optation? *Sociological Perspectives, 39*(2), 301-316.
Moore, C.
 1986 *The Mediation Process: Practical Strategies for Resolving Conflict*. San Francisco, CA: Jossey-Bass.
Moore, S.A.
 1996 Defining "successful" environmental dispute resolution: Case studies from public land planning in the United States and Australia. *Environmental Impact Assessment Review, 16*, 151-169.
Morgan, K.M., DeKay, M.L., Fischbeck, P.S., Morgan, M.G., Fischoff, B., and Florig, H.K.
 2001 A deliberative method for ranking risks(II): Evaluation of validity and agreement among risk managers. *Risk Analysis, 21*(5), 923-938.
Morgan, M.G., and Henrion, M.
 1990 *Uncertainty: A Guide to Dealing with Uncertainty in Quantitative Risk and Policy Analysis*. New York: Cambridge University Press.
Morgan, M.G., and Keith, D.W.
 1995 Subjective judgments by climate experts. *Environmental Science and Technology, 29*(10), 468A-476A.
Morgan, M.G., Cantor, R., Clark, W.C., Fisher, A., Jacoby, H.D., Janetos, A.C., Kinzig, A.P., Melillo, J., Street, R.B., and Wilbanks, T.J.
 2005 Learning from the U.S. National Assessment of Climate Change Impacts. *Environmental Science and Technology, 39*(23), 9023-9032.
Moroe, K.R.
 2001 Morality and sense of self: The importance of identity and categorization for moral action. *American Journal or Political Science, 45*(3), 491-507.
Morone, J.A.
 1990 *The Democratic Wish: Popular Participation and the Limits of American Government*. New York: Basic Books.
Morone, J.A., and Kilbreth, E.H.
 2003 Power to the people? Restoring citizen participation. *Journal of Health Politics, Policy and Law, 28*(2, 3), 271-288.
Moscovici, S., and Zavalloni, M.
 1969 The group as a polarizer of attitudes. *Journal of Personality and Social Psychology, 12*, 125-135.

Moser, S.C.
2005 *Impacts of Climate Change in the United States. The First U.S. National Assessment of the Potential Consequences of Climate Variability and Change: A Guide Through the Process.* Available: http://www.climatehotmap.org/impacts/assessment.html [accessed June 2008].

Moss, R., and Schneider, S.H.
2000 Uncertainties in the IPCC TAR: Recommendations to lead authors for more consistent assessment and reporting. In R. Pachauri, T. Taniguchi, and K. Tanaka (Eds.), *Guidance Papers on the Cross-Cutting Issues of the Third Assessment Report of the IPCC* (pp. 33-51). Geneva, Switzerland: World Meteorological Organization.

Mossberger, K., Tolbert, C., and Stansbury, M.
2003 *Virtual Inequality: Beyond the Digital Divide.* Washington, DC: Georgetown University Press.

Mouffe, C.
1999 Deliberative democracy or agonistic pluralism? *Social Research, 66,* 745-758.

Moynihan, D.P.
1969 *Maximum Feasible Misunderstanding: Community Action in the War on Poverty.* New York: Free Press.

Murphree, D.W., Wright, S.A., and Ebaugh, H.R.
1996 Toxic waste siting and community resistance: How cooptation of local citizen opposition failed. *Sociological Perspectives, 39*(4), 447-463.

Murphree, M.W.
1991 *Communities as Resource Management Institutions.* (Gatekeeper series #36.) London, England: International Institute for Environment and Development. Available: http://www.iied.org/pubs/pdfs/G01147.pdf [accessed June 2008].

Mutz, D.C.
2002a The consequences of cross-cutting networks for political participation. *American Journal of Political Science, 46*(4), 838-855.
2002b Cross-cutting social networks: Testing democratic theory in practice. *American Political Science Review, 96*(1), 111-126.

Myers, D.G., and Lamm, H.
1976 The group polarization phenomenon. *Psychological Bulletin, 83,* 602-662.

National Academy of Engineering
2004 *Accident Precursor Analysis and Management: Reducing Quantitative Risk Through Diligence.* J.R. Phimister, V.M. Bier, and H.C. Kunreuther (Eds.). Washington, DC: The National Academies Press.

National Environmental Justice Advisory Committee
1996 *The Model Plan for Public Participation.* (EPA #300-K-00-001.) Washington, DC: U.S. Environmental Protection Agency.

National Park Service Division of Park Planning and Special Studies
1997 *National Parks and Their Neighbors: Lessons from the Field on Building Partnerships with Local Communities.* Tucson, AZ: Sonoran Institute.

National Research Council
1983 *Risk Assessment in the Federal Government: Managing the Process.* Committee on the Institutional Means for Assessment of Risks to Public Health, Commission on Life Sciences. Washington, DC: National Academy Press.
1986 *Proceedings of the Conference on Common Property Resource Management.* Panel on Common Property Resource Management. Washington, DC: National Academy Press.

REFERENCES

1989 *Improving Risk Communication.* Committee on Risk Perception and Communication, Commission on Physical Sciences, Mathematics, and Resources. Washington, DC: National Academy Press.

1994 *Science and Judgment of Risk Assessment.* Committee on Risk Assessment of Hazardous Air Pollutants, Commission on Life Sciences. Washington, DC: National Academy Press.

1996 *Understanding Risk: Informing Decisions in a Democratic Society.* P.C. Stern and H.V. Fineberg (Eds.). Committee on Risk Characterization, Commission on Behavioral and Social Sciences and Education. Washington, DC: National Academy Press.

1999a *Making Climate Forecasts Matter.* P.C. Stern and W.E. Easterling (Eds.). Panel on the Human Dimensions of Seasonal-to-Interannual Climate Variability, Commission on Behavioral and Social Sciences and Education. Washington, DC: National Academy Press.

1999b *Perspectives on Biodiversity: Valuing Its Role in an Everchanging World.* Committee on Noneconomic and Economic Value of Biodiversity. Board on Biology, Commission on Life Sciences.Washington, DC: National Academy Press.

2000 *Disposition of High-Level Waste and Spent Nuclear Fuel: The Continuing Societal and Technical Challenges.* Committee on Disposition of High-Level Radioactive Waste Through Geological Isolation. Board on Radioactive Waste Management, Division on Earth and Life Studies. Washington, DC: National Academy Press.

2002a *The Drama of the Commons.* Committee on the Human Dimensions of Global Change. E. Ostrom, T. Dietz, N. Dolsak, P. C. Stern, S. Stonich, and E. Weber (Eds.). Division of Behavioral and Social Sciencs and Education. Washington, DC: National Academy Press.

2002b *Estimating the Public Health Benefits of Proposed Air Pollution Regulations.* Committee on Estimating the Health-Risk-Reduction Benefits of Proposed Air Pollution Regulations. Board on Environmental Studies and Toxicology. Washington, DC: The National Academies Press.

2003 *One Step at a Time: The Staged Development of Geologic Repositories for High-Level Radioactive Waste.* Committee on Principles and Operational Strategies for Staged Repository Systems. Board on Radioactive Waste Management, Division on Earth and Life Studies. Washington, DC: The National Academies Press.

2004 *Valuing Ecosystem Services: Towards Better Environmental Decision-Making.* Committee on Assessing and Valuing the Services of Aquatic and Related Terrestrial Ecosystems. Water Science and Technology Board, Division on Earth and Life Studies. Washington, DC: The National Academies Press.

2005a *Decision Making for the Environment: Social and Behavioral Science Research Priorities.* G.D. Brewer and P.C. Stern (Eds.). Panel on Social and Behavioral Science Research Priorities for Environmental Decision Making. Committee on the Human Dimensions of Global Change. Division of Behavioral and Social Sciences and Education. Washington, DC: The National Academies Press.

2005b *Thinking Strategically: The Appropriate Use of Metrics for the Climate Change Science Program.* Committee on Metrics for Global Change Research. Board on Atmospheric Sciences and Climate, Division on Earth and Life Studies. Washington, DC: The National Academies Press.

2007a *Analysis of Global Change Assessments: Lessons Learned.* Committee on Analysis of Global Change Assessments. Board on Atmospheric Sciences and Climate, Division on Earth and Life Studies. Washington, DC: The National Academies Press.

2007b *Scientific Review of the Proposed Risk Assessment Bulletin from the Office of Management and Budget.* Committee to Review the OMB Risk Assessment Bulletin. Board on Environmental Studies and Toxicology, Division on Earth and Life Studies. Washington, DC: The National Academies Press.

2007c *Models in Environmental Regulatory Decision Making.* Committee on Models in the Regulatory Decision Process. Board on Environmental Studies and Toxicology, Division on Earth and Life Studies Washington, DC: The National Academies Press.

2008 *Research and Networks for Decision Support in the NOAA Sectoral Applications Research Program.* Panel on Design Issues for the NOAA Sector Applications Research Program. H.L. Ingram and P.C. Stern (Eds.). Committee on the Human Dimensions of Global Change, Division of Behavioral and Social Sciences and Education. Washington, DC: The National Academies Press.

Newton, K.
1997 Social capital and democracy. *American Behavioral Scientist, 40*(5), 575-586.

Nicholson, M.
1991 Negotiation, agreement and conflict resolution: The role of rational approaches and their criticism. In R. Väyrynen (Ed.), *New Directions in Conflict Theory: Conflict Resolution and Conflict Transformation* (Chapter 3). Newbury Park, CA: Sage.

Niemeyer, S., and Spash, C.
2001 Environmental valuation analysis, public deliberation, and their pragmatic synthesis: A critical appraisal. *Environment and Planning, 19,* 567-585.

Niemi, R.G.
1969 Majority decision-making with partial unidimensionality. *American Political Science Review, 63*(2, June), 488-497.

Nonet, P.
1980 The legitimation of purposive decisions. *California Law Review, 68,* 263-300.

Nordenstam, B., and Vaughan, E.
1991 Farmworkers and pesticide exposure: Perceived risk and self-protective behavior. In B.J. Garrick and W.C. Gekler (Eds.), *The Analysis, Communication and Perception of Risk.* New York: Plenum Press.

North, D.W.
1968 A tutorial introduction to decision theory. *IEEE Transactions on Systems Science and Cybernetics, 4,* 105-115.

1995 Limitations, definitions, principles, and methods of risk analysis. *Scientific and Technical Review, Office International des Epizooties, 14*(4), 913-923.

2003 Reflections on the Red/mis-read book, 20 years after. *Human and Ecological Risk Assessment, 9*(5), 1145-1154.

North, D.W., and Balson, W.E.
1985 Risk assessment and acid rain policy: A decision framework that includes uncertainty. In P. Mandelbaum (Ed.), *Acid Rain: Economic Assessment.* New York: Plenum Press.

North, D.W., and Merkhofer, M.W.
1976 A methodology for analyzing emission control strategies. *Computers and Operations Research, 3,* 185-207.

North, D.W., and Renn, O.
2005 *Decision Analytic Tools and Participatory Decision Processes.* Paper prepared for the Panel on Public Participation in Environmental Assessment and Decision Making, March, National Academy of Sciences, Washington, DC.

North, D.W., Selker, F., and Guardino, T.
  2002 The value of research on health effects of ingested inorganic arsenic. In D.J. Paustenbach (Ed.), *Human and Ecological Risk Assessment: Theory and Practice* (2nd ed.). New York: Wiley.

Nye, J.S., Jr., Zelikow, P.D., and King, D.C. (Eds.).
  1997 *Why People Don't Trust Government.* Cambridge, MA: Harvard University Press.

O'Conner, K.M.
  1994 *Negotiation Teams: The Impact of Accountability in Representation Structure on Negotiator Cognition and Performance.* Eugene, OR: International Association of Conflict Management.

Office of Management and Budget and President's Council on Environmental Quality
  2005 *Memorandum on Environmental Conflict Resolution.* Available: http://www.whitehouse.gov/ceq/joint-statement.pdf [accessed June 2008].

Office of Technology Assessment
  1992 Public involvement in forest planning. In *Forest Service Planning: Accommodating Uses, Producing Outputs, and Sustaining Ecosystems* (Ch. 5, pp. 77-108). Darby, PA: Diane.

Okrent, D.
  1998 Risk perception and risk management: On knowledge, resource allocation and equity. *Reliability Engineering and Systems Safety, 59,* 17-25.

O'Leary, R., and Bingham, L.B.
  2003 *The Promise and Performance of Environmental Conflict Resolution.* Washington, DC: Resources for the Future Press.

O'Leary, R.R., and Summers, S.
  2001 Lessons learned from two decades of alternative dispute resolution programs and processes at the U.S. Environmental Protection Agency. *Public Administration Review, 61,* 682-692.

Olson, M.
  1965 *The Logic of Collective Action.* Cambridge, MA: Harvard University Press.
  1984 *Participatory Pluralism: Political Participation and Influence in the United States and Sweden.* Chicago: Nelson-Hall.

Omenn, G.S.
  2003 On the significance of "The Red Book" in the evolution of risk assessment and risk management. *Human and Ecological Risk Assessment, 9,* 1155-1167.

Organisation for Economic Co-Operation and Development
  2001 *Engaging Citizens in Policy-making: Information, Consultation and Public Participation.* (OECD Public Management Policy Brief, PUMA Policy Brief No. 10.) Available: http://www.oecd.org/dataoecd/24/34/2384040.pdf [accessed June 2008].

O'Rourke, D., and Macey, G.
  2003 Community environmental policing: Assessing new strategies of public participation in environmental regulation. *Journal of Policy Analysis and Management,* 22(30), 383-414. Available: http://nature.berkeley.edu/orourke/PDF/CEP-JPAM.pdf [accessed August 2008].

Osterman, P.
  2002 *Gathering Power.* Boston, MA: Beacon Press.

Ostrom, E.
  1990 *Governing the Commons: The Evolution of Institutions for Collective Action.* New York: Cambridge University Press.
  1998 A behavioral approach to the rational choice theory of collective action, *American Political Science Review,* 92(1), 1-22.

O'Toole, L.J., and Meier, K.J.
 2004 Desperately seeking Selznick: Cooptation and the dark side of public management in networks. *Public Administration Review,* 64(6), 681-693.
Ozawa, C.P.
 1991 *Recasting Science: Consensual Procedures in Public Policy Making.* Boulder, CO: Westview Press.
Parkinson, J.
 2006 *Deliberating in the Real World: Problems of Legitimacy in Deliberative Democracy.* New York: Oxford University Press.
Passy, F.
 2003 Social networks matter. But how? In M. Diani and D. McAdam (Eds.), *Social Movements and Networks: Rational Approaches to Collective Action* (pp. 21-48). Oxford, England: Oxford University Press.
Pateman, C.
 1970 *Participation and Democratic Theory.* Cambridge, England: Cambridge University Press.
Patil, S.R., and Frey, H.C.
 2004 Comparison of sensitivity analysis methods based on applications to a food safety risk assessment model. *Risk Analysis,* 24(3), 573-586.
Payne, J.W., Bettman, J.R., and Johnson, E.J.
 1992 Behavioral decision research: A constructive processing perspective. *Annual Review of Psychology,* 43, 87-131.
Payne, J.W., Bettman, J.R., and Schkade, D.A.
 1999 Measuring constructed preferences: Towards a building code. *Journal of Risk and Uncertainty,* 19, 243-270.
Peelle, E., Schweitzer, M., Munro, J., Carnes, S., and Wolfe, A.
 1996 *Factors Favorable to Public Participation Success.* (Paper prepared for the U.S. Department of Energy.) Presented at the Annual Conference of the National Association of Environmental Professionals: Practical Environmental Directions: A Changing Agenda, June 2-6, Houston, TX. Available: http://www.osti.gov/energycitations/servlets/purl/228492-EoYgiS/webviewable/228492.pdf [accessed June 2008].
Pellizoni, L.
 2001 The myth of the best argument: Power deliberation and reasons. *British Journal of Sociology,* 51(1), 59-86.
 2003 Uncertainty and participatory democracy. *Environmental Values,* 12, 195-224.
Pellow, D.N.
 1999 Framing emerging environmental movement tactics: Mobilizing consensus, demobilizing conflict. *Sociological Forum,* 14, 659-683.
 2002 *Garbage Wars: The Struggle for Environmental Justice in Chicago.* Cambridge, MA: MIT Press.
Perri 6
 1996 The morality of managing risk: Paternalism, prevention and precaution, and the limits of proceduralism. *Journal of Risk Research,* 3(2), 135-165. Available: http://www.hsmc.bham.ac.uk/staff/staffdetails/6p/pdfs/P6%20Morality%20of%20managing%20risk.pdf [accessed June 2008].
Perry, J.L.
 2000 Bringing society in: Toward a theory of public-service motivation. *Journal of Public Administration Research and Theory,* 10(2), 471-489.
Peters, E., and Slovic, P.
 1996 The role of affect and worldviews as orienting dispositions in the perception and acceptance of nuclear power. *Journal of Applied Social Psychology,* 26, 1427-1453.

Peterson, C., and Stunkard, A.J.
  1989 Personal control and health promotion. *Social Science and Medicine,* 28, 819-828.
Pidgeon, N.F., Poortinga, W., Rowe, G., Horlick-Jones, T., Walls, J., and O'Riordan, T.
  2005 Using surveys in public participation processes for risk decision making: The case of the 2003 British GM nation? Public Debate. *Risk Analysis,* 25(2), 467-479.
Pierce Colfer, C.J.
  2005 *The Complex Forest: Communities, Uncertainty, and Adaptive Collaborative Management.* Washington, DC: Resources for the Future.
Pierson, P.
  2000 Increasing returns, path dependence, and the study of politics. *American Political Science Review,* 94(2), 251-267.
Pinkerton, E.
  1994 Local fisheries co-management: A review of international experiences and their implications for British Columbia salmon management. *Canadian Journal of Fisheries and Aquatic Sciences,* 51(10), 2363-2378.
Piven, F.F., and Cloward, R.A.
  1971 *Regulating the Poor: The Functions of Public Welfare.* New York: Pantheon Books.
Plumlee, J.P., Starling, J.D., and Kramer, K.W.
  1985 Citizen participation in water quality planning. *Administration & Society,* 16(4), 455-473.
Policy Consensus Initiative
  1999 *A Practical Guide to Consensus.* Portland, OR: Author.
Pomeroy, R.S., and Rivera-Guieb, R.
  2006 *Fishery Co-Management: A Practical Handbook.* Wallingford, England: CABI.
Posavac, E.J.
  1991 *Program Evaluation: Methods and Case Studies.* Upper Saddle River, NJ: Prentice-Hall.
Powell, Bingham, G., Jr.
  1986 American voter turnout in comparative perspective. *American Political Science Review,* 80(1), 17-43.
Presidential/Congressional Commission on Risk Assessment and Risk Management
  1997a *Framework for Environmental Health Risk Management* (vol. 1, final report). Washington, DC: Author. Available: http://www.riskworld.com/Nreports/1997/risk-rpt/pdf/EPAJAN.PDF [accessed June 2008].
  1997b *Risk Assessment and Risk Management in Regulatory Decision-Making* (vol. 2, final report). Washington, DC: Author. Available: http://www.riskworld.com/Nreports/1997/risk-rpt/volume2/pdf/v2epa.PDF [accessed June 2008].
Price, V., and Capella, J.
  2001 Online deliberation and its influence: The electronic dialogue project in campaign 2000. *IT and Society,* 1(1), 303-328. Available: http://www.stanford.edu/group/siqss/itandsociety/v01i01/v01i01a20.pdf [accessed June 2008].
Price, V., Goldthwaite, D., Cappella, J.N., and Romantan, A.
  2003 *Online Discussion, Civic Engagement, and Social Trust.* Paper presented at the 2nd Annual Pre-APSA Conference on Political Communication, Conference on Mass Communication and Civil Engagement, Georgetown University, Washington, DC.
Price, V., Nir, L., and Cappella, J.
  2002 Does disagreement contribute to more deliberative opinion? *Political Communication,* 19, 95-112.

Pritzker, D.M., and Dalton, D.S.
   1990   *Negotiated Rulemaking Sourcebook.* Washington, DC: Administrative Conference of the United States.
Proctor, J.D.
   1998   Environmental values and popular conflict over environmental management: A comparative analysis of public comments on the Clinton forest plan. *Environmental Management,* 22, 347-358.
Pruitt, D., and Carnevale, P.J.
   1993   *Negotiation in Social Conflict.* Pacific Grove, CA: Brooks-Cole.
Pruitt, D., and Rubin, J.
   1986   *Social Conflict: Escalation, Stalemate, and Settlement.* New York: McGraw-Hill.
Pultzer, E., Maney, A., and O'Connor, R.E.
   1998   Ideology and elites' perceptions of the safety of new technologies. *American Journal of Political Science,* 42, 190-209.
Putnam, R.D.
   1993   *Making Democracy Work: Civic Traditions in Modern Italy.* Princeton, NJ: Princeton University Press.
   2000   *Bowling Alone: The Collapse and Revival of American Community.* New York: Simon and Schuster.
Pyle, W.
   2005   *Collective Action and Post-Communist Enterprise: The Economic Logic of Russia's Business Associations.* (WDI Working Paper #794.) Ann Arbor, MI: William Davidson Institute.
Quinn, R.E., and Rohrbaugh, J.W.
   1983   A spatial model of effectiveness criteria: Towards a competing values approach to organizational analysis. *Management Science,* 29(3), 363-377.
Ragin, C.C.
   1987   *The Comparative Method: Moving Beyond Qualitative and Quantitative Strategies.* Berkeley: University of California Press.
Ragin, C.C., and Becker, H.S. (Eds.)
   1992   *What Is a Case?: Exploring the Foundations of Social Inquiry.* New York: Cambridge University Press.
Raiffa, H.
   1968   *Decision Analysis: Introductory Lectures on Choices Under Uncertainty.* Reading, PA: Addison-Wesley.
   1994   *The Art and Science of Negotiation* (12th ed.). New York: Cambridge University Press.
   2005   *The Art and Science of Negotiation: How to Resolve Conflicts and Get the Best out of Bargaining.* Cambridge, MA: Harvard University Press.
   2007   *Negotiation Analysis: The Science and Art of Collaborative Decision Making.* Cambridge, MA: Harvard University Press.
Rauschmayer, F., and Wittmer, H.
   2006   Evaluating deliberative and analytical methods for the resolution of environmental conflicts. *Land Use Policy,* 23, 108-122.
Raven, B.H., and Rubin, J.Z.
   1983   *Social Psychology* (2nd ed.). New York: Wiley.
Rawls, J.
   1971   *A Theory of Justice.* Cambridge, MA: Belknap.
Reagan, M., and Fedor-Thurman, V.
   1987   Public participation: Reflections on the California energy policy experience. In J. DeSario and S. Langton (Eds.), *Citizen Participation in Public Decision Making* (pp. 89-113). Westport, CT: Greenwood Press.

Reid, W.V., Mooney, H.A., Cropper, A., Capistrano, D., Carpenter, S.R., Chopra, K., Dasgupta, P., Dietz, T., Duraiappah, A.K., Hassan, R., Kasperson, R., Leemans, R., May, R.M., McMichael, T., Pingali, P., Samper, C., Sholes, R., Watson, R.T., Zakri, A.H., Shidong, Z., Ash, N.J., Bennett, E., Kumar, P., Lee, M.J., Raudsepp-Hearne, C., Simons, H., Thonell, J., and Zurek, M.B.
  2005 *Ecosystems and Human Well-Being: Synthesis*. Washington, DC: Island Press.
Renn, O.
  1999 A model for an analytic-deliberative process in risk management. *Environmental Science and Technology*, 33, 3049-3055.
  2004 The challenge of integrating deliberation and expertise: Participation and discourse in risk management. In T.L. McDaniels and M.J. Small (Eds.), *Risk Analysis and Society: An Interdisciplinary Characterization of the Field* (pp. 289-366). Cambridge, England: Cambridge University Press.
  2005 *Risk Governance: Towards an Integrative Approach*. Geneva, Switzerland: International Risk Governance Council.
  2008 *Risk Governance: Coping with Uncertainty in a Complex World*. London, England: Earthscan.
Renn, O., and Schweizer, P.
  In press New models of citizen participation in Germany and France. In O. Gabriel (Ed.), *Political Governance Revisited: A Comparative Approach*. Bordeaux, France: Lawrence Erlbaum Associates.
Renn, O., and Walker, K.
  2008 Lessons learned and a way forward. In O. Renn and K. Walker (Eds.), *Global Risk Governance. Concept and Practice of Using the IRGC Framework* (pp. 331-367). Dordrecht, the Netherlands: Springer.
Renn, O., Webler, T., and Wiedemann, P. (Eds.)
  1995 *Fairness and Competence in Citizen Participation: Evaluating Models for Environmental Discourse*. Boston, MA: Kluwer Academic.
Renn, O., Webler, T., Rakel, H., Johnson, B., and Dienel, P.
  1993 A three-step procedure for public participation in decision making. *Policy Sciences*, 26, 189-214.
Revkin, A.C.
  2004 Bush's science aide rejects claims of distorted facts. *New York Times* (April 3), p. A9.
Rich, R.C., Edelstein, M., Hallman, W.K., and Wandersman, A.H.
  1995 Citizen participation and empowerment: The case of local environmental hazards. *American Journal of Community Psychology*, 23, 657-675.
Richardson, H.S.
  2002 *Democratic Autonomy: Public Reasoning about the Ends of Policy*. New York: Oxford University Press.
Rittel, H.W.J., and Webber, M.M.
  1973 Dilemmas in a general theory of planning. *Policy Sciences*, 4, 155-169.
Rivera, F.G., and Erlich, J.L.
  1998 *Community Organizing in a Diverse Society* (3rd ed.). Boston, MA: Allyn and Bacon.
Rivers, L.I.
  2006 *Risk Perception and Decision-Making in Minority and Marginalized Communities*. Columbus: Ohio State University.
Rivers, L.I., and Arvai, J.
  2007 Win some, lose some: The effect of chronic losses on decision making under risk. *Journal of Risk Research*, 10(8), 1085-1099.

Roberts, N.
2004 Public deliberation in an age of direct citizen participation. *American Review of Public Administration*, 34(4), 315-353.

Roch, I.
1997 Evaluation der 3. Phase des Bürgerbeteiligungsverfahrens in der Region Nordschwarzwald. (Research Report No. 71.) Stuttgart, Germany: Akademie für Technikfolgenabschätzung.

Rohrmann, B.
1992 The evaluation of risk communication effectiveness. *Acta Psychologica*, 81, 169-192.
1999 *Risk Perception Research: Review and Documentation.* (Studies in Risk Communication No. 68.) Juelich, Germany: Research Center Juelich-MUT.

Rohrmann, B., and Renn, O.
2000 Introduction. In O. Renn and B. Rohrmann (Eds.), *Cross-Cultural Risk Perception* (pp. 5-32). Boston, MA: Kluwer.

Rosa, E.
1998 Meta-theoretical foundations for post-normal risk. *Journal of Risk Research*, 1, 15-44.

Rosa, E.A., Matsuda, N., and Kleinhesselink, R.R.
2000 The cognitive architecture of risk: Pancultural unity or cultural shaping? In O. Renn and B. Rohrmann (Eds.), *Comparative Risk Perception* (pp. 185-210). Dordrecht, the Netherlands: Kluwer Academic.

Rosa, E.A., Renn, O., and McCright, A.
2007 *The Risk Society: Theoretical Frames and Management Challenges of Post-Modernity.* Department of Sociology, Washington State University, Pullman, WA.

Rose-Ackerman, S.
1994 Consensus versus incentives: A skeptical look at regulatory negotiation. *Duke Law Journal*, 43(6), 1206-1220, Twenty-Fifth Annual Administrative Law Issue.

Rosenbaum, N.
1978 Citizen participation and democratic theory. In S. Langton (Ed.), *Citizen Participation in America* (pp. 43-54). Lexington, MA: Lexington Books.

Rosener, J.B.
1981 User-oriented evaluation: A new way to view citizen participation. *Journal of Applied Behavioral Studies*, 17, 583-596.
1982 Making bureaucracy responsive: A study of the impacts of citizen participation and staff recommendations on regulatory decision making, *Public Administration Review*, 42, 339-345.

Rosenstone, S.J., and Hansen, J.M.
1993 *Mobilization, Participation, and Democracy in America.* New York: Macmillan.

Ross, M.H.
1993 *The Management of Conflict: Interpretations and Interests in Comparative Perspective.* New Haven, CT: Yale University Press.

Rossi, J.
1997 Participation run amok: The costs of mass participation for deliberative agency decision making. *Northwestern University Law Review*, 92, 173-249.

Rowe, G., and Frewer, L.J.
2000 Public participation methods: A framework for evaluation. *Science, Technology, and Human Values*, 25(1), 3-29. Available: http://sth.sagepub.com/cgi/reprint/25/1/3 [accessed June 2008].
2004 Evaluating public-participation exercises: A research agenda. *Science, Technology, and Human Values*, 29(4), 512-556.

Rowe, G., Marsh, R., and Frewer, L.J.
   2004   Evaluation of a deliberative conference. *Science, Technology, and Human Values,* 29(1), 88-121.
Saaty, T.L.
   1990   *Multicriteria Decision Making: The Analytic Hierarchy Process, Planning Priority Setting, Resource Allocation.* Berlin, Germany: RWS.
Sabatier, P.A., and Jenkins-Smith, H. (Eds.)
   1993   *Policy Change and Learning: An Advocacy Coalition Approach.* Boulder, CO: Westview Press.
Sabatier, P.A., and Weible, C.M.
   2007   The advocacy coalition framework: Innovation and clarification. In P.A. Sabatier (Ed.), *Theories of the Policy Process* (pp. 189-222). Boulder, CO: Westview Press.
Sabatier, P.A., Focht, W., Lubell, M., Trachtenberg, Z., Vedlitz, A., and Matlock, M.
   2005   *Swimming Upstream: Collaborative Approaches to Watershed Management.* Cambridge, MA: MIT Press.
Sabel, C.F., Fung, A., and Karkkainen, B.C.
   2000   Beyond Backyard Environmentalism. Boston, MA: Beacon Press.
Saegert, S., Thompson, J.P., and Warren, M.R.
   2002   *Social Capital and Poor Communities.* New York: Russell Sage Foundation.
Sager, T.
   1994   *Communicative Planning Theory.* Aldershot, England: Avebury.
Salisbury, R.
   1969   An exchange theory of interest groups. *Midwest Journal of Political Science,* 13(1), 1-32.
Sally, D.
   1995   Conversation and cooperation in social dilemmas: A meta-analysis of experiments from 1958 to 1992. *Rationality and Society,* 7, 58-92.
Sampson, R.J., Raudenbush, S., and Earls, F.
   1997   Neighborhoods and violent crime: A multi-level study of collective efficacy. *Science,* 277(5328), 918-924.
Sanders, L.
   1997   Against deliberation. *Political Theory,* 25(3), 347-376.
Saunders, H.H.
   1999   *A Public Peace Process: Sustained Dialogue to Transform Ethnic and Racial Conflicts.* New York: St. Martin's Press.
Savage, L.J.
   1954   *The Foundations of Statistics.* New York: Wiley.
Scherer, K.R.
   1997   The role of culture in emotion-antecedent appraisal. *Journal of Personality and Social Psychology,* 73, 902-922.
Schier, S.E.
   2000   *By Invitation Only: The Rise of Exclusive Politics in the United States.* Pittsburgh: University of Pittsburgh Press.
Schkade, D., Sunstein, C.R., and Kahneman, D.
   2000   Deliberating about dollars: The severity shift. *Columbia Law Review,* 100, 1139-1175.
Schlozman, K.L., and Tierney, J.
   1986   *Organized Interests and American Democracy.* New York: Harper and Row.
Schneider, A., and Ingram, H.
   1997   *Policy Design for Democracy.* Lawrence: University Press of Kansas.

Schneider, M., Scholz, J.T., Lubell, M., Mindruta, D., and Edwardsen, M.
  2003 Building consensual institutions: Networks and the National Estuary Program. *American Journal of Political Science*, 47, 143-158.
Schoenbrod, D.
  1983 Limits and dangers of environmental mediation: A review essay. *New York University Law Review*, 58(December), 1453-1476.
Schudson, M.
  1997 Why conversation is not the soul of democracy. *Critical Studies of Mass Communication*, 14, 297-309.
Schuett, M., Selin, S., and Carr, D.
  2001 Making it work: Keys to successful collaboration in natural resource management. *Environmental Management*, 27(4), 587-593.
Schultz-Hardt, S., Frey, D., Luthgens, C., and Moscovici, S.
  2000 Biased information search in group decision making. *Journal of Personality and Social Psychology*, 78, 655-669.
Schuman, S.
  2005 *The IAF Handbook of Group Facilitation: Best Practices from the Leading Organization in Facilitation*. San Francisco: Jossey-Bass.
Schumpeter, J.A.
  1942 *Capitalism, Socialism and Democracy*. New York: Harper and Row.
Schwartz, R.
  2002 *The Skilled Facilitator*. San Francisco: Jossey-Bass.
Schwarz, M., and Thompson, M.
  1990 *Divided We Stand: Re-defining Politics, Technology and Social Choice*. Philadelphia: University of Pennsylvania Press.
Sclove, R.
  1995 *Democracy and Technology*. New York: Guilford Press.
Scott, W.R.
  1992 *Organizations: Rational, Natural, and Open Systems*. Englewood Cliffs, NJ: Prentice-Hall.
Selin, S., and Chavez, D.
  1994 Characteristics of successful tourism partnerships: A multiple case study design (characteristics of successful tourism partnerships). *Journal of Park & Recreation Administration*, 12(2), 51-62.
Selin, S., and Myers, N.
  1995 Correlates of partnership effectiveness: The coalition for unified recreation in Eastern Sierra. *Journal of Park & Recreation Administration*, 13(4), 37-46.
Selin, S., Schuett, M., and Carr, D.
  2000 Modeling stakeholder perceptions of collaboration initiative effectiveness. *Society and Natural Resources*, 13, 735-745.
Selznick, P.A.
  1949 *TVA and the Grassroots: A Study in the Sociology of Formal Organization*. Berkeley: University of California Press.
Sen, A.
  1999 *Development as Freedom*. New York: Random House.
Shah, D.V., Domke, D., and Wackman, D.B.
  1996 To thine own self be true: Values, framing and voter decision making strategies. *Communications Research*, 23, 509-560.
Shannon, M.A.
  1987 Forest planning: Learning with people. In M.L. Miller, R.P. Gale, and P.J. Brown (Eds.), *Social Science in Natural Resource Management Systems* (pp. 233-252). Boulder, CO: Westview Press.

1991 *Building Public Decisions: Learning Through Planning*. Washington, DC: Office of Technology Assessment.
2003 The Northwest Forest Plan as a learning process: A call for new institutions bridging science and politics. In K. Arabase and J. Bowersox (Eds.), *Forest Futures: Science, Politics and Policy for the Next Century* (pp. 256-279). New York: Rowman and Littlefield.

Shannon, M.A., and Antypas, A.R.
1996 Civic science is democracy in action. *Northwest Science, 70*(1), 66-69.

Shannon, P., and Walker, P.
2006 *Partnerships and Control: Lessons from a Research Program into Strategies for Deliberative Governance*. Refereed paper presented to the Governments and Communities in Partnership Conference, Centre for Public Policy, September, University of Melbourne.

Shapiro, I.
1996 Elements of democratic justice. *Political Theory, 24*(4), 579-619.
1999 Enough of deliberation: Politics is about interests and power. In S. Macedo (Ed.), *Deliberative Politics: Essays on Democracy and Disagreement* (pp. 28-38). New York: Oxford University Press.
2003 *The State of Democratic Theory*. Princeton, NJ: Princeton University Press.

Sherington, M.V.
1997 Participatory research methods: Implementation, effectiveness, and institutional context. *Agricultural Systems, 55*(2), 195-216.

Shindler, B., and Neburka, J.
1997 Public participation in forest planning: 8 attributes of success. *Journal of Forestry, 95*(1), 17-19.

Shutkin, W.A.
2000 *The Land That Could Be: Environmentalism and Democracy in the Twenty-First Century*. Cambridge, MA: MIT Press.

Siegel-Jacobs, K., and Yates, J.F.
1996 Effects of procedural and outcome accountability on judgment quality. *Organizational and Human Decision Processes, 65*, 1-17.

Siegrist, M., Earle, T.C., and Gutscher, H. (Eds.)
2007 *Trust in Cooperative Risk Management: Uncertainty and Skepticism in the Public Mind*. London, England: Earthscan.

Silva, C.L., Jenkins-Smith, H.C., and Barke, R.P.
2007 Reconciling scientists' beliefs about radiation risks and social norms: Explaining preferred radiation protection standards. *Risk Analysis, 27*(3), 755-773.

Simmel, G.
1955 Conflict. In K.H. Wolff (Ed.), *Conflict and the Web of Group Affiliations* (pp. 3-17). New York: Free Press.

Simonson, I., and Staw, B.M.
1992 Deescalation strategies: A comparison of techniques for reducing commitment to losing courses of action. *Journal of Applied Psychology, 77*, 419-426.

Sirianni, C., and Friedland, L.
2001 *Civic Innovation in America*. Berkeley: University of California Press.

Sjöberg, L.
1999 Risk perception in Western Europe. *Ambio, 28*, 543-549.

Skocpol, T.
1996 Unraveling from above. *American Prospect, 25*, 20-25.

Slimak, M.W., and Dietz, T.
2006 Personal values, beliefs and ecological risk perception. *Risk Analysis, 26*, 1689-1705.

Slovic, P.
1987 Perception of risk. *Science, 236,* 280-285.
1992 Perception of risk: Reflections on the psychometric paradigm. In D. Golding and S. Krimsky (Eds.), *Theories of Risk* (pp. 117-152). London, England: Praeger.
1995 The construction of preference. *American Psychologist, 50,* 364-371.
1999 Trust, emotion, sex, politics, and science: Surveying the risk-assessment battlefield. *Risk Analysis, 19,* 689-701.
2000 *The Perception of Risk.* London, England: Earthscan.

Slovic, P., Fischhoff, B., and Lichtenstein, S.
1979 Rating the risks. *Environment, 21,* 14-20, 30, 36-39.
1980 Facts vs. fears: Understanding perceived risk. In R. Schwing and W.A. Albers (Eds.), *Societal Risk Assessment: How Safe is Safe Enough?* New York: Plenum Press.
1985 Characterizing perceived risk. In R.W. Kates, C. Hohenemser, and J.X. Kasperson (Eds.), *Perilous Progress: Managing the Hazards of Technology* (pp. 91-125). Boulder, CO: Westview Press.
1986 The psychometric study of risk perception. In V.T. Covello, J. Menkes, and J. Mumpower (Eds.), *Risk Evaluation and Management* (pp. 3-24). New York: Praeger.

Slovic, P., Kraus, N., Lappe, H., and Major, L.
1991 Risk perception of prescription drugs: Report on a survey in Canada. *Canadian Journal of Public Health, 82,* S15-S20.

Slovic, P., Malmfors, T., Krewski, D., Mertz, C.K., Neil, N., and Bartlett, S.
1995 Intuitive toxicology. II. Expert and lay judgments of chemical risks in Canada. *Risk Analysis, 15,* 661-675.

Smith, V.K., and Desvouges, W.H.
1986 *Measuring Water Quality Benefits.* Boston, MA: Kluwer.

Snow, D.A., Rochford, E.B., Jr., Worden, S.K., and Benford, R.D.
1986 Frame alignment processes, micromobilization, and movement participation. *American Sociological Review, 51,* 464-481.

Society of Professionals in Dispute Resolution
1992 *Competencies for Mediators of Complex Public Disputes.* Washington, DC: Author.

Sommarstrom, S., and Huntington, C.
1999 *An Evaluation of Selected Watershed Councils in the Pacific Northwest and Northern California.* (Report prepared for Trout Unlimited and Pacific Rivers Council.) Eugene, OR: Pacific Rivers Council.

Spetzler, C.S., and Staël von Holstein, C.-A.
1975 Probability Encoding in Decision Analysis. *Management Science, 22,* 340-352.

SPIDR
1997 *Best Practices for Government Agencies: Guidelines for Using Collaborative Agreement-Seeking Processes.* Report and Recommendations of the SPIDR Environment/Public Disputes Sector Critical Issues Committee, adopted by the SPIDR Board, January. Washington, DC: Society of Professionals in Dispute Resolution. Available: http://law.gsu.edu/cncr/pdf/papers/BestPracticesforGovtAgenices.pdf [accessed September 2008].

Stanovich, K.E., and West, R.F.
2000 Individual differences in reasoning: Implications for the rationality debate? [with commentaries]. *Behavioral and Brain Sciences, 23,* 645-726.

Starr, C.
1969 Societal benefit versus technological risk. *Science, 236,* 280-285.

Steelman, T., and Ascher, W.
 1997 Public involvement methods in natural resource policy making: Advantages, disadvantages and tradeoffs. *Policy Sciences, 30,* 71-90.
Sterman, J.D., and Sweeney, L.B.
 2002 Cloudy skies: Assessing public understanding of global warming. *Systems Dynamics Review, 18,* 207-240.
Stern, P.C.
 1991 Learning through conflict: A realistic approach to risk communication. *Policy Sciences, 24,* 99-119.
 2003 *Toward a Conceptual Framework for the Public Participation Study.* Paper prepared for the National Research Council Panel on Public Participation in Environmental Assessment and Decision Making, April, National Academy of Sciences, Washington, DC.
 2005a Deliberative methods for understanding environmental systems. *BioScience, 55,* 976-982.
 2005b *Implications of Research on Small-Group Processes for Environmental Public Participation.* Paper prepared for the National Research Council Panel on Public Participation in Environmental Assessment and Decision Making, February. Available: http://www7.nationalacademies.org/hdgc/Tab%20_14%20Implications.pdf [accessed June 2008].
Stern, P.C., and Druckman, D.
 2000 Evaluating interventions in history: The case of international conflict resolution. *International Studies Review, 2*(1), 33-63.
Stern, P.C., Dietz, T., Abel, T., Guagnano, G.A., and Kalof, L.
 1999 A value-belief-norm theory of support for social movements: The case for environmentalism. *Human Ecology Review, 6,* 81-97.
Stern, P.C., Dietz, T., and Black, J.S.
 1986 Support for environmental protection: The role of moral norms. *Population and Environment, 8,* 204-222.
Stern, P.C., Dietz, T., and Ostrom, E.
 2002 Research on the commons: Lessons for environmental resource managers. *Environmental Practice, 4*(2), 61-64.
Stewart, J., Kendall, E., and Coote, A.
 1994 *Citizen Juries.* London, England: Institute for Public Research.
Stewart, R.
 1975 The reformation of American administrative law. *Harvard Law Review, 88,* 1667-1813.
Stewart, T.R., Dennis, R.L., and Ely, D.W.
 1984 Citizen participation and judgment in policy analysis: A case of urban air quality policy. *Policy Sciences, 17,* 67-87.
Stirling, A.
 2004 Opening up or closing down: Analysis, participation and power in the social appraisal of technology. In M. Leach, I. Scoones, and B. Wynne (Eds.). *Science and Citizens: Globalization and the Challenge of Engagement.* London, England: Zed Books.
 2006 Analysis, participation and power: Justification and closure in participatory multi-criteria analysis. *Land Use Policy, 23,* 95-107.
 2008 "Opening up" and "closing down": Power, participation, and pluralism in the social appraisal of technology. *Science, Technology, and Human Values, 33*(2), 262-294.
Stone, D.
 2002 *Policy Paradox: The Art of Political Decision Making.* New York: Norton.

Strange, J.H.
 1972 Citizen participation in community action and Model Cities Programs. *Public Administration Review, 32,* 655-669.
Sunstein, C.R.
 1997 *Free Markets and Social Justice.* New York: Oxford University Press.
 2001 *Designing Democracy: What Constitutions Do.* Oxford, England: Oxford University Press.
 2003 The law of group polarization. In J.S. Fishkin and P. Laslett (Eds.), *Debating Deliberative Democracy* (pp. 80-101). Malden, MA: Blackwell.
 2006 *Infotopia: How Many Minds Produce Knowledge.* Oxford, England: Oxford University Press.
Susskind, L.E., and Cruikshank, J.
 1987 *Breaking the Impasse: Consensual Approaches to Resolving Public Disputes.* New York: Basic Books.
Susskind, L.E., and Field, P.
 1996 *Dealing with an Angry Public: The Mutual Gains Approach to Resolving Disputes.* New York: The Free Press.
Susskind, L.E., Bacow, L., and Wheeler, M.
 1983 *Resolving Environmental Regulatory Disputes.* Cambridge, England: Schenkman.
Susskind, L.E., McKearnan, S., Thomas-Larmer, J., and the Consensus Building Institute
 1999 *Negotiating Environmental Agreements: How to Avoid Escalating Confrontation, Needless Costs, and Unnecessary Litigation.* New York: Island Press.
Susskind, L.E., Thomas-Larmer, J., and Levy, P.
 1999 *The Consensus Building Handbook: A Comprehensive Guide to Reaching Agreement.* Thousand Oaks, CA: Sage.
Svedsäter, H.
 2003 Economic valuation of the environment: How citizens make sense of contingent valuation questions. *Land Economics, 79,* 88-109.
Sweeney, R.L.
 2004 *Environmental Risk Decision-Making: Risk Professionals and Their Use of Analytic-Deliberative Processes.* Fairfax, VA: George Mason University.
Syme, G.J., and Sadler, B.S.
 1994 Evaluation of public involvement in water resources planning. *Evaluation Review, 18,* 523-542.
Tani, S.N.
 1978 *Decision Analysis of the Synthetic Fuel Commercialization Program.* National Computer Conference Proceedings, *American Federation of Information Processing Societies, 47,* 23-29. (Reprinted in R.A. Howard and J.E. Matheson (Eds.), *Readings on the Principles and Applications of Decision Analysis.* Menlo Park, CA: Strategic Decisions Group.)
Tarrow, S.
 1998 *Power in Movement.* New York: Cambridge University Press.
Taylor, D.E.
 2000 The rise of the environmental justice paradigm: Injustice framing and the social construction of environmental discourses. *American Behavioral Scientist, 43,* 508-580.
Taylor, M., and Singleton, S.
 1993 The communal resource: Transaction costs and the solution of collective action problems. *Politics and Society, 21,* 195-214.
Tetlock, P.E., and Belkin, A. (Eds.)
 1996 *Counterfactual Thought Experiments in World Politics.* Princeton, NJ: Princeton University Press.

Thibaut, J., and Walker, L.
   1975   *Procedural Justice: A Psychological Analysis.* Hillsdale, NJ: Lawrence Erlbaum Associates.
Thomas, J.C.
   1995   *Public Participation in Public Decisions: New Skills and Strategies for Public Managers.* San Francisco: Jossey-Bass.
Thompson, L., and Gonzalez, R.
   1997   Environmental disputes. In M. Bazerman, D.M. Messick, and K.A. Wade-Benzadi (Eds.), *Environmental Ethics and Behavior* (pp. 75-103). San Francisco: New Lexington Press.
Triandis, H.
   2000   Culture and conflict. *International Journal of Psychology, 35,* 145-152.
Tuler, S.
   2003   *Relationships Between Process, Context, and Outcomes: Review of Findings from Multi-Case Study Literature (N≥5).* Paper prepared for the National Research Council Panel on Public Participation in Environmental Assessment and Decision Making, April, National Academy of Sciences, Washington, DC.
Tuler, S., and Webler, T.
   1995   Process evaluation for discursive decision making in environmental and risk policy. *Human Ecological Review, 2,* 62-74.
   1999   Voices from the forest: Participants and planners evaluate a public policy making process. *Society and Natural Resources, 12*(5), 437-453.
Tversky, A., and Kahneman, D.
   1972   Judgment under uncertainty: Heuristics and biases. *Science, 185,* 1124-1131.
   1981   The framing of decisions and the psychology of choice. *Science, 211,* 453-458.
Tversky, A., Slovic, P., and Kahneman, D.
   1990   The causes of preference reversal. *American Economic Review, 80,* 204-217.
Tyler, T.R.
   2000   Social justice: Outcome and procedure. *International Journal of Psychology, 35*(2), 117-125.
Tyler, T.R., and Blader, S.
   2000   *Cooperation in Groups: Procedural Justice, Social Identity, and Behavioral Engagement.* Philadelphia: Psychology Press.
Tyler, T.R., and Lynd, E.A.
   1992   A relational model of authority in groups. *Advances in Experimental Social Psychology, 25,* 115-191.
United Nations Economic Commission for Europe
   1998   *Convention on Access to Information, Public Participation in Decision-Making and Access to Justice in Environmental Matters.* Aarhus, Denmark: Author. http://www.unece.org/env/pp/documents/cep43e.pdf [accessed June 2008].
U.S. Department of Interior, Bureau of Land Management
   1998   *Winning for the Public: A Strategy for Licensing and Relicensing Dams.* Washington, DC: Author.
U.S. Environmental Protection Agency
   1986   Guidelines for carcinogenic risk assessment. (EPA/630/R-00/004.) *Federal Register,* 51(185), 33992-34003. Available: http://www.epa.gov/ncea/raf/car2sab/guidelines_1986.pdf [accessed June 2008].
   1992   *Community Relations in Superfund: A Handbook.* Washington, DC: U.S. Government Printing Office.
   1998   *Stakeholder Involvement Action Plan.* Washington, DC: Author.

2000a *Engaging the American People: A Review of EPA's Public Participation Policy and Regulations with Recommendations for Action.* (EPA 240-R-00-005.) Washington, DC: Author. Available: http://www.epa.gov/stakeholders/pdf/eap_report.pdf [accessed June 2008].

2000b *Guide on Consultation and Collaboration with Indian Tribal Governments and the Public Participation of Indigenous Groups and Tribal Members in Environmental Decision Making.* (Prepared by the National Environmental Justice Advisory Council Indigenous Peoples Subcommittee.) Washington, DC: Author. Available: http://www.lm.doe.gov/env_justice/pdf/ips_consultation_guide.pdf [accessed June 2008].

2001 *Stakeholder Involvement & Public Participation at the U.S. EPA: Lessons Learned, Barriers, & Innovative Approaches.* (EPA-100-R-00-040.) Washington, DC: Author. Available: http://www.epa.gov/stakeholders/pdf/sipp.pdf [accessed June 2008].

2005 Guidelines for carcinogenic risk assessment. (EPA/630/P-03/001B.) *Federal Register,* 70(66), 17765-17817. Available: http://www.epa.gov/IRIS/cancer032505.pdf [accessed June 2008].

U.S. Environmental Protection Agency Science Advisory Board

2001 *Improved Science-Based Environmental Stakeholder Processes.* (EPA-SAB-EC-COM-01-006.) Washington, DC: Author. Available: http://www.epa.gov/sab/pdf/eccm01006.pdf [accessed June 2008].

U.S. National Assessment Synthesis Team

2000 *Climate Change Impacts on the United States: The Potential Consequences of Climate Variability and Change.* Cambridge, England: Cambridge University Press.

U.S.D.A. Forest Service

1997 *Sustaining the People's Lands: Recommendations for Stewardship of the National Forests and Grasslands into the Next Century.* Washington, DC: Author.

2000 *Collaborative Stewardship Within the Forest Service: Findings and Recommendations from the National Collaborative Stewardship Team.* Available: http://www.partnershipresourcecenter.org/resources/publications/docs/Report_National_Collaborative_Stewardship_Team.doc [accessed June 2008].

2002 *The Process Predicament: How Statutory, Regulatory, and Administrative Factors Affect National Forest Management.* Washington, DC: Author.

U.S.D.A. Forest Service Inventory and Monitoring Institute and Business Genetics

2001 *Report Abstract: Reflecting Complexity & Impact of Laws on a USDA Forest Service Project.* Fort Collins, CO: Author.

U.S.D.A. Office of General Counsel Natural Resources Division

2002 *Overview of Forest Planning and Project Level Decision Making.* Washington, DC: Author.

Van Asselt, M.B.A.

2000 *Perspectives on Uncertainty and Risk.* Boston, MA: Kluwer.

Van de Wetering, S.B.

2006 *The Legal Framework for Cooperative Conversation.* (Collaborative Governance Report #1.) Missoula: Public Policy Research Institute, University of Montana.

Van den Belt, M.

2004 *Mediating Modeling: A Systems Dynamics Approach to Environmental Consensus Building.* Washington, DC: Island Press.

Van den Bos, K.

2001 Uncertainty management: The influence of uncertainty salience on reactions to perceived procedural fairness. *Journal of Personality and Social Psychology, 80,* 931-941.

Van den Bos, K., and Lind, E.A.
2002 Uncertainty management by means of fairness judgments. *Advances in Experimental Social Psychology*, 34, 1-60.
van den Daele, W.
1992 Scientific evidence and the regulation of technical risks: Twenty years of demythologizing the experts. In N. Stehr and R.V. Ericson (Eds.), *The Culture and Power of Knowledge. Inquiries into Contemporary Societies* (pp. 323-340). Berlin, Germany: de Gruyter.
Van Horn, C.E.
1988 *Breaking the Environmental Gridlock*. New Brunswick, NJ: The Eagleton Institute of Politics, Rutgers University.
Vaughan, E.
1993 Individual and cultural differences in adaptation to environmental risks. *American Psychologist*, 48, 673-680.
Vaughan, E., and Nordenstam, B.
1991 The perception of environmental risks among ethnically diverse groups in the United States. *Journal of Cross-Cultural Psychology*, 22, 29-60.
Vaughan, E., and Seifert, M.
1992 Variability in the framing of risk issues. *Journal of Social Issues*, 48, 119-135.
Ventriss, C., and Kuentzel, W.
2005 Critical theory and the role of citizen involvement in environmental decision making: A re-examination. *International Journal of Organization Theory and Behavior*, 8(4), 519-539. Available: http://www.uvm.edu/envnr/welcome/gradpages/pdf_files/Ventriss_Kuentzel.pdf [accessed August 2008].
Verba, S., and Nie, N.H.
1972 *Participation in America*. New York: Harper and Row.
Verba, S., Schlozman, K., and Brady, H.
1994 *Voice and Equality: Civic Voluntarism in American Politics*. Cambridge, MA: Harvard University Press.
Verba, S., Schlozman, K.L., Brady, H.E., and Nie, N.H.
1993 Race, ethnicity, and political resources: Participation in the United States. *British Journal of Political Science*, 23, 453-497.
Viscusi, W.K.
1998 *Rational Risk Policy. The 1996 Arne Ryde Memorial Lectures*. New York: Oxford University Press.
von Winterfeldt, D., and Edwards, W.
1986 *Decision Analysis and Behavioral Research*. Cambridge, England: Cambridge University Press.
Vorwerk, V., and Kämper, E.
1997 *Evaluation der 3. Phase des Bürgerbeteiligungsverfahrens in der Region Nordschwarzwald*. (Working Report No. 70.) Stuttgart, Germany: Center of Technology Assessment.
Wade, R.
1994 *Village Republics: Economic Conditions for Collective Action in South India*. San Francisco: ICP Press. (Original work published in 1988, Cambridge University Press.)
Walinsky, A.
1969 Review: Daniel Moynihan's *Maximum Feasible Misunderstanding: Community Action in the War on Poverty*. *New York Times*, October 5. Available: http://www.nytimes.com/books/98/10/04/specials/moynihan-community.html [accessed August 2008].

Waller, T.
1995 Knowledge, power and environmental policy: Expertise, the lay public and water management in the Western United States. *The Environmental Professional*, 7, 153-166.

Wallsten, T.S., and Budescu, D.V.
1983 Encoding subjective probabilities: A psychological and psychometric review. *Management Science*, 29, 151-173.

Walters, L.C., Aydelotte, J., and Miller, J.
2000 Putting more public in policy analysis. *Public Administration Review*, 60(4), 349-359.

Walters, L.C., Balint, P.J., Desai, A., and Stewart, R.E.
2003 *Risk and Uncertainty in Management of the Sierra Nevada National Forests.* USDA Forest Service, Pacific Southwest Region. Available: http://64.233.169.104/ search?q=cache:nUmyDULYquYJ:gunston.doit.gmu.edu/snfpa_risk/Finalreport. doc+Risk+and+Uncertainty+in+Management+of+the+Sierra+Nevada+National+ Forests&hl=en&ct=clnk&cd=1&gl=us [accessed September 2008].

Walton, R., and McKersie, R.
1965 *A Behavioral Theory of Labor Negotiations.* New York: McGraw-Hill.

Warren, M.
1992 Democratic theory and self-transformation. *American Political Science Review*, 86, 8-23.

Warren, M.E., and Pearse, H.
2008 *Designing Deliberative Democracy: The British Columbia Citizens' Assembly.* New York: Cambridge University Press

Warren, M.R.
2001 *Dry Bones Rattling: Community Building to Revitalize American Democracy.* Princeton, NJ: Princeton University Press.

Watson, M., Bulkeley, H., and Hudson, R.
2004 *Vertical and Horizontal Integration in the Governance of UK Municipal Waste Policy.* Paper presented at the IDHP Berlin Conference on the Human Dimensions of Global Environmental Change, Greening of Policies: Interlinkages and Policy Integration, Dec. 3-4, Freie Universität, Berlin.

Weber, E.P.
2003 *Bringing Society Back In: Grassroots Ecosystem Management, Accountability, and Sustainable Communities.* Cambridge, MA: MIT Press.

Weber, E.U., and Hsee, C.
1998 Cross-cultural differences in risk perception but cross-cultural similarities in attitudes towards perceived risk. *Management Science*, 44, 1205-1217.

Webler, T.
1995 "Right" discourse in citizen participation: An evaluative yardstick. In O. Renn, T. Webler, and P. Wiedemann (Eds.), *Fairness and Competence in Citizen Participation: Evaluating Models for Environmental Discourse* (pp. 33-84). Boston, MA: Kluwer Academic.

1999 The craft and theory of public participation: A dialectical process. *Journal of Risk Research*, 2(1), 55-71.

Webler, T., and Renn, O.
1995 A brief primer on participation: Philosophy and practice. In O. Renn, T. Webler, and P. Wiedemann (Eds.). *Fairness and Competence in Citizen Participation: Evaluating Models for Environmental Discourse* (pp. 17-33). Boston, MA: Kluwer Academic.

Webler, T., and Tuler, S.
- 2008 Organizing a deliberative participatory process: What does the science say? In S. Odugbemi and T. Jacobson (Eds.), *Governance Reform Under Real-World Conditions: Citizens, Stakeholders, and Voice* (pp. 125-160). Washington, DC: World Bank.

Webler, T., Rakel, H., Renn, O., and Johnson, B.
- 1995 Eliciting and classifying concerns: A methodological critique. *Risk Analysis, 15*, 421-436.

Webler, T., Tuler, S., and Krueger, R.
- 2001 What is a good public participation process? Five perspectives from the public. *Environmental Management, 27*(3), 435-450.

Wehr, P.
- 1979 *Conflict Regulation*. Boulder, CO: Westview Press.

Weidner, H.
- 1993 *Mediation as a Policy for Resolving Environmental Disputes with Special References to Germany*. (Manuscript of the Series "Mediationsverfahren im Umweltschutz.") Berlin, Germany: Science Center.

Welp, M., and Stoll-Kleemann, S.
- 2006 Integrative theory of reflexive dialogues. In S. Stoll-Kleemann and M. Welp (Eds.), *Stakeholder Dialogues in Natural Resources Management: Theory and Practice* (pp. 43-78). Heidelberg, Germany: Springer.

Welp, M., de la Vega-Leinert, A.C., Stoll-Kleemann, S., and Furstenau, C.
- 2007 Science-based stakeholder dialogues in climate change research. In S. Stoll-Klemmann and M. Welp (Eds.), *Stakeholder Dialogues in Natural Resources Management* (pp. 213-240). Heidelberg, Germany: Springer.

Wengert, N.
- 1976 Citizen participation: Practice in search of a theory. *Natural Resources Journal, 16*(1), 23-40.

Western Center for Environmental Decision-Making
- 1997 *Public Involvement in Comparative Risk Projects: Principles and Best Practices: A Sourcebook for Project Managers*. Boulder, CO: Meridian West Institute.

Wilbanks, T.J.
- 2003 Geographic scaling issues in integrated assessments of climate change. In J. Rotmans and D. Rothman (Eds.), *Scaling Issues in Integrated Assessment* (pp. 5-34). Linne, the Netherlands: Swets and Zeitlinger.
- 2006 Stakeholder involvement in local smart growth: Needs and challenges. In M. Ruth (Ed.), *Smart Growth and Climate Change: Regional Development, Infrastructure and Adaptation*. Northampton, MA: Edward Elgar.

Williams, B.A., and Matheny, A.R.
- 1995 *Democracy, Dialogue, and Environmental Disputes*. New Haven, CT: Yale University Press.

Williams, E., and Ellefson, P.
- 1996 *Natural Resource Partnerships: Factors Leading to Cooperative Success in the Management of Landscape Level Ecosystems Involving Mixed Ownership*. (Staff Paper Series Number 113.) St. Paul: Department of Forest Resources, College of Natural Resources and the Agricultural Experiment Station, University of Minnesota.

Willis, H.H., and DeKay, M.L.
- 2007 The roles of group membership, beliefs, and norms in ecological risk perception. *Risk Analysis, 27*(5), 1365-1380.

Willis, H.H., DeKay, M.L., Fischhoff, B., and Morgan, M.G.
2005 Aggregate, disaggregate, and hybrid analyses of ecological risk perceptions. *Risk Analysis, 25*, 405-428.

Willis, H.H., DeKay, M.L., Morgan, M.G., Florig, H.K., and Fishbeck, P.S.
2004 Ecological risk ranking: Development and evaluation of a method for improving public participation in environmental decision making. *Risk Analysis, 24,* 363-378.

Wilson, J.
2002 Scientific uncertainty, complex systems, and the design of common-pool institutions. In National Research Council, *The Drama of the Commons* (pp. 327-359). Committee on the Human Dimensions of Global Change. E. Ostrom, T. Dietz, N. Dolsak, P.C. Stern, S. Stonich, and E. Weber (Eds.). Division of Behavioral and Social Sciences and Education. Washington, DC: National Academy Press.

Winquist, J.R., and Larson, J.R., Jr.
1998 Information polling: When it impacts group decision making. *Journal of Personality and Social Psychology, 74,* 371-377.

Wittenbaum, G.M., Hubbell, A.P., and Zuckerman, C.
1999 Mutual enhancement toward an understanding of the collective preference for shared information. *Journal of Personality and Social Psychology, 77,* 967-978.

Wondolleck, J.M.
1988 *Public Lands Conflict and Resolution: Managing National Forest Disputes.* New York: Plenum Press.

Wondolleck, J.M., and Ryan, C.M.
1999 What hat do I wear now?: An examination of agency roles in collaborative processes. *Negotiation Journal, 15*(2), 117-134.

Wondolleck, J.M., and Yaffee, S.L.
1994 *Building Bridges Across Agency Boundaries: In Search of Excellence in the United States Forest Service.* (Research report to the USDA Forest Service Pacific Northwest Research Station, School of Natural Resources and Environment, University of Michigan, Ann Arbor.)

1997 *Sustaining the Success of Collaborative Partnerships: Revisiting the Building Bridges Cases.* (Research report submitted to the USDA Forest Service Pacific Northwest Research Station.) Available: http://www.snre.umich.edu/ecomgt/collaboration/Sustaining_Success.pdf [accessed June 2008].

2000 *Making Collaboration Work: Lessons from Innovation in Natural Resource Management.* Washington, DC: Island Press.

Woolcock, M.
1998 Social capital and economic development: Towards theoretical synthesis and policy framework. *Theory and Society, 27*(2), 151-208.

World Bank
1996 *World Bank Participation Sourcebook.* Washington, DC: Author.

Wynne, B.
1989 Sheepfarming after Chernobyl: A case study in communicating scientific information. *Environment, 31*(2), 10-15, 33-39.

1992 Uncertainty and environmental learning. *Global Environmental Change, 2,* 111-127.

1995 Public understanding of science. In S. Jasanoff, G.E. Markle, J.C. Petersen, and T. Pinch (Eds.), *Handbook of Science and Technology Studies* (pp. 361-388). London, England: Sage.

2005 Risk as Globalising "Democratic Discourse": Framing Subjects and Citizens. In M. Leach, I. Scoones, and B. Wynne (Eds.), *Science and Citizens* (pp. 66-82). London, England: Zed Books.

Yaffee, S.L., Wondolleck, J.M., and Lippman, S.
1997 *Factors That Promote and Constrain Bridging: A Summary and Analysis of the Literature*. Ecosystem Management Initiative, School of Natural Resources and Environment, University of Michigan. Ann Arbor. Available: http://www.snre.umich.edu/ecomgt/collaboration/Factors_that_Promote_and_Contrain_Bridging.pdf [accessed June 2008].

Yang, K., and Callahan, K.
2007 Citizen involvement efforts and bureaucratic responsiveness: Participatory values, stakeholder pressures, and administrative practicality. *Public Administration Review*, 67(2), 249-264.

Yearley, S.
2000 Making systematic sense of public discontents with expert knowledge: Two analytic approaches and a case study. *Public Understanding of Science*, 9, 105-122.

Yosie, T.F., and Herbst, T.D.
1998 *Using Stakeholder Processes in Environmental Decisionmaking: An Evaluation of Lessons Learned, Key Issues, and Future Challenge*. Washington, DC: Ruder Finn. Available: http://gdrc.org/decision/nr98ab01.pdf [accessed June 2008].

Yost, N.C.
1979 New NEPA regulations stress cooperation rather than conflict. *Environmental Consensus*, 2(l), March.

Young, I.M.
2000 *Inclusion and Democracy*. New York: Oxford University Press.

Young, O.R.
2002 *The Institutional Dimensions of Environmental Change: Fit, Interplay, and Scale*. Cambridge, MA: MIT Press.

Young, P.
1993 *Equity in Theory and Practice*. Princeton, NJ: Princeton University Press.

Zarger, R.
2003 *Practitioner Perspectives on Successful Public Participation in Environmental Decisions*. Paper prepared for the National Research Council Panel on Public Participation in Environmental Assessment and Decision Making, May, National Academy of Sciences, Washington, DC.

Zavestoski, S., McCormick, S., and Brown, P.
2004 Gender, embodiment, and disease: Environmental breast cancer activists' challenges to science, the biomedical model, and policy. *Science as Culture*, 13(4), 563-586.

Zeckhauser, R., and Viscusi, K.W.
1996 The risk management dilemma. In H. Kunreuther and P. Slovic (Eds.), *Annals of the American Academy of Political and Social Science, Special Issue, Challenges in Risk Assessment and Risk Management* (pp. 144-155). Thousand Oaks: Sage.

Appendix

Biographical Sketches of Panel Members and Staff

Thomas Dietz *(Chair)* is professor of sociology and of crop and soil sciences, director of the Environmental Science and Policy Program, and assistant vice president for environmental research at Michigan State University. His research interests include the role of deliberation in environmental decision making, the human dimensions of global environmental change and cultural evolution. He is a fellow of the American Association for the Advancement of Science, a Danforth fellow, and past president of the Society for Human Ecology. He is the recipient of the distinguished contribution award from the Section on Environment, Technology, and Society of the American Sociological Association and of the Sustainability Science Award of the Ecological Society of America. He holds a bachelor's degree in general studies from Kent State University and a Ph.D. in ecology from the University of California at Davis.

Gail Bingham is president of RESOLVE and has been a practicing mediator for 30 years with a focus on the environment and natural resources. She has served as a mediator for a variety of local, state and federal agencies and private parties on such diverse subjects as the economic implications of proposed climate change legislation, geologic sequestration of carbon dioxide, regulatory policy under the Safe Drinking Water Act, national wetlands policy, watershed management and pollutant policy, children's health protection, allocation of water rights, hydroelectric relicensing, chemicals policy, hazardous waste management, and community land use and infrastructure issues. She also is the author of several publications, including *Resolving Environmental Disputes: A Decade of Experience*, *When the*

Sparks Fly: Building Consensus When the Science Is Contested, and *Seeking Solutions: Alternative Dispute Resolution and Western Water Issues.* She was the 2006 recipient of the Mary Parker Follett Award from the Association for Conflict Resolution. She received a B.S. degree from Huxley College of Environmental Studies in Washington State and did graduate work in environmental planning at the University of California at Berkeley.

Jennifer Brewer *(Program Officer)* is now an assistant professor in the Department of Geography and the Institute for Coastal Science and Policy of East Carolina University. Her research investigates models of environmental governance, especially in the areas of marine resources and climate change. Prior to her work on this study at the National Research Council, she worked in the areas of environmental policy, natural resource management, and international voluntary service. She held a fellowship in the U.S. House of Representatives, positions on the staff and board of Volunteers for Peace, and staff and consulting positions with nonprofit and governmental organizations involved in fisheries and coastal resources. She has a B.A. degree with high honors from the University of Michigan, an M.S. degree in marine policy from the University of Maine, and a Ph.D. in human geography from Clark University.

Caron Chess is a professor in the Department of Human Ecology at Rutgers University. She conducts research on the evaluation of public participation and the impact of organizational factors on public participation and risk communication. She has served as the president of the Society for Risk Analysis and she currently sits on the editorial board of *Risk Analysis* and the boards of two journals of environmental communication. In addition to publishing in academic journals, she has also authored publications that are used widely by government and industry practitioners, including *Communicating with the Public: Ten Questions Environmental Managers Should Ask* and *Improving Dialogue with Communities: A Short Guide to Government Risk Communication*, which has been translated into three languages. Prior to her academic career, she coordinated environmental programs for state government and environmental organizations and played a central role in the campaign for the country's first public access right-to-know law. She received an M.S. degree from the University of Michigan and a Ph.D. degree from the State University of New York College of Environmental Science and Forestry.

Michael L. DeKay is an associate professor in the Department of Psychology at Ohio State University. Previously, he was an associate professor in the Department of Engineering and Public Policy and the H. John Heinz III School of Public Policy and Management at Carnegie Mellon University.

His research concerns judgment and decision making, particularly in the environmental and medical domains. With colleagues at Carnegie Mellon, he developed and assessed a deliberative method for ranking health, safety, and environmental risks, with specific attention to the validity and replicability of the resulting rankings. His current projects involve precautionary reasoning, distortion of outcome and probability information in risky decisions, and the appropriateness of aggregating outcomes across repeated decisions. He has authored or coauthored numerous journal articles and book chapters, including many articles in *Risk Analysis* and *Medical Decision Making*. He holds a B.S. in chemistry from Caltech (1985), an M.S. in chemistry from Cornell (1987), and an M.A. and a Ph.D. in social psychology from the University of Colorado at Boulder.

Jeanne M. Fox is president of the New Jersey Board of Public Utilities and serves as a member of the governor's cabinet. She also serves on several committees of the National Association of Regulatory Commissioners and on the advisory council to the board of directors and the executive committee of the Electric Power Research Institute. She is chair of the National Council on Electricity Policy. Previously, she was regional administrator of the U.S. Environmental Protection Agency with responsibility for New Jersey, New York, Puerto Rico, and the U.S. Virgin Islands, and she also served as commissioner of the New Jersey Department of Environmental Protection and Energy and as New Jersey's commissioner on the interstate Delaware River Basin Commission. She has also been a visiting distinguished lecturer at the Bloustein School of Planning and Public Policy at Rutgers University and a visiting lecturer in public and international affairs at the Woodrow Wilson School of Public and International Affairs at Princeton University. She received a bachelor's degree from Douglass College and a J.D. degree from Rutgers University School of Law.

Steven C. Lewis is president and principal scientist of Integrative Policy & Science, Inc. (IPSi), which provides consulting services in general toxicology, qualitative and quantitative assessment of risk from environmental hazards, science policy, and legislative/regulatory affairs. Prior to founding IPSi, he held various positions at Exxon-Mobil, including manager of the petroleum and synthetic fuels group. His research and safety assessment activities focused on potential health risks from exposure to chemical carcinogens, toxicants to the nervous system, and chemical hazards to reproductive health, and he also had responsibility for public and community affairs, including management of a multi-stakeholder process to address concerns of rural Alaskans after the Exxon Valdez oil spill. He is a diplomate of the American Board of Toxicology and has served on the editorial boards of five scientific journals. He is an adjunct professor at the University of Medicine and Dentistry of New Jersey and the Robert Wood Johnson Medical School and a senior fellow at the University of Texas at Dallas. He holds a B.A. degree in chemistry from Indiana University and a Ph.D. degree in toxicology from the Indiana University School of Medicine.

Gregory B. Markus is a professor of political science at the University of Michigan and a research professor in the Center for Political Studies at the university's Institute for Social Research. His research, teaching, and public work focus on political participation and urban politics, primarily in the United States. He has worked for more than 25 years with organizations at the local, state, national, and international levels that build the capacities of individuals and communities to devise and implement practical strategies to address public issues. He previously served as vice president of MOSES, a community organizing project based in Detroit, and he is board chair of the Harriet Tubman Center for Community Organizing, also based in Detroit. He is a past recipient of the sociopsychological prize of the American Association for the Advancement of Science and the Amoco Award for excellence in teaching. He holds a Ph.D. degree in political science from the University of Michigan.

D. Warner North is president and principal scientist of the consulting firm NorthWorks, Inc., and a consulting professor in the Department of Management Science and Engineering at Stanford University. Over his career, he has carried out applications of decision analysis and risk analysis for electric utilities in the United States and Mexico, for the petroleum and chemical industries, and for government agencies with responsibility for energy and environmental protection. He has served as a member and consultant to the Science Advisory Board of the U.S. Environmental Protection Agency since 1978, and he previously served as a member of the U.S. Nuclear Waste Technical Review Board. He received a B.S. degree in physics from Yale University and a Ph.D. degree in operations research from Stanford University.

Ortwin Renn serves as full professor and chair of environmental sociology at Stuttgart University. He directs the Interdisciplinary Research Unit for Risk Governance and Sustainable Technology Development at the University of Stuttgart and DIALOGIK, a nonprofit research institute for the investigation of communication and participation processes in environmental policy making. He is also the elected deputy dean of the Economics and Social Science Department and acting director of the Institute of Social Sciences at the University of Stuttgart. He work focuses primarily on risk governance, political participation, and technology assessment. He is a member of the Scientific and Technical Council of the International Risk Governance Council in Geneva and the European Academy of Science and Arts, and he serves on the senate of the Berlin-Brandenburg Academy of Sciences and on the governing board of the German National Academy of Technology and Engineering. He also chairs the State Sustainability Commission. His is a recipient of an honorary doctorate from the Swiss Institute of Technology and the distinguished achievement award of the Society for Risk Analysis. He holds a doctoral degree in sociology and social psychology from the University of Cologne.

Margaret A. Shannon is the associate dean for undergraduate education and faculty development and professor in the Rubenstein School of Environment and Natural Resources at the University of Vermont. She is also a professor in honor on the faculty of forest and environmental sciences at the University of Freiburg, where she teaches international environmental governance and supervises doctoral students. Previous academic appointments were at the Buffalo Law School, State University of New York (SUNY); the Maxwell School of Citizenship and Public Affairs at Syracuse University; the College of Forestry of the University of Washington; the College of Environmental Science and Forestry of SUNY Syracuse; and at the Lewis and Clark Law School. Her research focuses on the emergence of a participatory approach that actively engages people and organizations in creating new modes of environmental governance. She received B.A. degrees in anthropology and sociology from the University of Montana and M.S. and Ph.D. degrees in natural resource management, policy, and sociology from the School of Renewable Natural Resources at the University of California at Berkeley.

Paul C. Stern *(Study Director)* is a principal staff officer at the National Research Council/National Academy of Sciences and director of its standing Committee on the Human Dimensions of Global Change. His research interests include the determinants of environmentally significant behavior, particularly at the individual level; participatory processes for informing environmental decision making; and the governance of environmental resources and risks. He is coauthor of the textbook *Environmental Problems and Human Behavior* (2nd ed., 2002); coeditor of numerous National Research Council publications, including *Decision Making for the Environment: Social and Behavioral Science Priorities*, *The Drama of the Commons*, *Making Climate Forecasts Matter*, *Understanding Risk*, and *Energy Use: The Human Dimension*. His coauthored article in *Science*, "The Struggle to Govern the Commons," won the 2005 Sustainability Science Award from the Ecological Society of America. He is a fellow of the American Association for the Advancement of Science and the American Psychological Association. He holds a B.A. degree from Amherst College and M.A. and Ph.D. degrees from Clark University, all in psychology.

Seth Tuler *(Consultant)* has been a senior researcher at SERI since its founding in 1995. His research interests are focused on the human dimensions of natural resource management and environmental remediation, including public participation and risk communication. He seeks to apply insights emerging from research to practical applications in a wide range of policy arenas, including the clean-up of contaminated sites, marine oil spill response, fisheries management regulations, worker and public safety in national parks, and wildland fire management. He has also been involved with a variety of projects to facilitate environmental health education, training, and public participation with community residents affected by contamination from U.S. nuclear weapons production and related facilities. He is a member of the Board of Scientific Advisors' Subcommittee for the National Center for Environmental Research of the U.S. Environmental Protection Agency.

Elaine Vaughan is professor emerita and research professor of psychology in the Department of Psychology and Social Behavior at the University of California at Irvine. Her research interests include risk communication, public understanding and use of scientific risk information, cultural values and beliefs and their influence on psychological responses to risk and uncertainty, risk perceptions of culturally and socioeconomically diverse populations, and measurement issues related to research that targets such social groups. She has designed and conducted numerous studies on community reactions to both conventional and nontraditional or extreme risk events with an emphasis on the effects of uncertain and evolving information on responses. She has published numerous scientific articles on these topics. She has served on numerous national and state committees, including the joint White House-Congressional Advisory Board on Veterans' Dose Reconstruction, the University of California's Scientific Advisory Panel on the Disposal of Low Level Radioactive Waste, and California's Project on Comparative Risk Policy. She received a B.A. degree in psychology from the University of California at Los Angeles and M.A. and Ph.D. degrees in psychology from Stanford University.

Thomas J. Wilbanks is a corporate research fellow at the Oak Ridge National Laboratory and leads the laboratory's Global Change and Developing Country Programs. A past president of the Association of American Geographers, he conducts research on such issues as sustainable development, energy and environmental technology and policy, responses to global climate change, and the role of geographical scale in all of these regards. His recently coedited books include *Global Change and Local Places*, *Geographical Dimensions of Terrorism*, and *Bridging Scales and Knowledge Systems: Linking Global Science and Local Knowledge*. For the Inter-

APPENDIX

governmental Panel on Climate Change, he is the coordinating lead author for the Fourth Assessment Report, Working Group II, Chapter 7 (industry, settlement, and society); for the U.S. Climate Change Science Program, he is the coordinating lead author for the Synthesis and Assessment Product (SAP 4.5) on the effects of climate change on energy production and use in the United States, and lead author for the section of another SAP (4.6) on effects of global change on human health and welfare and human systems.